Bart de Bruyn

Near Polygons

Birkhäuser Verlag
Basel • Boston • Berlin

Author:

Bart De Bruyn
Ghent University
Department of Pure Mathematics and
Computer Algebra
Galglaan 2
B-9000 Gent
Belgium
e-mail: bdb@cage.ugent.be

2000 Mathematical Subject Classification 51E20, 05B25, 51A50, 51E20, 05C12, 05C75

A CIP catalogue record for this book is available from the
Library of Congress, Washington D.C., USA

Bibliographic information published by Die Deutsche Bibliothek
Die Deutsche Bibliothek lists this publication in the Deutsche National-
bibliografie; detailed bibliographic data is available in the Internet at
<http://dnb.ddb.de>.

ISBN 3-7643-7552-3 Birkhäuser Verlag, Basel – Boston – Berlin

Part of Springer Science+Business Media
Cover design: Birgit Blohmann, Zürich, Switzerland
Printed on acid-free paper produced from chlorine-free pulp. TCF ∞
Printed in Germany
ISBN 10: 3-7643-7552-3 e-ISBN: 3-7643-7553-1
ISBN 13: 978-3-7643-7552-2

9 8 7 6 5 4 3 2 1 www.birkhauser.ch

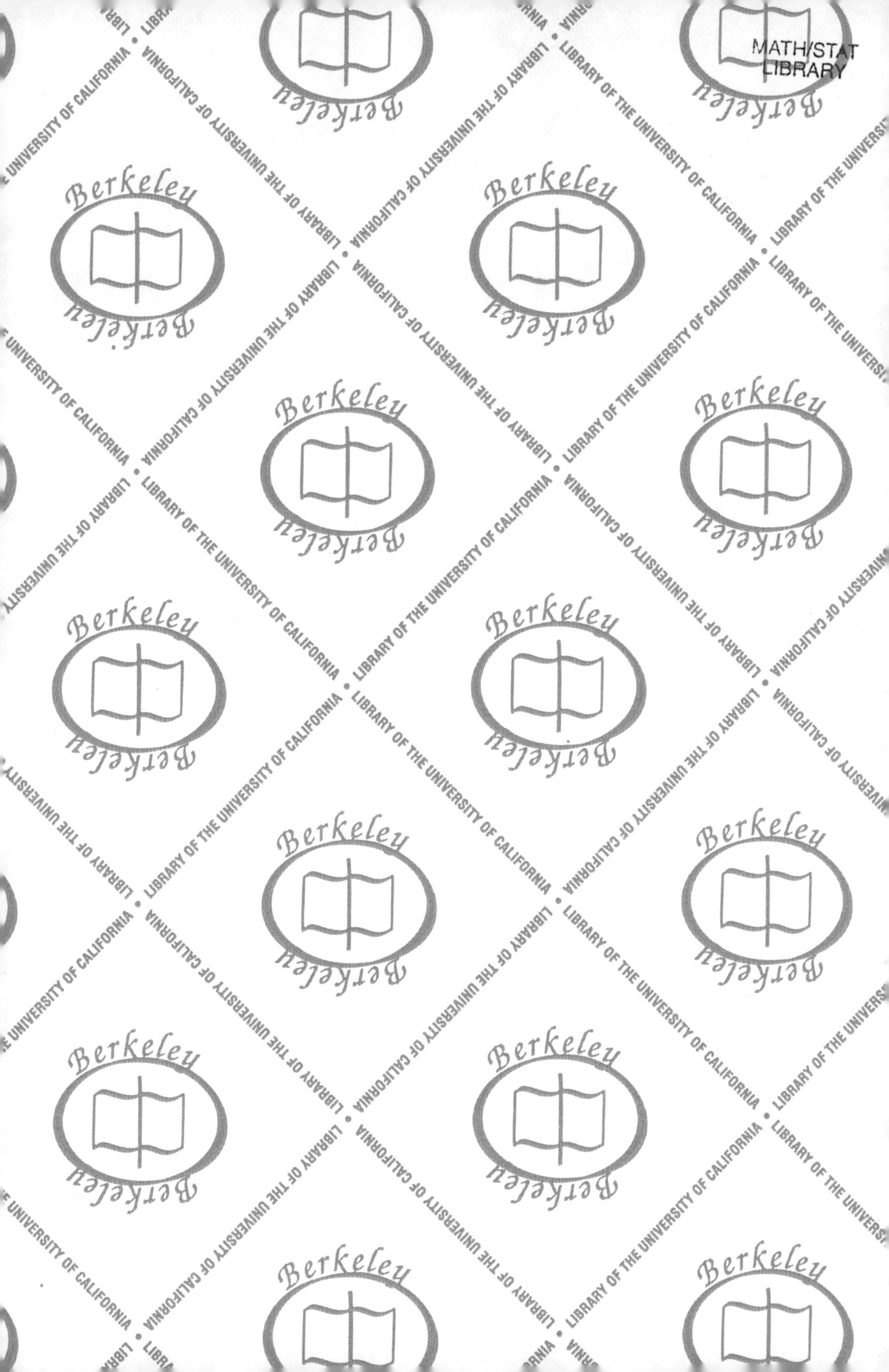

Frontiers in Mathematics

Contents

Preface

In this book, we intend to give an extensive treatment of the basic theory of general near polygons. The subject of near polygons has been around for about 25 years now. Excellent handbooks have appeared on certain important subclasses of near polygons like generalized quadrangles ([82]) and generalized polygons ([100]), but no book has ever occurred dealing with the topic of general near polygons. Although generalized polygons and especially generalized quadrangles are indispensable to the study of near polygons, we do not aim at giving a profound study of these incidence structures. In fact, this book can be seen as complementary to the two above-mentioned books.

Although generalized quadrangles and generalized polygons were intensively studied since they were introduced by Tits in his celebrated paper on triality ([96]), the terminology near polygon first occurred in a paper in 1980. In [91], Shult and Yanushka showed the connection between the so-called tetrahedrally closed line-systems in Euclidean spaces and a class of point-line geometries which they called *near polygons*. In [91], also some very fundamental results regarding the geometric structure of near polygons were obtained, like the existence of quads, a result which was later generalized by Brouwer and Wilbrink [16] who showed that any dense near polygon has convex subpolygons of any feasible diameter. The paper [16] gives for the first time a profound study of dense near polygons. Other important papers on near polygons from the 1980s and the beginning of the 1990s deal with dual polar spaces, the classification of regular near polygons in terms of their parameters and the classification of the slim dense near hexagons. The subject of near polygons has regained interest in the last years. Important new contributions to the theory were the theory of glued near polygons, the theory of valuations and important breakthrough results regarding the classification of dense near polygons with three and four points on every line. These new contributions will be discussed extensively in this book.

This book essentially consists of two main parts. In the first part of the book, which consists of the first five chapters, we develop the basic theory of near polygons. In Chapters 2, 3 and 4, we study three classes of near polygons: the dense, the regular and the glued near polygons. Our treatment of the dense and glued near polygons is rather complete. The treatment of the regular near polygons is concise and results are not always accompanied with proofs. More detailed information on regular near polygons can be found in the book *Distance-regular graphs* [13] by Brouwer, Cohen and Neumaier. In that book regular near polygons are considered as one of the main classes of distance-regular graphs. In Chapter 5, we discuss the notion of *valuation of a near polygon* which is a very important tool for classifying near polygons.

The second main part of this book, which consists of Chapters 6–9, discusses the problem of classifying all slim dense near polygons. These are dense near polygons with three points on every line. There is a conjecture which states that all these near polygons can be obtained by applying the direct product and the glueing construction to a list of near polygons which consists of five infinite classes and three exceptional near hexagons. In Chapter 6, we will discuss all the near polygons of that list, thereby providing information on their convex subpolygons, spreads of symmetry and valuations. The first main result with respect to the above-mentioned classification was obtained by Brouwer, Cohen, Hall and Wilbrink [12] who succeeded in classifying all slim dense near hexagons. Their proof relies on Fisher's theory of groups generated by 3-transpositions [72] and Buekenhout's geometric interpretation of that theory [18]. In Chapter 7, we will give a purely combinatorial proof of the classification of the slim dense near hexagons. In Chapter 8, we classify all slim dense near polygons with a so-called nice chain of convex subpolygons. All known slim dense near polygons, except for the ones with a so-called $\mathbb{E}_1$- or $\mathbb{E}_2$-hex, have such a chain of subpolygons. In Chapter 9, we will give a complete classification of all slim dense near octagons. In order to obtain the classification results mentioned in Chapters 7, 8 and 9, we must invoke almost the whole theory of near polygons which we developed in the earlier chapters. It is our sincere hope that the techniques provided in Chapters 1–9 will ultimately lead to a complete classification of all slim dense near polygons.

In the final chapter we will discuss nondense slim near hexagons and in the appendix we will give an overview of what is known on the classification of dense near polygons with four points on each line.

The study of near polygons is rather interesting and some of the tools which we developed for dealing with general near polygons seem to have nice applications to some specific classes of near polygons. We give two examples.

Admissible triples were originally introduced for the study of glued near hexagons. One of the advantages of these structures was that they allowed a unified construction for several classes of generalized quadrangles ([29]). Recently, these structures have also been used to obtain some interesting characterization results regarding the symplectic generalized quadrangle $W(q)$ ([49], [55]).

Several of the near polygon techniques which we will discuss throughout this book turn out to be useful for the study of certain important substructures of dual polar spaces. E.g., the study of valuations of dual polar spaces has led to two new classes of hyperplanes in dual polar spaces ([61]). Hyperplanes of dual polar spaces are often near polygons themselves and/or contain near polygons as substructures, see e.g. [57].

Finally, I want to thank all my co-authors; it was a pleasure to work with them on the subject of near polygons. I list them in alphabetical order: R. J. Blok, I. Cardinali, F. De Clerck, U. Meierfrankenfeld, A. Pasini, S. E. Payne, H. Pralle, K. Thas and P. Vandecasteele. Spccial thanks goes to my former Ph.D. student P. Vandecasteele who was involved in realizing several of the results to be discussed in this book. A large part of this book was written while the author was visiting Antonio Pasini at the University of Siena in the first half of 2005. During this visit, the author enjoyed his and several other people's warm hospitality. I am also very grateful to the Fund for Scientific Research – Flanders (Belgium) for financial support during the last eight years.

Ghent, December 2005 Bart De Bruyn

Chapter 1

Introduction

1.1 Definition of near polygon

Let $\Gamma = (V, E)$ be a simple undirected graph without loops. The adjacency relation in Γ will usually be denoted as $\sim$. A *clique* of Γ is a set of mutually adjacent vertices. A clique is called *maximal* if it is not properly contained in another clique. We will denote the distance between two vertices x and y of Γ by $\mathrm{d}(x, y)$. If X_1 and X_2 are two nonempty sets of vertices, then we denote by $\mathrm{d}(X_1, X_2)$ the minimal distance between a vertex of X_1 and a vertex of X_2. If X_1 is a singleton $\{x_1\}$, then we will also write $\mathrm{d}(x_1, X_2)$ instead of $\mathrm{d}(\{x_1\}, X_2)$. For every $i \in \mathbb{N}$ and every nonempty set X of vertices, we denote by $\Gamma_i(X)$ the set of all vertices y for which $\mathrm{d}(y, X) = i$. If X is a singleton $\{x\}$, then we also write $\Gamma_i(x)$ instead of $\Gamma_i(\{x\})$.

A *near* $2d$*-gon* is a connected graph of finite diameter d with the property that for every vertex x and every maximal clique M there exists a unique vertex x' in M nearest to x. A near 0-gon consists of one vertex and a near 2-gon is just a complete graph with at least two vertices.

A *(point-line) incidence structure* is a triple $\mathcal{S} = (\mathcal{P}, \mathcal{L}, \mathrm{I})$, with $\mathcal{P}$ a nonempty set whose elements are called *points*, $\mathcal{L}$ a possibly empty set whose elements are called *lines* and I a subset of $\mathcal{P} \times \mathcal{L}$, called the *incidence relation*. If $(p, L) \in \mathrm{I}$, then we say that p is incident with L, that p is contained in L, that L contains p, etc… A point-line incidence structure is called a *partial linear space* (respectively a *linear space*) if every line is incident with at least two points and if every two different points are incident with at most (respectively exactly) one line. The *point graph* or *collinearity graph* of a point-line incidence structure $\mathcal{S}$ is the graph whose vertices are the points of $\mathcal{S}$ with two different points adjacent whenever they are *collinear*, i.e. whenever there exists a line incident with these points.

There is a bijective correspondence between the class of near polygons and a class of partial linear spaces. If a graph Γ is a near polygon, then the point-line incidence structure with points, respectively lines, the vertices, respectively maximal cliques, of Γ (natural incidence) is a partial linear space $\mathcal{S}$. The graph Γ can easily be retrieved from $\mathcal{S}$: Γ is the point graph of $\mathcal{S}$. Because of this bijective correspondence, we will call the partial linear spaces which correspond with near polygons also near polygons. In the sequel we will always adopt the geometric point of view. A near 0-gon is then a point and a near 2-gon a line. In the sequel we will denote the line with $s+1$ points by $\mathbb{L}_{s+1}$. If x is a point and L a line of a near polygon, then we denote by $\pi_L(x)$ the unique point of L nearest to x. The diameter of a near polygon $\mathcal{S}$ is the diameter of its point graph and will be denoted by $\mathrm{diam}(\mathcal{S})$.

A near $2d$-gon, $d \geq 2$, is called a *generalized $2d$-gon* if $|\Gamma_1(y) \cap \Gamma_{i-1}(x)| = 1$ for every $i \in \{1, \ldots, d-1\}$ and every two points x and y at distance i from each other. The generalized quadrangles (GQ's) are precisely the near quadrangles. A generalized $2d$-gon is called *degenerate* if it does not contain ordinary $2d$-gons as subgeometries. For more background information on generalized quadrangles and generalized polygons, we refer to [82], [95] and [100].

1.2 Genesis

Consider the n-dimensional Euclidean space with origin O. A *line system of type* $(a_1, a_2, \ldots, a_k)$ is a set A of lines through O with the property that $|\cos\alpha| \in \{a_1, a_2, \ldots, a_k\}$ for any two different lines of A at angle α. A *system of vectors of type* $(a_1, a_2, \ldots, a_k)$ is a set Σ of vectors satisfying:

(i) $\Sigma = -\Sigma := \{-x | x \in \Sigma\}$;

(ii) the norm (x, x) of x is a nonzero constant c for all $x \in \Sigma$;

(iii) if $x, y \in \Sigma$ and $x \neq \pm y$, then $|\frac{(x,y)}{c}| \in \{a_1, a_2, \ldots, a_k\}$.

Line systems of type $(a_1, \ldots, a_k)$ are clearly equivalent to systems of vectors of type $(a_1, \ldots, a_k)$. In [91], E. Shult and A. Yanushka discussed line systems of type $(0, \frac{1}{3})$ and showed that they were related to a class of point-line incidence structures which they called *near polygons*. The angle α for which $\cos\alpha = -\frac{1}{3}$ is the angle subtended by two chords drawn from the barycenter of a tetrahedron to two of its corner vertices. A line system of type $(0, \frac{1}{3})$ is called *tetrahedrally closed* if it has 0, 1, 2 or 4 lines in common with every set of four lines obtained by connecting the barycenter of an arbitrary tetrahedron centered at O with its four corner vertices. A system of vectors of type $(0, \frac{1}{3})$ is called a *tetrahedrally closed* $(0, \frac{1}{3})$ *system of vectors* if the corresponding line system is tetrahedrally closed.

Let Σ be a tetrahedrally closed $(0, \frac{1}{3})$ system of vectors with norm 3. Fix a vector $u \in \Sigma$ and let $\Sigma_{-1}(u)$ be the set of vectors of Σ having inner product -1 with u. If $\Sigma_{-1}(u) \neq \emptyset$, then we can define the following incidence system $\mathcal{S}$: the point set is equal to $\Sigma_{-1}(u)$; a triplet of three vectors $\{y_1, y_2, y_3\} \subseteq \Sigma_{-1}(u)$ is a line if and only if u, y_1, y_2, y_3 define the four vertices of a tetrahedron centered at O.

Theorem 1.1 ([91, Proposition 3.10]). *If for every two vectors of $\Sigma_{-1}(u)$ having inner product 1, there exists at least one vector of $\Sigma_{-1}(u)$ having inner product -1 with both vectors, then $\mathcal{S}$ is a near 2d-gon with $d \leq 3$. Moreover, every line of $\mathcal{S}$ is incident with precisely three points.*

1.3 Near polygons with an order

A near $2n$-gon, $n \geq 1$, is said to have *order* (s,t) if every line is incident with precisely $s+1$ points and every point is incident with precisely $t+1$ lines. Here, s and t are allowed to be infinite. If $s=t$, then we also say that the near polygon has order s.

Theorem 1.2. *Let $\mathcal{S} = (\mathcal{P}, \mathcal{L}, \mathrm{I})$ be a finite near $2n$-gon, $n \geq 1$, of order (s,t) and let x be a point of $\mathcal{S}$, then $\sum_{y \in \mathcal{P}} (-\frac{1}{s})^{\mathrm{d}(x,y)} = 0$.*

Proof. Since every line L of $\mathcal{S}$ contains a unique point nearest to x, the sum $\sum_{y \mathrm{I} L} (-\frac{1}{s})^{\mathrm{d}(x,y)}$ is equal to 0. Hence,

$$0 = \sum_{L \in \mathcal{L}} \sum_{y \mathrm{I} L} (-\frac{1}{s})^{\mathrm{d}(x,y)} = \sum_{y \in \mathcal{P}} \sum_{L \mathrm{I} y} (-\frac{1}{s})^{\mathrm{d}(x,y)} = (t+1) \cdot \sum_{y \in \mathcal{P}} (-\frac{1}{s})^{\mathrm{d}(x,y)},$$

which proves the theorem. □

1.4 Parallel lines

The notion of parallel lines was introduced in [16].

Theorem 1.3. *If K and L are two lines of a near polygon, then precisely one of the following cases occurs.*

(a) *There exists a unique point k^* on K and a unique point l^* on L such that $d(k,l) = d(k,k^*) + d(k^*,l^*) + d(l^*,l)$ for every point k on K and every point l on L.*

(b) *For every point k on K, there exists a unique point l on L such that $d(k,l) = d(K,L)$.*

Proof. We will make use of the following obvious observation, see also Lemma 1.8.

> If u and v are two collinear points such that $\mathrm{d}(u,L) = \mathrm{d}(v,L) - 1$, then $\pi_L(u) = \pi_L(v)$.

We distinguish the following two possibilities.

(a) All points $\pi_L(u)$, $u \in K$, are equal, say to l^*. Let k^* denote the unique point of K nearest to l^*. Then for all points $k \in K$ and $l \in L$, we have $\mathrm{d}(k,l) = \mathrm{d}(k,l^*) + \mathrm{d}(l^*,l) = \mathrm{d}(k,k^*) + \mathrm{d}(k^*,l^*) + \mathrm{d}(l^*,l)$.

(b) There exist points $u_1, u_2 \in K$ for which $\pi_L(u_1) \neq \pi_L(u_2)$. By the above observation, $\mathrm{d}(u_1, L) = \mathrm{d}(u_2, L)$. Also by the above observation, every point $u \in K$ has distance precisely $\mathrm{d}(u_1, L)$ from L (otherwise $\pi_L(u_1) = \pi_L(u) = \pi_L(u_2)$). So, $\mathrm{d}(K, L) = \mathrm{d}(u, L)$ for every point u of K. In a similar way one proves that every point v of L has distance $\mathrm{d}(K, L)$ from K. □

Definition. If case (b) of Theorem 1.3 occurs, then the lines K and L are called *parallel* and we will write $K \parallel L$.

1.5 Substructures

A nonempty set X of points of a partial linear space $\mathcal{S} = (\mathcal{P}, \mathcal{L}, \mathrm{I})$ is called a *subspace* if every line meeting X in at least two points is completely contained in X. A subspace X of $\mathcal{S}$ is called *geodetically closed* or *convex* if every point on a shortest path between two points of X is also contained in X. Obviously, the intersection of two (convex) subspaces is again a (convex) subspace. Having a subspace X, we can define a subgeometry $\mathcal{S}_X$ of $\mathcal{S}$ by considering only those points and lines of $\mathcal{S}$ which are completely contained in X.

If X is a convex subspace of a near polygon $\mathcal{S}$, then $\mathcal{S}_X$ clearly is a sub-near-polygon of $\mathcal{S}$. In the sequel, we will use the word "subpolygon" as a shortening of "sub-near-polygon". If the convex subpolygon $\mathcal{S}_X$ is a nondegenerate generalized quadrangle, then X (and often also $\mathcal{S}_X$) will be called a *quad*. For every point x of a near polygon $\mathcal{S}$, we can define the following point-line incidence structure $\mathcal{L}(\mathcal{S}, x)$: the points of $\mathcal{L}(\mathcal{S}, x)$ are the lines through x, the lines of $\mathcal{L}(\mathcal{S}, x)$ are the quads through x and incidence is containment. $\mathcal{L}(\mathcal{S}, x)$ is called *the local space at* x. The *modified local space* $\mathcal{ML}(\mathcal{S}, x)$ at a point x is obtained from $\mathcal{L}(\mathcal{S}, x)$ by deleting all lines of size 2.

Theorem 1.4. *$\mathcal{L}(\mathcal{S}, x)$ is a partial linear space.*

Proof. Suppose that two different lines L_1 and L_2 through x are contained in two different quads Q_1 and Q_2. Without loss of generality, we may suppose that Q_1 contains a point u not contained in Q_2. If all points of Q_2 are collinear with x, then Q_2 has no ordinary subquadrangles and hence is degenerate, a contradiction. Hence, there exists a point x' in Q_2 noncollinear with x. Let x_i, $i \in \{1, 2\}$, denote the unique point of L_i collinear with x'. The point x' belongs to Q_1 since it is on a shortest path between $x_1 \in Q_1$ and $x_2 \in Q_1$. Hence, also the line $x'x_2$ belongs to Q_1. Let u', respectively u'', denote the unique point of xx_1, respectively $x'x_2$, nearest to u. Since Q_2 is convex and $u', u'' \in Q_2$, also u belongs to Q_2, a contradiction. □

Every nonempty set X of points is contained in a unique minimal convex subpolygon $\mathcal{C}(X)$, namely the intersection of all convex subpolygons through X. We define $\mathcal{C}(\emptyset) := \emptyset$. If $X_1, \ldots, X_k$ are sets of points, then $\mathcal{C}(X_1 \cup \cdots \cup X_k)$ is also denoted by $\mathcal{C}(X_1, \ldots, X_k)$. If one of the arguments of $\mathcal{C}$ is a singleton $\{x\}$, we will often omit the braces and write $\mathcal{C}(\cdots, x, \cdots)$ instead of $\mathcal{C}(\cdots, \{x\}, \cdots)$.

A point x of a near polygon is called *classical* with respect to a convex subpolygon F if there exists a (necessarily unique) point $\pi_F(x) \in F$ such that $\mathrm{d}(x, y) = \mathrm{d}(x, \pi_F(x)) + \mathrm{d}(\pi_F(x), y)$ for every point y of F. We call $\pi_F(x)$ the *projection* of x on F. If x is a point of a near polygon and L a line, then x is always classical with respect to L.

Theorem 1.5. *If x is a point of a near polygon and F is a convex subpolygon such that $d(x, F) \leq 1$, then x is classical with respect to F.*

Proof. Obviously, the theorem holds if $x \in F$. So, suppose $x \notin F$. Since F is convex, there exists a unique point x' in F collinear with x. Suppose there exists a point y in F such that $\mathrm{d}(y, x) \neq 1 + \mathrm{d}(y, x')$. Then $\mathrm{d}(y, x) \leq \mathrm{d}(y, x')$. Hence, the unique point z of the line xx' nearest to y lies at distance $\mathrm{d}(y, x') - 1$ from y. So, z is contained on a shortest path between the points y and x' of F. This would imply that $z \in F$ and $x'z \subseteq F$, contradicting $x \notin F$. □

A convex subpolygon F of $\mathcal{S}$ is called *classical* in $\mathcal{S}$ if every point of $\mathcal{S}$ is classical with respect to F. This means that F is *gated* in the sense of [69]. A convex subpolygon F of $\mathcal{S}$ is called *big* in $\mathcal{S}$ if $F \neq \mathcal{S}$ and if every point of $\mathcal{S} \setminus F$ is collinear with a (necessarily unique) point of F. By Theorem 1.5, every big convex subpolygon of $\mathcal{S}$ is classical in $\mathcal{S}$.

Theorem 1.6. *The intersection of two classical convex subpolygons F_1 and F_2 is either empty or again a classical convex subpolygon.*

Proof. For an arbitrary point x of $\mathcal{S}$, let x' denote the unique point in F_1 nearest to x and let x'' denote the unique point of F_2 nearest to x'. Suppose that $F_1 \cap F_2 \neq \emptyset$ and let u denote an arbitrary point of $F_1 \cap F_2$. Since F_2 is classical, x'' is contained on a shortest path between $x' \in F_1$ and $u \in F_1$. Since F_1 is convex, $x'' \in F_1$ and hence $x'' \in F_1 \cap F_2$. Now, let y denote an arbitrary point of $F_1 \cap F_2$. Then $\mathrm{d}(x, y) = \mathrm{d}(x, x') + \mathrm{d}(x', y) = \mathrm{d}(x, x') + \mathrm{d}(x', x'') + \mathrm{d}(x'', y) = \mathrm{d}(x, x'') + \mathrm{d}(x'', y)$. Hence x is classical with respect to $F_1 \cap F_2$. Since x was arbitrary, $F_1 \cap F_2$ is classical. □

Theorem 1.7. *Let F be a big convex subpolygon of a near polygon $\mathcal{S}$. Then every convex subpolygon F' which meets F either is contained in F or intersects F in a big convex subpolygon of F'.*

Proof. Suppose that F' is not contained in F and let x denote a common point of F and F'. Let y denote an arbitrary point of F' not contained in $F \cap F'$ and let y' denote the unique point of F collinear with y. Since F is classical, $\mathrm{d}(y, x) = \mathrm{d}(y, y') + \mathrm{d}(y', x)$. Hence, the point y', which is contained on a shortest path between the points x and y of F', also belongs to F'. So, every point of F' not contained in $F \cap F'$ is collinear with a unique point of $F \cap F'$. This proves the theorem. □

Lemma 1.8. *Let F be a convex subpolygon of a near polygon $\mathcal{S}$ which is classical in $\mathcal{S}$. If x_1 and x_2 are collinear points of $\mathcal{S}$ such that $d(x_1, F) = d(x_2, F) - 1$, then $\pi_F(x_1) = \pi_F(x_2)$.*

Proof. The point $\pi_F(x_1)$ has distance at most $\mathrm{d}(x_1, \pi_F(x_1)) + \mathrm{d}(x_1, x_2) = \mathrm{d}(x_1, F) + 1 = \mathrm{d}(x_2, F)$ from x_2 and hence coincides with $\pi_F(x_2)$. □

Theorem 1.9. *Let K be a line of a near polygon $\mathcal{S}$ and let F denote a convex subpolygon of $\mathcal{S}$ which is classical in $\mathcal{S}$. Then one of the following holds.*

- *Every point of K has the same distance from F. In this case $\pi_F(K) := \{\pi_F(x) \,|\, x \in K\}$ is a line of F parallel with K.*
- *There exists a unique point on K nearest to F. In this case all points $\pi_F(x)$, $x \in K$, are equal.*

Proof. Suppose that all points $\pi_F(x)$, $x \in K$, are equal, to u say. Then there exists a unique point on K nearest to F, namely the unique point of K nearest to u. Suppose therefore that there exist points $x_1, x_2 \in K$ such that $\pi_F(x_1) \neq \pi_F(x_2)$. By Lemma 1.8, $\mathrm{d}(x_1, F) = \mathrm{d}(x_2, F)$. Put $i := \mathrm{d}(x_1, F)$. Since $\mathrm{d}(\pi_F(x_1), \pi_F(x_2)) = \mathrm{d}(x_1, \pi_F(x_2)) - \mathrm{d}(x_1, \pi_F(x_1)) \leq \mathrm{d}(x_1, x_2) + \mathrm{d}(x_2, \pi_F(x_2)) - \mathrm{d}(x_1, \pi_F(x_1)) = 1$, $\pi_F(x_1)$ and $\pi_F(x_2)$ are contained in a line K'. If u is a point of K different from x_1 and x_2, then u has distance at most $i+1$ from the points $\pi_F(x_1)$ and $\pi_F(x_2)$ of K'. Hence there exists a point u' on K' at distance at most i from u. By Lemma 1.8, $\mathrm{d}(u, F) = i$ and $\pi_F(u) = u'$. This proves that $\pi_F(K) \subseteq K'$ and that every point of K has the same distance i from F. Suppose now that there exists a point u' in $K' \setminus \pi_F(K)$. Then u' has distance at most $i+1$ from at least two points of K and hence distance at most i from a point u of K, showing that $u' = \pi_F(u)$, a contradiction. □

In Section 1.4 we defined the notion of parallel lines. We will now generalize this notion to arbitrary subpolygons. Two convex subpolygons F_1 and F_2 of a near polygon are called *parallel* (notation: $F_1 \| F_2$) if the following holds for every $i \in \{1, 2\}$ and every point x of F_i:

- x is classical with respect to F_{3-i};
- $\mathrm{d}(x, \pi_{F_{3-i}}(x)) = \mathrm{d}(F_1, F_2)$.

By Theorem 1.5 every two disjoint big convex subpolygons of a near polygon are parallel.

Theorem 1.10. *Let F_1 and F_2 be two parallel convex subpolygons of $\mathcal{S}$. Then the map $\pi_{i,3-i} : F_i \to F_{3-i}$, $i \in \{1, 2\}$, which maps a point x of F_i to the unique point of F_{3-i} nearest to x, is an isomorphism. Moreover, $\pi_{2,1} = \pi_{1,2}^{-1}$.*

Proof. (a) We first show that $\pi_{1,2}$ and $\pi_{2,1}$ are bijections and that $\pi_{2,1} = \pi_{1,2}^{-1}$. Let x denote an arbitrary point of F_1. Since $\mathrm{d}(x, \pi_{1,2}(x)) = \mathrm{d}(F_1, F_2)$, x is the unique point in F_1 nearest to $\pi_{1,2}(x)$. Hence, $\pi_{1,2}$ is injective and $x = \pi_{2,1} \circ \pi_{1,2}(x)$. Since x was arbitrary, $\pi_{2,1}$ is surjective. By symmetry, $\pi_{1,2}$ and $\pi_{2,1}$ are bijections and $\pi_{2,1} = \pi_{1,2}^{-1}$.

(b) Next, we show that $\pi_{1,2}$ and $\pi_{2,1}$ determine isomorphisms between F_1 and F_2. Let x and y denote two different collinear points of F_1. Since $\pi_{1,2}(y)$ is the unique point of F_2 at minimal distance $\mathrm{d}(F_1, F_2)$ from y, we have $\mathrm{d}(y, \pi_{1,2}(x)) \geq \mathrm{d}(F_1, F_2) + 1$. Now, $\mathrm{d}(y, \pi_{1,2}(x)) \leq \mathrm{d}(y, x) + \mathrm{d}(x, \pi_{1,2}(x)) \leq 1 + \mathrm{d}(F_1, F_2)$. So, $1 + \mathrm{d}(F_1, F_2) = \mathrm{d}(y, \pi_{1,2}(x)) = \mathrm{d}(y, \pi_{1,2}(y)) + \mathrm{d}(\pi_{1,2}(y), \pi_{1,2}(x)) = \mathrm{d}(F_1, F_2) + \mathrm{d}(\pi_{1,2}(y), \pi_{1,2}(x))$, namely $\mathrm{d}(\pi_{1,2}(x), \pi_{1,2}(y)) = 1$. So, $\pi_{1,2}$ determines an isomorphism between the point graphs of F_1 and F_2 and hence also between F_1 and F_2 themselves. □

Definition. The map $\pi_{i,3-i}$ defined in Theorem 1.10 is called the *projection from* F_i *to* F_{3-i}.

Definition. Let $\mathcal{S} = (\mathcal{P}, \mathcal{L}, \mathrm{I})$ be a near polygon with three points on each line and let F be a big convex subpolygon of $\mathcal{S}$. For every point x of F, we define $\mathcal{R}_F(x) := x$. If x is a point of $\mathcal{S}$ not contained in F, then we put $\mathcal{R}_F(x)$ equal to the unique point of the line $x\,\pi_F(x)$ different from x and $\pi_F(x)$. Obviously, π_F is a permutation of $\mathcal{P}$. We call $\mathcal{R}_F$ the *reflection about* F.

Theorem 1.11. *$\mathcal{R}_F$ is an automorphism of $\mathcal{S}$.*

Proof. It suffices to show that $\mathcal{R}_F$ maps two different collinear points x and y to collinear points $\mathcal{R}_F(x)$ and $\mathcal{R}_F(y)$. Obviously, this holds if the line xy meets F. Suppose therefore that xy is disjoint from F. By Theorem 1.9, $\pi_F(x)$ and $\pi_F(y)$ are collinear. Consider the lines $L_x := x\,\pi_F(x)$ and $L_y := y\pi_F(y)$. Since $\mathrm{d}(x, y) = 1$ and $\mathrm{d}(\pi_F(x), \pi_F(y)) = 1$, $L_x \parallel L_y$ and $\mathrm{d}(L_x, L_y) = 1$. Hence $\mathrm{d}(\mathcal{R}_F(x), \mathcal{R}_F(y)) = 1$. This proves the theorem. □

1.6 Product near polygons

For any two graphs $\Gamma_1 = (V_1, E_1)$ and $\Gamma_2 = (V_2, E_2)$, a new graph $\Gamma_1 \times \Gamma_2$ can be defined. The vertices of $\Gamma_1 \times \Gamma_2$ are the elements of the cartesian product $V_1 \times V_2$. Two vertices (x_1, x_2) and (y_1, y_2) of Γ are adjacent if and only if either ($x_1 = y_1$ and $x_2 \sim y_2$) or ($x_1 \sim y_1$ and $x_2 = y_2$). The graph $\Gamma_1 \times \Gamma_2$ is called the *direct product* of Γ_1 and Γ_2.

If $\mathcal{S}_1 = (\mathcal{P}_1, \mathcal{L}_1, \mathrm{I}_1)$ and $\mathcal{S}_2 = (\mathcal{P}_2, \mathcal{L}_2, \mathrm{I}_2)$ are two near polygons, then we can define the following incidence structure $\mathcal{S} = (\mathcal{P}, \mathcal{L}, \mathrm{I})$:

- $\mathcal{P} = \mathcal{P}_1 \times \mathcal{P}_2$;
- $\mathcal{L} = (\mathcal{P}_1 \times \mathcal{L}_2) \cup (\mathcal{L}_1 \times \mathcal{P}_2)$;

- the point (x, y) of $\mathcal{S}$ is incident with the line $(z, L) \in \mathcal{P}_1 \times \mathcal{L}_2$ if and only if $x = z$ and $y \mathrm{I}_2 L$, the point (x, y) of $\mathcal{S}$ is incident with the line $(M, u) \in \mathcal{L}_1 \times \mathcal{P}_2$ if and only if $x \mathrm{I}_1 M$ and $y = u$.

Let $\mathrm{d}_i(\cdot, \cdot)$, $i \in \{1, 2\}$, denote the distance in $\mathcal{S}_i$ and let $\mathrm{d}(\cdot, \cdot)$ denote the distance in $\mathcal{S}$. One easily verifies that $\mathrm{d}[(x_1, x_2), (y_1, y_2)] = \mathrm{d}_1(x_1, y_1) + \mathrm{d}_2(x_2, y_2)$ for all points (x_1, x_2) and (y_1, y_2) of $\mathcal{S}$. From this it follows that $\mathcal{S}$ is a near polygon. We denote $\mathcal{S}$ also by $\mathcal{S}_1 \times \mathcal{S}_2$ and call it the *direct product* of $\mathcal{S}_1$ and $\mathcal{S}_2$. The point graph of $\mathcal{S}_1 \times \mathcal{S}_2$ is isomorphic to the direct product of the point graphs of $\mathcal{S}_1$ and $\mathcal{S}_2$. Since $\mathcal{S}_1 \times \mathcal{S}_2 \cong \mathcal{S}_2 \times \mathcal{S}_1$ and $(\mathcal{S}_1 \times \mathcal{S}_2) \times \mathcal{S}_3 \cong \mathcal{S}_1 \times (\mathcal{S}_2 \times \mathcal{S}_3)$, also the direct product of $k \geq 3$ near polygons $\mathcal{S}_1, \ldots, \mathcal{S}_k$ is well-defined. If $\mathcal{S}_i$, $i \in \{1, 2\}$, is a near $2n_i$-gon, then $\mathcal{S}_1 \times \mathcal{S}_2$ is a near $2(n_1 + n_2)$-gon. A *product near polygon* is a near polygon which is the direct product of two near polygons with diameter at least 1. A *Hamming near $2n$-gon*, $n \geq 1$, is the direct product of n lines. An *$(n_1 \times n_2)$-grid* is the direct product of a line of size n_1 and a line of size n_2. Clearly, an $(n_1 \times n_2)$-grid is a generalized quadrangle. An $(n_1 \times n_2)$-grid is called *symmetrical*, respectively nonsymmetrical, if $n_1 = n_2$, respectively $n_1 \neq n_2$. The point-line dual of a (symmetrical, nonsymmetrical) grid is called a *(symmetrical, nonsymmetrical) dual grid.* The following result is due to Brouwer and Wilbrink [16].

Theorem 1.12. *Let $\mathcal{S} = (\mathcal{P}, \mathcal{L}, \mathrm{I})$ be a near polygon and let s be a possibly infinite number for which the following holds:*

- *$\mathcal{S}$ has a line of size $s + 1$;*
- *$\mathcal{S}$ has a line whose size is different from $s + 1$;*
- *if x and y are points at mutual distance 2 and if x and y have a common neighbour z such that the line xz has size $s + 1$ and the line zy has size different from $s + 1$, then x and y have at least two common neighbours.*

Then there exists two near polygons $\mathcal{S}_1$ and $\mathcal{S}_2$ such that

(i) *every line of $\mathcal{S}_1$ has size $s + 1$,*

(ii) *no line of $\mathcal{S}_2$ has size $s + 1$,*

(iii) *$\mathcal{S} \cong \mathcal{S}_1 \times \mathcal{S}_2$.*

Proof. Let $\mathcal{L}_1$ denote the set of all lines of $\mathcal{S}$ which are incident with precisely $s+1$ points and put $\mathcal{L}_2 := \mathcal{L} \setminus \mathcal{L}_1$. Let $\mathcal{A}_i$ denote the partial linear space with point set $\mathcal{P}$ and line set $\mathcal{L}_i$ (natural incidence). For every point x of $\mathcal{S}$, let $F_x^{(i)}$ $(i \in \{1, 2\})$ denote the connected component of $\mathcal{A}_i$ containing x. Every line through x is either contained in $F_x^{(1)}$ or $F_x^{(2)}$.

Definition. A path $(x_0, \ldots, x_k)$ is called *nice* if the following holds for a certain point x_l $(0 \leq l \leq k)$ of that path:

- for every $i \in \{0, \ldots, l - 1\}$, $x_i x_{i+1}$ has size $s + 1$,

- for every $i \in \{l, \ldots, k-1\}$, $x_i x_{i+1}$ has size different from $s+1$.

The unique point x_l with that property is called the *turning point* of the nice path.

Property. Among all shortest paths between two points of $\mathcal{S}$, there always exists one which is nice.

Proof. Let x and y denote two points of $\mathcal{S}$ at mutual distance 2 and let (x, z, y) denote a shortest path between x and y which is not nice. By our assumptions, there exists a common neighbour z' of x and y different from z. Now, $xz' \| zy$ and $z'y \| xz$. Since parallel lines contain the same number of points, the line xz' has size $s+1$ and the line $z'y$ has size different from $s+1$. As a consequence, (x, z', y) is a nice path.

Let x and y be two points of $\mathcal{S}$ such that $\mathrm{d}(x, y) \geq 3$. Starting from an arbitrary shortest path between x and y, we can obtain a nice shortest path between x and y by successively altering a subpath of length 2 which is not nice to a nice path of length 2. □

Definition. For every $i \in \{1, 2\}$, put $P_i := \{F_x^{(i)} \,|\, x \in \mathcal{S}\}$. Then P_i is a partition of the point set of $\mathcal{S}$.

Property. Every element of P_1 intersects every element of P_2 in at least one point.

Proof. Let $F_x^{(1)}$ denote an arbitrary element of P_1 and let $F_y^{(2)}$ denote an arbitrary element of P_2. Consider a nice path between x and y and let z denote the turning point of that path. Then z is a common point of $F_x^{(1)}$ and $F_y^{(2)}$. □

Property. Let $(x_0, \ldots, x_k)$, $x_k = x_0$, denote a closed path in $\mathcal{S}$, and let L_i, $i \in \{0, \ldots, k-1\}$, denote the line defined by x_i and x_{i+1}. Then for every $j \in \{0, \ldots, k-1\}$, there exists a $j' \in \{0, \ldots, k-1\} \setminus \{j\}$ such that $L_j \| L_{j'}$.

Proof. Without loss of generality, we may suppose that $j = 0$. Suppose that none of the lines $L_1, \ldots, L_{k-1}$ is parallel with L_0. Since L_i, $i \in \{1, \ldots, k-1\}$, is not parallel with L_0, $\pi_{L_0}(x_i) = \pi_{L_0}(x_{i+1})$. So, we must have that $\pi_{L_0}(x_1) = \pi_{L_0}(x_k)$ or that $x_1 = x_0$, a contradiction. □

Property. Every element of P_1 intersects every element of P_2 in a unique point.

Proof. Let G_1 denote an arbitrary element of P_1 and let G_2 denote an arbitrary element of P_2. Suppose that x and y are two different points of $G_1 \cap G_2$. Let γ_i, $i \in \{1, 2\}$, denote a path between x and y in the collinearity graph of G_i, let γ denote a shortest path between x and y and let L denote a line defined by two successive points of γ. Let $L_i \neq L$, $i \in \{1, 2\}$, be a line parallel with L containing two successive points u_i and v_i of the closed path between $\gamma + \gamma_i$ (see previous property). Since γ is a shortest path, it is impossible that both u_i and v_i lie on γ. So, u_i and v_i are points of γ_i. Hence, L_1 has size $s+1$ and L_2 has size different from $s+1$. Since the lines L_1 and L_2 are parallel with L, the three lines L, L_1 and L_2 must contain the same number of lines, a contradiction. So, G_1 and G_2 have precisely one point in common. □

Property. Every element α of $P_1 \cup P_2$ is a convex subpolygon of $\mathcal{S}$.

Proof. Suppose that α is an element of P_1. Consider a shortest path $\gamma = (x_0, \ldots, x_k)$ between two vertices x_0 and x_k of α. Let γ' denote a path in α connecting x_0 and x_k. By the proof of the previous property, there exist two successive points u_i and v_i in γ' such that $x_i x_{i+1} \| u_i v_i$ ($i \in \{0, \ldots, k-1\}$). So, every line $x_i x_{i+1}$, $i \in \{0, \ldots, k-1\}$, contains $s+1$ points and γ is completely contained in α. As a consequence, α is convex. In a similar way one shows that every element of P_2 is convex. □

Definition. Define now the following graph $\Gamma_i = (P_i, E_i)$. Two different elements α and β of P_i are adjacent if and only if α contains a point which is collinear with a point of β.

Property. If α and β are adjacent in Γ_i, $i \in \{1, 2\}$, then every point of α is collinear with a point of β.

Proof. Let x and y be two collinear points of α. We will show that if $\mathrm{d}(x, \beta) = 1$, then also $\mathrm{d}(y, \beta) = 1$. The property then follows from the connectedness of α. Let x' denote a point of β collinear with x. Then $\mathrm{d}(x', y) = 2$ and x is a common neighbour of x' and y. Moreover, one of the lines $x'x$, xy has size $s+1$, while the other line has size different from $s+1$. Hence, x and y' have a second common neighbour y'. Since $x'y' \| xy$, $x'y'$ has the same number of points as xy. Hence, $y' \in \beta$. This proves the property. □

Property. Let $i \in \{1, 2\}$. Then two vertices α and β of Γ_i are adjacent if and only if $\mathrm{d}(\alpha \cap \gamma, \beta \cap \gamma) = 1$ for every $\gamma \in P_{3-i}$.

Proof. This follows immediately from the previous properties. □

Corollary. The point graph of every element of P_{3-i}, $i \in \{1, 2\}$, is isomorphic to Γ_i. As a consequence, all elements of P_{3-i} are isomorphic (to $\mathcal{S}_{3-i}$, say).

Let Γ denote the point graph of $\mathcal{S}$. The following property completes the proof of Theorem 1.12.

Property. The graphs Γ and $\Gamma_1 \times \Gamma_2$ are isomorphic.

Proof. Consider the bijection $\theta : x \mapsto (F_x^{(1)}, F_x^{(2)})$ from the vertex set of Γ to the vertex set of $\Gamma_1 \times \Gamma_2$. We will show that θ is an isomorphism. Let x and x' denote two arbitrary vertices of Γ. Then

$$\begin{aligned} x \sim x' \quad &\Leftrightarrow \quad \exists i \in \{1,2\}, F_x^{(i)} = F_{x'}^{(i)} \text{ and } \mathrm{d}(F_x^{(1)} \cap F_x^{(2)}, F_{x'}^{(1)} \cap F_{x'}^{(2)}) = 1, \\ &\Leftrightarrow \quad \exists i \in \{1,2\}, F_x^{(i)} = F_{x'}^{(i)} \text{ and } F_x^{(3-i)} \sim F_{x'}^{(3-i)}, \\ &\Leftrightarrow \quad \theta(x) \text{ and } \theta(x') \text{ are adjacent.} \end{aligned}$$

□

Corollary 1.13. *Let $\mathcal{S}$ be a near polygon which satisfies the following properties:*

- *not every line of $\mathcal{S}$ is incident with the same number of points,*
- *if two points x and y at mutual distance 2 have a common neighbour z such that the lines xz and yz have different sizes, then x and y have at least two common neighbours.*

If $s_1 + 1, \ldots, s_k + 1$ denote the $k \geq 2$ different line sizes which occur in $\mathcal{S}$, then there exist near polygons $\mathcal{S}_1, \ldots, \mathcal{S}_k$ which satisfy the following properties:

- *every line of $\mathcal{S}_i$ ($i \in \{1, \ldots, k\}$) has size $s_i + 1$,*
- *$\mathcal{S} \cong \mathcal{S}_1 \times \cdots \times \mathcal{S}_k$.*

Proof. We will prove this by induction on k. By Theorem 1.12, the corollary holds if $k = 2$. So, suppose $k \geq 3$. Again by Theorem 1.12, there exist near polygons $\mathcal{S}_1$ and $\mathcal{S}'$ satisfying the following properties: (i) every line of $\mathcal{S}_1$ is incident with s_1+1 points; (ii) no line of $\mathcal{S}'$ is incident with s_1+1 points; (iii) $\mathcal{S} \cong \mathcal{S}_1 \times \mathcal{S}'$. Now, the product near polygon $\mathcal{S}_1 \times \mathcal{S}'$ contains a convex subpolygon F isomorphic to $\mathcal{S}'$. If x and y are two points of F at distance 2 from each other having a common neighbour z such that xz and yz have different sizes, then x and y have at least two common neighbours in $\mathcal{S}_1 \times \mathcal{S}'$ and hence also in F. So, the induction hypothesis can be applied to $\mathcal{S}'$. The corollary now readily follows. □

Theorem 1.14. *If $\mathcal{S}$ is a generalized quadrangle, then precisely one of the following holds:*

- *$\mathcal{S}$ is degenerate: there exists a point in $\mathcal{S}$ which is incident with all lines of $\mathcal{S}$;*
- *$\mathcal{S}$ is a nonsymmetrical grid;*
- *$\mathcal{S}$ is a nonsymmetrical dual grid;*
- *$\mathcal{S}$ has an order (s, t).*

Proof. Suppose first that every two points at distance 2 have at least two common neighbours. Let x denote an arbitrary point of $\mathcal{S}$. Suppose that there exists only one line through x. Since $\mathcal{S}$ has diameter 2, there exists a point $y \in \Gamma_2(x)$. If z_1 and z_2 denote two different common neighbours of x and y, then the lines xz_1 and xz_2 are different, contrary to our assumption. Hence, every point of $\mathcal{S}$ is incident with at least two lines. It follows that the point-line dual of $\mathcal{S}$ is also a generalized quadrangle. If not every line of $\mathcal{S}$ is incident with the same number of points, then by Theorem 1.12, $\mathcal{S}$ is a grid. If not every point of $\mathcal{S}$ is incident with the same number of lines, then again by Theorem 1.12, $\mathcal{S}$ is the point-line dual of a grid.

Suppose next that there exist two noncollinear points x and y which have a unique common neighbour z. We will show that every point of $\mathcal{S}$ is collinear with z. Suppose the contrary and let u be a point noncollinear with z. Let v denote the unique point of zy collinear with u. Then $v \neq z$ and $x \notin uv$. Let K denote the unique line through x intersecting uv. Then K contains a unique point collinear with y contradicting the fact that x and y have only one common neighbour. So, every point is collinear with z. It now easily follows that z is contained in every line. So, no ordinary subquadrangles exist and $\mathcal{S}$ is degenerate. This proves the theorem. $\square$

1.7 Existence of quads

In this section, we will prove one of the main results in the theory of near polygons. Theorem 1.20 which is due to Shult and Yanushka [91] gives sufficient conditions for the existence of quads through two points at distance 2 from each other.

Let $\mathcal{S}$ be a near polygon. Let a and b denote two points of $\mathcal{S}$ at distance 2 from each other and let c and d denote two common neighbours of a and b such that at least one of the lines ac, ad, bc, bd contains at least three points. Let $X(a, b, c, d)$ denote the set of points at distance at most 2 from a, b, c and d.

Lemma 1.15. *$X(a, b, c, d)$ is a subspace.*

Proof. The set $\Gamma_{\leq 2}(a)$ of points at distance at most 2 from a is a subspace. (If a line L contains at least two points of $\Gamma_{\leq 2}(a)$, then it has distance at most 1 from a and hence every point of L belongs to $\Gamma_{\leq 2}(a)$.) Hence, $\Gamma_{\leq 2}(a) \cap \Gamma_{\leq 2}(b) \cap \Gamma_{\leq 2}(c) \cap \Gamma_{\leq 2}(d)$ is, as an intersection of subspaces, again a subspace. $\square$

Lemma 1.16. *Suppose c' is a point of the line ac different from a and let b' denote the unique point of the line bd collinear with c'. Then $X(a, b, c, d) = X(a, b', c', d)$.*

Proof. The points at distance at most 2 from a and c are the points at distance at most 1 from the line ac. Similarly, the points at distance at most 2 from b and d are the points at distance at most 1 from the line bd. The lemma now easily follows. $\square$

Lemma 1.17. *If x_1 and x_2 are points of $X(a, b, c, d)$, then there exist points a', b', c', d' for which the following holds:*

(i) *$d(a', b') = 2$ and c' and d' are two common neighbours of a' and b';*

(ii) *at least one of the lines $a'c', a'd', b'c', b'd'$ contains at least three points;*

(iii) *$X(a', b', c', d') = X(a, b, c, d)$;*

(iv) *for every $i \in \{1, 2\}$, x_i has distance at most 1 from $\{a', b', c', d'\}$;*

(v) *if $x_i \in a'c' \cup c'b' \cup b'd' \cup d'a'$ for a certain $i \in \{1, 2\}$, then $x_i \in \{a', b', c', d'\}$.*

Proof. We can construct the points a', b', c' and d' by applying Lemma 1.16 a few times. Let $\tilde{x}_i$, $i \in \{1, 2\}$, denote a point of $ac \cup cb \cup bd \cup da$ nearest to x_i. By applying Lemma 1.16 at most two times, we can find points a'', b'', c'' and d'' satisfying (i), (ii), (iii), (iv) and $\{\tilde{x}_1, \tilde{x}_2\} \subseteq \{a'', b'', c'', d''\}$. If also condition (v) is satisfied, then we are done. If condition (v) is not satisfied, then we can find the required points a', b', c', d' by applying Lemma 1.16 again at most two times. □

Lemma 1.18. *Every two points x_1 and x_2 of $X(a,b,c,d)$ lie at distance at most 2 from each other.*

Proof. By Lemma 1.17, we may suppose that the following holds for every $i \in \{1,2\}$:

- x_i has distance at most 1 from $\{a,b,c,d\}$
- if $x_i \in ac \cup cb \cup bd \cup da$, then $x_i \in \{a,b,c,d\}$.

The lemma obviously holds if $\{x_1,x_2\} \cap \{a,b,c,d\} \neq \emptyset$. So, we may suppose that $\{x_1,x_2\} \cap \{a,b,c,d\} = \emptyset$. Let y_i, $i \in \{1,2\}$, denote a point of $\{a,b,c,d\}$ collinear with x_i. We distinguish between the following possibilities.

- $\mathrm{d}(y_1,y_2) = 0$.
  Then we have $\mathrm{d}(x_1,x_2) \leq \mathrm{d}(x_1,y_1) + \mathrm{d}(y_1,x_2) = 2$.
- $\mathrm{d}(y_1,y_2) = 1$.
  Without loss of generality, we may suppose that $y_1 = a$ and $y_2 = c$. The point b has distance 2 from a and distance at most 2 from x_1. Hence, there exists a unique point z_1 on the line ax_1 collinear with b. Similarly, there exists a unique point z_2 on the line cx_2 collinear with d. Since $ad \| bc \| az_1 \| bd \| ac$, each of the lines ad, bc, az_1, bd and ac contains at least three points. Let u denote an arbitrary point of az_1 different from a and z_1. Since $az_1 \| bd$, there exists a unique point u' on bd collinear with u. Obviously, $d \neq u' \neq b$. Since $cz_2 \| bd$, there exists a unique point u'' on cz_2 collinear with u'. Obviously, $c \neq u'' \neq z_2$. If the lines ax_1 and cx_2 were not parallel, then we would have $\mathrm{d}(u,u'') = \mathrm{d}(u,a) + \mathrm{d}(a,c) + d(c,u'') = 3$, contradicting $\mathrm{d}(u,u'') \leq \mathrm{d}(u,u') + \mathrm{d}(u',u'') = 2$. So, the lines ax_1 and cx_2 are parallel. This implies that $\mathrm{d}(x_1,x_2) \leq 2$.
- $\mathrm{d}(y_1,y_2) = 2$.
  The point y_1 has distance 2 from y_2 and distance at most 2 from x_2. Hence, there exists a unique point on the line y_2x_2 collinear with y_1. Similarly, there exists a unique point on the line x_1y_1 collinear with y_2. It follows that either $x_1y_1 \cap x_2y_2 \neq \emptyset$ or $x_1y_1 \| x_2y_2$. In any case, $\mathrm{d}(x_1,x_2) \leq 2$. □

Lemma 1.19. *The set $X(a,b,c,d)$ is a quad.*

Proof. Take a nonincident point-line pair (x,L) with $x \in X(a,b,c,d)$ and $L \subseteq X(a,b,c,d)$. Since every point of L has distance at most 2 from x, L contains a unique point x' collinear with x. This proves that $X(a,b,c,d)$ is a generalized quadrangle. Since $X(a,b,c,d)$ contains an ordinary quadrangle as subgeometry, it is nondegenerate. It remains to show that $X(a,b,c,d)$ is geodetically closed. Let a' and b' be arbitrary points of $X(a,b,c,d)$ at distance 2 from each other, let c' and d' denote two common neighbours of a' and b' contained in $X(a,b,c,d)$ and let e'

denote a third common neighbour of a' and b'. The points a, b, c, d and e' belong to the subquadrangle $X(a', b', c', d')$. So, $e' \in \Gamma_{\leq 2}(a) \cap \Gamma_{\leq 2}(b) \cap \Gamma_{\leq 2}(c) \cap \Gamma_{\leq 2}(d)$ by Lemma 1.18. This proves that $X(a, b, c, d)$ is geodetically closed. $\square$

Theorem 1.20. *Let a and b denote two points of a near polygon at distance 2 from each other. Let c and d denote two common neighbours of a and b such that at least one of the lines ac, ad, bc, bd contains at least three points. Then a and b are contained in a unique quad. This quad coincides with $\mathcal{C}(a, b)$ and consists of all points of the near polygon which have distance at most 2 from a, b, c and d.*

Proof. Let Q denote a quad through a and b. Since Q is convex, c and d are contained in Q. Any point of Q has distance at most 2 from a, b, c and d. Hence, $Q \subseteq X(a, b, c, d)$. Every point x of $X(a, b, c, d)$ has distance at most 1 from the lines ac and bd. Since x has distance at most 1 from two different points of Q, x must be contained in Q. So, $X(a, b, c, d) \subseteq Q$. Hence, $X(a, b, c, d)$ is the unique quad through a and b. $\square$

1.8 The point-quad and line-quad relations

In this section, we determine the possible relations between a point and a quad and a line and a quad. These possible relations were first described in [91] and [16].

Theorem 1.21. *Let Q be a generalized quadrangle which is not a dual grid. Then, for every point x of Q, $\Gamma_2(x)$ is connected and has diameter at most 3.*

Proof. Let y_1 and y_2 denote two different points of $\Gamma_2(x)$. We distinguish between the following possibilities.

- $\mathrm{d}(y_1, y_2) = 1$. Then y_1 and y_2 have distance 1 in $\Gamma_2(x)$.
- $\mathrm{d}(y_1, y_2) = 2$ and there exists a point in $\Gamma_1(y_1) \cap \Gamma_1(y_2)$ not collinear with x. Then y_1 and y_2 have distance 2 in $\Gamma_2(x)$.
- $\mathrm{d}(y_1, y_2) = 2$ and every point of $\Gamma_1(y_1) \cap \Gamma_1(y_2)$ is collinear with x. Let z_1 and z_2 denote two different points of $\Gamma_1(y_1) \cap \Gamma_1(y_2)$ and let y_2' denote an arbitrary point of the line $y_2 z_2$ different from y_2 and z_2. Then y_2' is not collinear with z_1. Hence, the unique point y_1' on $y_1 z_1$ collinear with y_2' is contained in $\Gamma_2(x)$. Now, the path y_1, y_1', y_2', y_2 is completely contained in $\Gamma_2(x)$. So, y_1 and y_2 have distance 3 in $\Gamma_2(x)$. $\square$

Definition. An *ovoid* of a generalized quadrangle is a set of points which intersects each line in a unique point. A *fan of ovoids* is a partition in ovoids. A *rosette of ovoids* is a set of ovoids through a point partitioning the set of points at distance 2 from that point.

If Q is a finite generalized quadrangle of order (s, t), then every ovoid of Q contains $st + 1$ points, every fan of ovoids contains $s + 1$ elements and every rosette of ovoids contains s elements.

Theorem 1.22. *Let x be a point and Q a quad of a near polygon. Then precisely one of the following cases occurs.*

(1) *The point x is classical with respect to Q. We will also say that the pair (x, Q) is classical.*

(2) *The points in Q nearest to x form an ovoid in Q. In this case we say that x is* ovoidal *with respect to Q, or that (x, Q) is ovoidal.*

(3) *Q is a dual grid and the set of points of Q nearest to x contains at least two points and is a proper subset of one of the two ovoids of Q. In this case (x, Q) is called* thin ovoidal.

Proof. Put $d(x, Q) = \delta$ and let O_x denote the set of points of Q at distance δ from x. No two points of O_x are collinear. So, if Q is a dual grid, then we have one of the cases (1), (2) or (3). We may therefore suppose that Q is not a dual grid. If every point of Q has distance at most $\delta + 1$ from x, then every line of Q contains precisely one point at distance δ from x. So, O_x is an ovoid and we have case (2). So, suppose there exists a point y in Q at distance $\delta + 2$ from x. Let z denote a point of Q at distance δ from x. Then every point of Q at distance 1 from z has distance $\delta + 1$ from x. Suppose that y' is a point of $\Gamma_2(z) \cap Q$ collinear with y and let y'' denote the unique point of the line yy' collinear with z. Since $d(x, y'') = \delta + 1$ and $d(x, y) = \delta + 2$, $d(x, y') = \delta + 2$. So, every point of $\Gamma_2(z) \cap Q$ collinear with y has distance $\delta + 2$ from x. By Theorem 1.21, it now easily follows that every point of $\Gamma_2(z) \cap Q$ has distance $\delta + 2$ from x. So, we have case (1). □

Let Q be a quad of a near polygon of diameter d and suppose that Q is not a dual grid. For every $i \in \mathbb{N}$, let $X_i(Q)$ denote the set of points at distance i from Q, let $X_{i,C}(Q)$ denote the set of points of $X_i(Q)$ which are classical with respect to Q and let $X_{i,O}(Q)$ denote the set of points of $X_i(Q)$ which are ovoidal with respect to Q. If no confusion is possible, we will write X_i, $X_{i,C}$ and $X_{i,O}$ instead of $X_i(Q)$, $X_{i,C}(Q)$ and $X_{i,O}(Q)$. Clearly, $X_0 = Q$, $X_{1,O} = \emptyset$, $X_{d-1,C} = \emptyset$ and $X_i = \emptyset$ if $i \geq d$. If $x \in X_{i,C}(Q)$, then $\pi_Q(x)$ denotes the unique point of Q nearest to x.

Theorem 1.23. *Let Q be a quad of a near polygon $\mathcal{S}$ and suppose that Q is not a dual grid. Let $i \in \mathbb{N}$ and let L be a line of $\mathcal{S}$.*

(1) *If L contains points of X_i and X_{i+1}, then $|L \cap X_i| = 1$.*

(2) *No point of $X_{i,O}$ is collinear with a point of $X_{i,C}$.*

(3) *If L is a line contained in $X_{i,C}$, then $\pi_Q(L)$ is a line of Q parallel with L.*

(4) *If L is a line contained in $X_{i,O}$, then the points of L determine a fan of ovoids in Q.*

(5) *If L contains points of $X_{i,O}$ and X_{i+1}, then L is contained in $X_{i,O} \cup X_{i+1,O}$. In this case, all points of L determine the same ovoid of Q.*

(6) *If L contains points of $X_{i,C}$ and $X_{i+1,C}$, then all points of L determine the same point in Q.*

(7) *If L contains points of $X_{i,C}$ and $X_{i+1,O}$, then the points of $L \cap X_{i+1,O}$ determine a rosette of ovoids in Q. The common point of all these ovoids is the point $\pi_Q(L \cap X_{i,C})$.*

Proof. (1) Suppose the contrary. Let x and y denote two different points of $L \cap X_i$ and let z denote a point of X_{i+1}. Every point of $\Gamma_i(x) \cap \Gamma_i(y) \cap Q$ must have distance $i-1$ from a unique point of L, which is impossible. Hence, $\Gamma_i(x) \cap Q$ and $\Gamma_i(y) \cap Q$ are disjoint. Since both $\Gamma_i(x) \cap Q$ and $\Gamma_i(y) \cap Q$ are contained in $\Gamma_{i+1}(z) \cap Q$, $\Gamma_{i+1}(z) \cap Q$ must be an ovoid, $\Gamma_i(x) \cap Q$ must be a point x', $\Gamma_i(y) \cap Q$ must be a point y' and $\mathrm{d}(x', y') = 2$. Now, x is classical with respect to Q and hence $\mathrm{d}(x, y') = \mathrm{d}(x, x') + \mathrm{d}(x', y') = i + 2$. On the other hand, $\mathrm{d}(x, y') \leq \mathrm{d}(x, y) + \mathrm{d}(y, y') \leq i + 1$. So, our assumption was wrong and $|L \cap X_i| = 1$.

(2) Suppose the contrary. Let $x \in X_{i,C}$ be collinear with $y \in X_{i,O}$. Let x' denote the unique point of Q nearest to x and let O be the ovoid $\Gamma_i(y) \cap Q$. Every point of $\Gamma_2(x') \cap Q$ has distance $i + 2$ from x and hence distance $i + 1$ from y. So, every point of the ovoid O is contained in $\{x'\} \cup (\Gamma_1(x') \cap Q)$. This is not possible, since Q is not a dual grid.

(3) Let x and y be two different points of L. If $\pi_Q(x) = \pi_Q(y)$, then $\pi_Q(x)$ has distance i from two different points of L and hence distance $i-1$ from a point of L, a contradiction. Hence, $\pi_Q(x) \neq \pi_Q(y)$. Now, $\mathrm{d}(x, \pi_Q(y)) = \mathrm{d}(x, \pi_Q(x)) + \mathrm{d}(\pi_Q(x), \pi_Q(y))$ and so $\mathrm{d}(\pi_Q(x), \pi_Q(y)) = \mathrm{d}(x, \pi_Q(y)) - \mathrm{d}(x, \pi_Q(x)) \leq \mathrm{d}(x, y) + \mathrm{d}(y, \pi_Q(y)) - \mathrm{d}(x, \pi_Q(x)) = \mathrm{d}(x, y) = 1$. Hence, the points $\pi_Q(z)$, $z \in L$, are mutually collinear. Hence, there exists a line L' in Q containing all points $\pi_Q(z)$, $z \in L$. Obviously, L is parallel with L and $\pi_Q(L) = L'$.

(4) For every point x of L, $O_x := \Gamma_i(x) \cap Q$ is an ovoid of Q. If y is a point of Q, then $\mathrm{d}(y, x) \in \{i, i+1\}$ for every point $x \in L$. Hence, L contains a unique point at distance i from y. It follows that the ovoids O_x, $x \in L$, partition the point set of Q.

(5) Let x denote the unique point of L contained in $X_{i,O}$ and let y denote an arbitrary point of $L \cap X_{i+1}$. The ovoid $\Gamma_i(x) \cap Q$ is contained in $\Gamma_{i+1}(y) \cap Q$. Hence, $y \in X_{i+1,O}$ and $\Gamma_{i+1}(y) \cap Q = \Gamma_i(x) \cap Q$. This proves the property.

(6) Let x denote the unique point of L contained in $X_{i,C}$ and let y denote an arbitrary point of $X_{i+1,C} \cap L$. The point $\pi_Q(x)$ has distance at most $i + 1$ from y. Hence, $\pi_Q(y) = \pi_Q(x)$. This proves the property.

(7) Let x denote the unique point of L contained in $X_{i,C}$. For every point $y \in L \cap X_{i+1,O}$, put $O_y := \Gamma_{i+1}(y) \cap Q$. Obviously, $\pi_Q(x) \in O_y$. Every point z of $Q \cap \Gamma_2(\pi_Q(x))$ lies at distance $i + 2$ from x and distance at most $i + 2$ from any point of $L \setminus \{x\}$. It follows that z is contained in a unique ovoid O_y, $y \in L \setminus \{x\}$. □

1.9 Some classes of near polygons

1.9.1 Thin and slim near polygons

A near polygon is called *thin* if every line is incident with precisely two points. A near polygon is called *slim* if every line is incident with precisely three points.

Theorem 1.24. *The class of thin near polygons coincides with the class of connected bipartite graphs of finite diameter.*

Proof. If Γ is a connected bipartite graph, then the maximal cliques of Γ are the edges of Γ. For every vertex x and every edge $E = \{y, z\}$, $d(x, y)$ and $d(x, z)$ have different parity and hence exactly one of the points y and z is nearest to x. This proves that Γ is the point graph of a thin near polygon $\mathcal{S}$.

Conversely, suppose that $\mathcal{S}$ is a thin near polygon and let Γ denote its point graph. If $x_0, x_1, \ldots, x_k$ denotes a path of length $k \geq 1$ in Γ, then since the line $\{x_i, x_{i+1}\}$, $i \in \{0, \ldots, k-1\}$, contains a unique point nearest to x_0, $d(x_0, x_i)$ and $d(x_0, x_{i+1})$ have different parity. So, $d(x_0, x_k)$ and k must have the same parity. Hence, every closed path in Γ has even length. This implies that Γ is a bipartite graph. □

1.9.2 Dense near polygons

A near polygon is called *dense* if every line is incident with at least three points and if every two points at distance 2 have at least two common neighbours. We will discuss dense near polygons in detail in Chapter 2. A large part of this book is devoted to the classification of the slim dense near polygons (Chapters 6, 7, 8 and 9). If x and y are two points of a dense near polygon at distance 2 from each other, then by Theorem 1.20, x and y are contained in a unique quad. We will generalize this property in Theorem 2.3.

1.9.3 Regular near polygons

A finite near $2d$-gon $\mathcal{S}$ ($d \geq 1$) is called *regular* if it has an order (s, t) and if there exist constants t_i, $i \in \{0, \ldots, d\}$, such that for every two points x and y at distance i, there are precisely $t_i + 1$ lines through y containing a (necessarily unique) point at distance $i - 1$ from x. Obviously, $t_0 = -1$, $t_1 = 0$ and $t_d = t$. The numbers s, t, t_i ($i \in \{0, \ldots, d\}$) are called the *parameters* of $\mathcal{S}$.

A finite graph Γ of diameter $d \geq 1$ is called *distance-regular* if there exist constants a_i, b_i, c_i ($i \in \{0, \ldots, d\}$) such that $|\Gamma_i(x) \cap \Gamma_1(y)| = a_i$, $|\Gamma_{i+1}(x) \cap \Gamma_1(y)| = b_i$ and $|\Gamma_{i-1}(x) \cap \Gamma_1(y)| = c_i$ for any two vertices x and y at distance i from each other. Obviously, $a_0 = c_0 = b_d = 0$ and Γ is regular with valency $k = b_0$. Also, $a_i + b_i + c_i = k$ for every $i \in \{0, \ldots, d\}$. The numbers a_i, b_i, c_i ($i \in \{0, \ldots, d\}$) are called the *parameters* of the distance-regular graph.

Theorem 1.25. *The regular near $2d$-gons, $d \geq 1$, are precisely those near $2d$-gons whose point graph is distance-regular.*

Proof. Suppose that $\mathcal{S}$ is a regular near $2d$-gon with parameters s, t and t_i ($i \in \{0, \ldots, d\}$). Then the point graph of $\mathcal{S}$ is regular with parameters $a_i = (s-1)(t_i + 1)$, $b_i = s(t - t_i)$ and $c_i = t_i + 1$ ($i \in \{0, \ldots, d\}$).

Conversely, suppose that $\mathcal{S}$ is a near $2d$-gon whose point graph is a distance-regular graph with parameters a_i, b_i, c_i ($i \in \{0, \ldots, d\}$). Then every line of $\mathcal{S}$ contains $s + 1 := a_1 + 2$ points. Since every point of $\mathcal{S}$ is collinear with b_0 other points, every point of $\mathcal{S}$ is incident with $t+1 := \frac{b_0}{s}$ lines. If x and y are two points of $\mathcal{S}$ at distance $i \in \{0, \ldots, d\}$ from each other, then c_i lines through y contain a (necessarily unique) point at distance $i - 1$ from x. This proves that $\mathcal{S}$ is a regular near polygon. □

We refer to [13] for more background information on distance-regular graphs. We will give a more extensive treatment of regular near polygons in Chapter 3.

1.9.4 Generalized polygons

A generalized $2n$-gon, $n \geq 2$, is a near $2n$-gon with the property that $|\Gamma_{i-1}(x) \cap \Gamma_1(y)| = 1$ for every $i \in \{1, \ldots, n-1\}$ and for every two points x and y at distance i from each other. A generalized $2n$-gon, $n \geq 2$, is called *thick* if every line is incident with at least three points and if every point is incident with at least three lines. A generalized $2n$-gon is called *degenerate* if it does not contain an ordinary $2n$-gon as subgeometry. A degenerate generalized quadrangle consists of a number of lines through a given point. The point-line dual of a nondegenerate generalized quadrangle is again a nondegenerate quadrangle.

Theorem 1.26 ([100, Corollary 1.5.3]). *Let $\mathcal{S}$ be a generalized $2n$-gon with the property that every line is incident with at least three points and that every point is incident with at least three lines. Then $\mathcal{S}$ has an order.*

Theorem 1.27 ([71]). *Let $\mathcal{S}$ be a finite generalized $2n$-gon of order (s,t). Then at least one of the following holds:*

(a) *$\mathcal{S}$ is an ordinary $2n$-gon.*

(b) *$\mathcal{S}$ is a generalized quadrangle.*

(c) *$\mathcal{S}$ is a generalized hexagon. In this case, st is a square if $s \neq 1 \neq t$.*

(d) *$\mathcal{S}$ is a generalized octagon. In this case, $2st$ is a square if $s \neq 1 \neq t$.*

(e) *$\mathcal{S}$ is a generalized dodecagon. In this case, $s = 1$ or $t = 1$.*

Theorem 1.28. *Let $\mathcal{S}$ be a finite generalized $2n$-gon of order (s,t) with $s, t, n \geq 2$. Then the following holds:*

- ([75]) *if $n = 2$, then $s \leq t^2$ and $t \leq s^2$ (Higman's bound);*

- ([74]) *if* $n = 3$, *then* $s \leq t^3$ *and* $t \leq s^3$ *(Inequality of Haemers and Roos);*
- ([75]) *if* $n = 4$, *then* $s \leq t^2$ *and* $t \leq s^2$ *(Higman's bound).*

If Q is a generalized quadrangle of order (s,t), then Q contains $(s+1)(st+1)$ points and $(t+1)(st+1)$ lines. We have the following restriction on the parameters.

Theorem 1.29 (e.g. Theorem 1.2.2 of [82]). *If Q is a finite generalized quadrangle of order (s,t), then $s+t$ divides $st(s+1)(t+1)$.*

Theorem 1.30. *If Q' is a subquadrangle of order (s,t') of a finite generalized quadrangle Q of order (s,t), $s \neq 1$, then either $t' = t$ or $t' \leq \frac{t}{s}$. If $t' = \frac{t}{s}$, then every line of Q meets Q'.*

Proof. Suppose that Q' is a proper subquadrangle of Q and let x denote an arbitrary point of Q not contained in Q'. The points in Q' collinear with x form an ovoid of Q'. From the $t+1$ lines through x, $st'+1$ meet Q'. The theorem now immediately follows. □

1.9.5 Dual polar spaces

In this section, we mean by a *projective space* an incidence structure with the property that any three noncollinear points generate a (possibly degenerate) projective plane. It is called *irreducible* when there are no lines of size 2; otherwise it is called *reducible.* Now, let $(\mathcal{P}_i)_{i \in I}$ be a family of irreducible projective spaces whose point sets are pairwise disjoint. Then the union $\mathcal{P}$ of their point sets carries the structure of a projective space whose lines are the lines on each $\mathcal{P}_i$ on the one hand and all pairs $\{x_i, y_j\}$ with $x_i \in \mathcal{P}_i, y_j \in \mathcal{P}_j, i \neq j$, on the other hand. $\mathcal{P}$ is called the *direct sum* of all the $\mathcal{P}_i$'s. It is known that a reducible projective space is the direct sum of irreducible projective spaces.

Definition. A *polar space of rank* $n \geq 1$ is a set together with a set of subsets, called *subspaces*, satisfying the following axioms.

(P1) Any proper subspace, together with the subspaces it contains, is a projective space of dimension at most $n-1$. This projective space may be reducible.

(P2) The intersection of two subspaces is again a subspace.

(P3) If L is a subspace of dimension $n-1$ and if p is a point outside L, then there is a unique subspace M through p such that $\dim(L \cap M) = n-2$; it contains all points q of L with the property that there is a one-dimensional subspace through p and q.

(P4) There exist two disjoint subspaces of dimension $n-1$.

Polar spaces of rank 2 are precisely the nondegenerate generalized quadrangles. With each polar space of rank $n \geq 1$, there is associated a *dual polar space* as follows. The points of the dual polar space are the maximal subspaces of the polar space (i.e. the subspaces of dimension $n-1$), the lines are the next-to-maximal subspaces (i.e. the subspaces of dimension $n-2$) and incidence is reverse containment. Each dual polar space of rank n is a near $2n$-gon by [21] or [91], see also Section 6.1. By convention, the unique near 0-gon is a dual polar space of rank 0.

If α is an $(n-1-i)$-dimensional subspace of a polar space Γ of rank n $(0 \leq i \leq n)$, then the set of all maximal subspaces through α defines a convex subspace of diameter i of the dual polar space $\mathcal{S}$ associated with Γ. Conversely, every convex subspace of $\mathcal{S}$ is obtained in this way. A dual polar space is called *thick* if all its quads (i.e. convex subspaces of diameter 2) are thick generalized quadrangles.

Given a number of polar spaces, many others can be constructed. Let $(\Gamma_i)_{i\in I}$ be a family of polar spaces defined on the sets $(P_i)_{i\in I}$ and with $(\mathcal{A}_i)_{i\in I}$ as a collection of subspaces. We suppose that all P_i's are mutually disjoint. A new polar space, called the *direct sum*, can then be constructed on the set $P = \bigcup_{i\in I} P_i$. A general form of a subspace is as follows: $\bigcup_{i\in I} \alpha_i$ where $\alpha_i \in \mathcal{A}_i$. A polar space is called *irreducible* if it is not isomorphic to a direct sum of at least two polar spaces of rank at least 1. Let Γ_1, Γ_2 be two polar spaces and let Γ_3 denote the direct sum of Γ_1 and Γ_2. If $\mathcal{S}_i$, $i \in \{1,2,3\}$, denotes the dual polar space related to Γ_i, then $\mathcal{S}_3 \cong \mathcal{S}_1 \times \mathcal{S}_2$.

Dual polar spaces can be characterized in a nice way.

Definition. A near polygon is called *classical* if it satisfies the following properties:

- every two points at distance 2 are contained in a unique quad,
- every point-quad pair is classical.

Theorem 1.31 ([21]). *The classical near polygons are precisely the dual polar spaces.*

By Tits' classification of polar spaces ([97]), every finite irreducible polar space of rank $n \geq 3$ without lines of size 2 is isomorphic to one of the following examples:

- $W(2n-1,q)$: the polar space with as subspaces the totally isotropic subspaces of a symplectic polarity in $\mathrm{PG}(2n-1,q)$;
- $Q(2n,q)$: the polar space with as subspaces the subspaces of $\mathrm{PG}(2n,q)$ lying on a given nonsingular quadric;
- $Q^-(2n+1,q)$: the polar space with as subspaces the subspaces of $\mathrm{PG}(2n+1,q)$ lying on a given nonsingular elliptic quadric;
- $Q^+(2n-1,q)$: the polar space with as subspaces the subspaces of $\mathrm{PG}(2n-1,q)$ lying on a given nonsingular hyperbolic quadric;
- $H(2n,q^2)$: the polar space with as subspaces the subspaces of $\mathrm{PG}(2n,q^2)$ lying on a given nonsingular hermitian variety;

- $H(2n-1,q^2)$: the polar space with as subspaces the subspaces of $\mathrm{PG}(2n-1,q^2)$ lying on a given nonsingular hermitian variety.

We denote the dual polar spaces arising from these polar spaces respectively by $DW(2n-1,q)$, $DQ(2n,q)$, $DQ^-(2n+1,q)$, $DQ^+(2n-1,q)$, $DH(2n,q^2)$ and $DH(2n-1,q^2)$. All these dual polar spaces are regular near $2n$-gons with parameters s and $t_i := \frac{t_2^i - t_2}{t_2 - 1}$, $i \in \{0,\ldots,n\}$.

dual polar space	(s,t_2)
$DW(2n-1,q)$	(q,q)
$DQ(2n,q)$	(q,q)
$DQ^-(2n+1,q)$	(q^2,q)
$DQ^+(2n-1,q)$	$(1,q)$
$DH(2n,q^2)$	(q^3,q^2)
$DH(2n-1,q^2)$	(q,q^2)

The dual polar spaces $DW(2n-1,q)$ and $DQ(2n,q)$ are isomorphic if and only if q is even.

By the above classification it follows that every classical slim near polygon of diameter at least 1 is isomorphic to the direct product of $k \geq 1$ elements of the set $\{\mathbb{L}_3\} \cup \{DQ(2n,2) \,|\, n \geq 2\} \cup \{DH(2n-1,4) \,|\, n \geq 2\}$.

1.10 Generalized quadrangles of order $(2,t)$

In this section, we will determine all (possibly infinite) generalized quadrangles of order $(2,t)$. We determine all ovoids of these generalized quadrangles and derive some properties which we will need later.

1.10.1 Examples

We now list some examples of generalized quadrangles of order $(2,t)$.

- The (3×3)-grid, which is the direct product of two lines of size 3, is a generalized quadrangle of order $(2,1)$.
- (i) The dual polar space $DQ(4,2)$ is a generalized quadrangle of order 2.
  (ii) The polar space $Q(4,2)$, see Section 1.9.5, is a generalized quadrangle of order 2.
  (iii) The polar space $W(2) := W(3,2)$ is a generalized quadrangle of order 2.
  (iv) For any set X of size 6, we can define a generalized quadrangle Q_X of order 2 in the following way, see Sylvester [92]. The points of Q_X are the elements of $\binom{X}{2}$, the lines of Q_X are the partitions of X in three subsets of size 2 and the incidence relation is containment.

  As we will see in Theorem 1.35, the four above-mentioned examples of generalized quadrangles of order 2 are isomorphic.

- (i) The dual polar space $DH(3,4)$ is a generalized quadrangle of order $(2,4)$. (ii) The polar space $Q(5,2) := Q^-(5,2)$ is a generalized quadrangle of order $(2,4)$.

As we will see in Theorem 1.37, the two above-mentioned examples of generalized quadrangles of order $(2,4)$ are isomorphic.

1.10.2 Possible orders

Theorem 1.32. *Let Q be a (possibly infinite) generalized quadrangle in which each line is incident with precisely three points. Then Q is finite and its order is equal to either $(2,1)$, $(2,2)$ or $(2,4)$.*

Proof. Let x be a given point of Q. For every point $y \in \Gamma_2(x)$, we define $A(y) := \Gamma_1(x) \cap \Gamma_1(y)$ and $\overline{A(y)} := \Gamma_1(x) \setminus A(y)$. If y and y' are two collinear points of $\Gamma_2(x)$, then $A(y) \cap A(y')$ consists of the unique point of yy' collinear with x. By Theorem 1.21, the diameter of $\Gamma_2(x)$ is at most 3. So, if y and y' are two points of $\Gamma_2(x)$, we have one of the following possibilities:

- $y = y'$. Then $|A(y) \cap \overline{A(y')}| = 0$.
- $y \sim y'$. Then $|A(y) \cap A(y')| = 1$.
- y and y' have distance 2 in $\Gamma_2(x)$. Let y'' denote a point of $\Gamma_2(x)$ collinear with y and y'. Let L_1 and L_2 denote the lines through x meeting yy'' and $y''y'$, respectively, and let x_i, $i \in \{1,2\}$, denote the point of L_i collinear with y. Then $A(y) \cap \overline{A(y')} = \{x_1, x_2\}$ and hence $|A(y) \cap \overline{A(y')}| = 2$.
- y and y' have distance 3 in $\Gamma_2(x)$. Let y'' and y''' denote points of $\Gamma_2(x)$ such that $y \sim y'' \sim y''' \sim y'$. Let L_1, L_2 and L_3 denote the lines through x meeting yy'', $y''y'''$ and $y'''y'$, respectively, and let x_i, $i \in \{1,2,3\}$, denote the point of L_i collinear with y. Then $A(y) \cap A(y') \subseteq \{x_1, x_2, x_3\}$ and hence $|A(y) \cap A(y')| \leq 3$.

Hence, we always have that $|A(y) \cap A(y')| \leq 3$ or $|A(y) \cap \overline{A(y')}| \leq 2$ for all $y, y' \in \Gamma_2(x)$.

Now, suppose t is infinite. Let u_0 and u_1 be two distinct collinear points of $\Gamma_2(x)$ and let L_1 denote the unique line through x meeting u_0u_1. We will now define in an inductive way a path $u_0, u_1, u_2, u_3, \ldots$ in $\Gamma_2(x)$. For every $i \geq 2$, let L_i denote a line through x different from the lines $L_1, \ldots, L_{i-1}$ and let u_i denote the unique point of $\Gamma_1(u_{i-1}) \cap \Gamma_2(x)$ such that the line $u_{i-1}u_i$ meets L_i. For every $i \in \mathbb{N} \setminus \{0\}$, let x_i denote the unique point of L_i collinear with u_0. One easily shows by induction that $A(u_0) \cap \overline{A(u_{2i})} = \{x_1, \ldots, x_{2i}\}$ and $A(u_0) \cap A(u_{2i+1}) = \{x_1, \ldots, x_{2i+1}\}$ for every $i \in \mathbb{N}$.

So, if t would not be finite, then there would exist points u and v in $\Gamma_2(x)$ such that $|A(u) \cap \overline{A(v)}| = 4$ and $|A(u) \cap A(v)|$ is infinite. This is impossible. Hence, t is finite. From Theorems 1.28 and 1.29 it now follows that $t \in \{1,2,4\}$. $\square$

1.10.3 Generalized quadrangles of order $(2, 1)$

Lemma 1.33. *Let Q be a generalized quadrangle of order $(2, t)$, $t \in \{1, 2, 4\}$, then every two disjoint lines are contained in a subquadrangle of order $(2, 1)$.*

Proof. Let K and L denote two disjoint lines of Q, let M_1, M_2 and M_3 denote the three lines of Q meeting K and L and let x_i, $i \in \{1, 2, 3\}$, denote the unique point of M_i not contained on K and L. Then it is easily seen that any two points of $\{x_1, x_2, x_3\}$ are collinear. So, $\{x_1, x_2, x_3\}$ is a line. This proves the lemma. □

Corollary 1.34. *There exists a unique generalized quadrangle of order $(2, 1)$, namely the (3×3)-grid.*

In the sequel of this section, we will call a subquadrangle of order $(2, 1)$ briefly a subgrid.

1.10.4 Generalized quadrangles of order 2

Theorem 1.35. *There exists a unique generalized quadrangle of order 2.*

Proof. Since $W(2)$ is a generalized quadrangle of order 2, there exists at least one such GQ. Let Q denote a GQ of order 2, then Q contains a subgrid G by Lemma 1.33. If x is a point of Q not contained in G, then the points in G collinear with x form an ovoid of G. It is now easily seen that there exists a bijective correspondence between the six points of Q not contained in G and the six ovoids of G. So, we have given a description of the points of Q in terms of the points, lines and ovoids of a (3×3)-grid G. The same is possible for the lines of Q as we will show now. The generalized quadrangle Q has 15 lines, 6 lines are contained in G and the other 9 lines intersect G in a unique point. Every point x of G is contained in a unique line L_x not contained in G and the two points of L_x different from x correspond with the two ovoids of G through x. The theorem now readily follows. □

1.10.5 Generalized quadrangles of order $(2, 4)$

A proof for the uniqueness of the generalized quadrangle of order $(2, 4)$ was independently obtained by several people ([67], [73], [86], [89], [93]).

Lemma 1.36. *Let Q be a generalized quadrangle of order $(2, 4)$, let G be a subgrid of Q and let x be a point of Q not contained in G. Then Q has a unique subquadrangle isomorphic to $W(2)$ containing G and x. As a consequence, G is contained in three subquadrangles isomorphic to $W(2)$.*

Proof. Let x_1, x_2 and x_3 denote the three points of G collinear with x and let y_i, $i \in \{1, 2, 3\}$, denote the unique third point of the line xx_i. Let $\{a^1_{12}, a^1_{13}, a^1_{23}\}$ and $\{a^2_{12}, a^2_{13}, a^2_{23}\}$ denote the two ovoids of G disjoint from $\{x_1, x_2, x_3\}$. Without loss of generality, we may suppose that $x_j \sim a^i_{jk} \sim x_k$ for every $i \in \{1, 2\}$ and all $j, k \in \{1, 2, 3\}$ with $j \neq k$. Now, let $i \in \{1, 2\}$ and let $\{j, k, l\} = \{1, 2, 3\}$. Since a^i_{jk}

is not collinear with x and x_l, it is collinear with y_l and we denote the unique third point of the line through a^i_{jk} and y_l by b^i_{jk}. We will now show that $b^i_{12} = b^i_{13} = b^i_{23}$ for every $i \in \{1,2,3\}$. Consider the two lines $a^i_{12}y_3$ and $a^i_{13}y_2$. Since $a^i_{12} \not\sim a^i_{13}$, $a^i_{12} \not\sim y_2$, $y_3 \not\sim a^i_{13}$ and $y_3 \not\sim y_2$, the point b^i_{12} is collinear with a^i_{13} and y_2 and hence belongs to the line $a^i_{13}y_2$. It follows that $b^i_{12} = b^i_{13}$. Hence, there exists a point b_i such that $b_i = b^i_{12} = b^i_{13} = b^i_{23}$. The lemma now easily follows. The unique subquadrangle of order 2 through x and G consists of the 9 points of G, the points $x, y_1, y_2, y_3, b_1, b_2$, the six lines of G, the lines xx_i ($i \in \{1,2,3\}$) and the lines $a^i_{jk}b_i$ ($i \in \{1,2\}$, $j,k \in \{1,2,3\}$, $j \neq k$). □

Theorem 1.37. *There exists a unique generalized quadrangle of order* $(2,4)$.

Proof. Let G denote a subgrid of Q and let R_1, R_2 and R_3 denote the three subquadrangles through G which are isomorphic to $W(2)$. With every point x of Q not contained in G, we associate the pair (O_x, i_x) where O_x is the ovoid $\Gamma_1(x) \cap G$ of G and $i_x \in \{1,2,3\}$ such that R_{i_x} contains x. There are 18 pairs (O,i) where O is an ovoid of G and $i \in \{1,2,3\}$, and these pairs are in bijective correspondence with the 18 points of Q not contained in G. Just as in the proof of Theorem 1.35, the lines which intersect G can easily be described in terms of the points, lines and ovoids of the grid G and the elements of the set $\{1,2,3\}$. The same holds for the lines of Q disjoint with G. Any such line looks like $\{(O_1,i_1),(O_2,i_2),(O_3,i_3)\}$, where $\{O_1,O_2,O_3\}$ is a fan of ovoids of G and $\{i_1,i_2,i_3\} = \{1,2,3\}$. This proves the theorem. □

Lemma 1.38. *A subgrid* G_1 *of* $Q(5,2)$ *defines a unique partition* $\{G_1,G_2,G_3\}$ *of* $Q(5,2)$ *into three subgrids.*

Proof. For a point x of $Q := Q(5,2)$, let $x^\perp$ denote the set of points of Q collinear with x. Call two vertices $x, y \in Q \setminus G_1$ equivalent if $x^\perp \cap G_1$ and $y^\perp \cap G_1$ are equal or disjoint. There are two equivalence classes C_2 and C_3 each containing 9 points. A point $x \in C_i$ is contained in three lines meeting G_1 and two lines which are entirely contained in C_i. So, each C_i contains $\frac{9 \cdot 2}{3} = 6$ lines. Clearly, a grid G_i is formed by the 9 points and 6 lines in C_i. The uniqueness of $\{G_1,G_2,G_3\}$ is also obvious. □

Remark. In the sequel the generalized quadrangles $\mathbb{L}_3 \times \mathbb{L}_3$, $W(2)$ and $Q(5,2)$ will often occur as quads in other near polygons. We will refer to them as *grid-quads*, $W(2)$*-quads* and $Q(5,2)$*-quads*, respectively.

1.10.6 Ovoids in generalized quadrangles of order $(2,t)$

For all generalized quadrangles of order $(2,t)$, we list all ovoids, fans of ovoids and rosettes of ovoids.

- Let Q be the generalized quadrangle of order $(2,1)$. So, Q is isomorphic to the (3×3)-grid. Clearly, Q has 6 ovoids, 2 fans of ovoids and 9 rosettes of ovoids.

- Let Q be the generalized quadrangle of order 2. We consider Sylvester's model for $W(2)$. So, the points of Q are the subsets of size 2 of a certain set X of size 6. If $x \in X$, then the five subsets of size 2 containing x define an ovoid O_x of Q. Since every two different points of an ovoid correspond with subsets of size 2 which intersect in a point, it is easily seen that every ovoid of Q is of the form O_x for a certain point x in X. So, Q has six ovoids and any two different ovoids intersect in a unique point. There are 15 rosettes of ovoids but no fan of ovoids.

- Let Q be the generalized quadrangle of order $(2,4)$. We will now show that Q has no ovoids. Suppose the contrary and let O denote an ovoid of Q. Let G denote a subgrid of Q and let R_1, R_2 and R_3 denote the three subquadrangles through G which are isomorphic to $W(2)$. The ovoid O intersects each of the subquadrangles G, R_1, R_2 and R_3 in an ovoid of the respective quadrangles. So, $|G \cap O| = 3$, $|G \cap R_1| = |G \cap R_2| = |G \cap R_3| = 5$. The ovoid in R_i, $i \in \{1,2,3\}$, is completely determined by $O \cap G$. The two points of $O \cap R_i$ which are not contained in G are the two points x of $R_i \setminus G$ for which $\Gamma_1(x) \cap G$ is an ovoid of G disjoint from $O \cap G$. Let x denote a point of R_1 for which $\Gamma_1(x) \cap G = O \cap G$ and let $\{x, y_1, y_2\}$ denote a line through x not intersecting G. Then $\{\Gamma_1(x) \cap G, \Gamma_1(y_1) \cap G, \Gamma_1(y_2) \cap G\}$ is a fan of ovoids of G. Since $\Gamma_1(y_i) \cap G$, $i \in \{1,2\}$, is disjoint from $O \cap G$, the point y_i belongs to O. So, $|L \cap O| = 2$, a contradiction.

Lemma 1.39. *Let O be an ovoid of the generalized quadrangle $W(2)$. Then there exists a cycle of length 5 in $W(2) \setminus O$.*

Proof. We take Sylvester's model for $W(2)$ using the set $X = \{1,2,3,4,5,6\}$. Without loss of generality, we may suppose that $O = \{\{1,2\},\{1,3\},\{1,4\},\{1,5\},\{1,6\}\}$. Then the cycle $\{2,3\},\{4,5\},\{3,6\},\{2,4\},\{5,6\},\{2,3\}$ satisfies all required properties. □

Theorem 1.40. *Let x be a point of the generalized quadrangle $Q(5,2)$. Then $\Gamma_2(x)$ contains a cycle of length 5.*

Proof. Let R be a subquadrangle of $Q(5,2)$ isomorphic to $W(2)$ and not containing x. Such a subquadrangle exists by Lemma 1.36. The points in R collinear with x form an ovoid O of R. By Lemma 1.39, there exists a path of length 5 completely contained in $R \setminus O$. This path is completely contained in $\Gamma_2(x)$. □

Chapter 2

Dense near polygons

2.1 Main results

A near polygon is called *dense* if every line is incident with at least three points and if every two points at distance 2 have at least two common neighbours. The aim of this chapter is to summarize the various nice properties which are satisfied by dense near polygons. From Theorem 1.20, we immediately have:

Theorem 2.1. *Let $\mathcal{S}$ be a dense near polygon. Then*

(a) *every two points at distance 2 are contained in a unique quad;*

(b) *every two intersecting lines are contained in a unique quad;*

(c) *every local space is linear.*

Using the existence of quads, one can easily show the following.

Theorem 2.2. *If $\mathcal{S}$ is a dense near polygon, then every point of $\mathcal{S}$ is incident with the same number of lines.*

Proof. If Q is a quad of $\mathcal{S}$, then by Theorem 1.14, every point of Q is contained in the same number of lines, say $t_Q + 1$. Now, take two collinear points x and y. The number of lines through x is equal to $1 + \sum t_Q$, where the summation ranges over all quads through xy. The number $1 + \sum t_Q$ is also the total number of lines through y. Hence, any two collinear points of $\mathcal{S}$ are contained in the same number of lines. The theorem now follows by connectedness of $\mathcal{S}$. □

If $\mathcal{S}$ is a dense near polygon, then we denote by $t_{\mathcal{S}} + 1$ the constant number of lines through a point. In Section 2.2 we will prove the following Theorems 2.3, 2.4 and 2.5. These results are due to Brouwer and Wilbrink [16].

Theorem 2.3. *If x and y are two points of a dense near polygon $\mathcal{S}$, then $\mathcal{C}(x,y)$ is the unique convex sub-$[2 \cdot d(x,y)]$-gon through x and y.*

Convex subhexagons of a dense near polygon are also called *hexes*. Using Theorem 2.3, we will show the following result.

Theorem 2.4. *Let $\mathcal{S}$ be a dense near polygon, let H be a convex sub-2δ-gon of $\mathcal{S}$ and let L be a line of $\mathcal{S}$ intersecting H in a unique point. Then $\mathcal{C}(H, L)$ is a convex sub-$2(\delta+1)$-gon.*

With every dense near $2n$-gon, $n \geq 2$, there is associated a rank-n-geometry. We refer to Pasini [79] for more background information on rank-n-geometries.

Theorem 2.5. *Let $\mathcal{S}$ be a dense near $2d$-gon, $d \geq 2$. Let Δ be the rank d geometry whose elements of type $i \in \{1, \ldots, d\}$ are the convex sub-$2(i-1)$-gons of $\mathcal{S}$ (natural incidence). Then Δ belongs to the following diagram.*

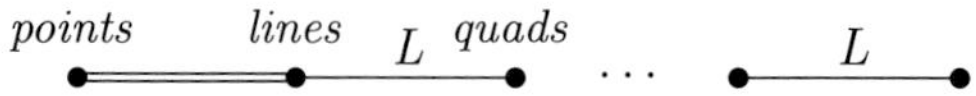

points *lines* L *quads* ... L

Definition. Let $\mathcal{S}$ be a dense near $2n$-gon, $n \geq 3$, and let x be a point of $\mathcal{S}$. Let Δ be the rank n geometry associated with $\mathcal{S}$, see Theorem 2.5. Then the objects of Δ which are incident with x form a rank $n-1$ geometry $\mathcal{G}(\mathcal{S}, x)$, called the *local geometry at the point x*. $\mathcal{G}(\mathcal{S}, x)$ is the so-called *residue* of x ([79]).

In this chapter, we will also prove the following results.

Theorem 2.6 (Section 2.3). *Let $\mathcal{S}$ be a finite dense near $2d$-gon. Then $|\Gamma_i(x)|$ only depends on $i \in \mathbb{N}$ and not on the chosen point x of $\mathcal{S}$.*

Theorem 2.7 (Section 2.4). *If x is a point of a dense near $2d$-gon, then (the subgraph induced by) $\Gamma_d(x)$ is connected and has diameter at most $\lfloor \frac{3d}{2} \rfloor$.*

In this chapter, we will also consider the special case of dense near polygons with three points on each line. For these near polygons, we will derive upper bounds for the constant number of lines through a point (Section 2.5) and we will prove a theorem in Section 2.6 which will be very useful later when we will classify slim dense near polygons which contain a given big convex subpolygon.

2.2 The existence of convex subpolygons

In this section we prove the existence of convex subpolygons in dense near polygons. We follow more or less the approach of [16].

Let $\mathcal{S} = (\mathcal{P}, \mathcal{L}, \mathrm{I})$ be a dense near $2d$-gon with $d \geq 2$. For all two points x and y of $\mathcal{S}$, let $S(x, y)$ denote the set of lines through x containing a point at distance $\mathrm{d}(x, y) - 1$ from y. If $\mathrm{d}(x, y) = 0$, then $S(x, y) = \emptyset$. If $\mathrm{d}(x, y) = 1$, then $S(x, y) = \{xy\}$. If $\mathrm{d}(x, y) = d$, then $S(x, y) = \mathcal{L}(\mathcal{S}, x)$.

Lemma 2.8. *For every two different points x and y of $\mathcal{S}$, $S(x, y)$ is a subspace of $\mathcal{L}(\mathcal{S}, x)$.*

Proof. Let L_1 and L_2 denote two different lines through x belonging to $S(x, y)$ and let Q denote the unique quad through L_1 and L_2. Since Q contains at least one point at distance $\mathrm{d}(x, y)$ from y and at least two points at distance $\mathrm{d}(x, y) - 1$ from y, one of the following possibilities occurs:

(a) y is ovoidal with respect to Q and $\mathrm{d}(y, Q) = \mathrm{d}(x, y) - 1$;

(b) y is classical with respect to Q and $\mathrm{d}(x, \pi_Q(y)) = 2$.

In any case each line through x contained in Q contains a point at distance $\mathrm{d}(x, y) - 1$ from y. This proves indeed that $S(x, y)$ is a subspace of $\mathcal{L}(\mathcal{S}, x)$. □

Lemma 2.9. *If x, y and y' are points such that $d(y, y') = 1$ and $d(x, y) = d(x, y')$, then $S(x, y) = S(x, y')$.*

Proof. Put $i := \mathrm{d}(x, y) \geq 1$. Let y'' denote the unique point of yy' at distance $i - 1$ from x. By symmetry it suffices to show that $S(x, y) \subseteq S(x, y')$. Let L denote an arbitrary line through x belonging to $S(x, y)$ and let x' denote the unique point of L nearest to y''. We distinguish two possibilities:

(a) $x' \neq x$. Then $\mathrm{d}(y'', x') = i - 2$, $\mathrm{d}(y, x') = \mathrm{d}(y', x') = i - 1$. So, L also belongs to $S(x, y')$.

(b) $x' = x$. Then the projections of y and y'' on L are different. So, L is parallel with yy''. Hence, $\mathrm{d}(y', L) = \mathrm{d}(y, L) = i - 1$, proving that $L \in S(x, y')$. □

Lemma 2.10. *If x, y and z are points of $\mathcal{S}$ such that $d(y, z) = 1$ and $d(x, z) = d(x, y) + 1$, then $S(x, y)$ is a proper subspace $S(x, z)$.*

Proof. Every point of $\Gamma_1(x)$ at distance $\mathrm{d}(x, y) - 1$ from y has distance $\mathrm{d}(x, z) - 1$ from z. This proves that $S(x, y) \subseteq S(x, z)$. We will now show that $S(x, y) \neq S(x, z)$. We prove this by induction on the distance $i := \mathrm{d}(x, y)$. Obviously, this holds if $i \in \{0, 1\}$. So, suppose that $i \geq 2$. Let x' denote a point collinear with x at distance $i - 1$ from y. By the induction hypothesis, there exists a line $L \in S(x', z) \setminus S(x', y)$. The line L is different from $x'x$ and hence $Q := \mathcal{C}(x'x, L)$ is a quad. The point z is classical with respect to Q and $\mathrm{d}(z, Q) = i - 1$. There are three possibilities for the relation of y with respect to Q.

(i) The point y is ovoidal with respect to Q. Then $\mathrm{d}(y, Q) = i - 1$. This is however impossible since no point of $X_{i-1,O}(Q)$ is collinear with a point of $X_{i-1,C}(Q)$, see Theorem 1.23.

(ii) The point y is classical with respect to Q and $\mathrm{d}(y, Q) = i - 2$. Then we would have that $\pi_Q(y) = \pi_Q(z)$, contradicting $L \notin S(x', y)$.

(iii) The point y is classical with respect to Q and $\mathrm{d}(y, Q) = i - 1$. Then $\pi_Q(y) = x'$ and every line of Q through x different from xx' belongs to $S(x, z) \setminus S(x, y)$. □

Lemma 2.11. *Let x, a, b and c be points of $\mathcal{S}$ such that $d(a,b) = d(b,c) = 1$, $d(x,a) = i$ and $d(x,b) = d(x,c) = i-1$ for a certain $i \in \mathbb{N} \setminus \{0\}$. Then any common neighbour d of a and c different from b lies at distance i from x.*

Proof. The line ab contains a unique point nearest to x. Hence, $c \notin ab$. So, $\mathrm{d}(a,c) = 2$ and there exists a unique quad Q through a and c. Let d' denote the unique point of the line bc nearest to x. Then $\mathrm{d}(x,d') = i-2$. Since Q contains points at distance $i-2$, $i-1$ and i from x, x is classical with respect to Q. If d denotes any common neighbour of a and c different from b. Then $\mathrm{d}(x,d) = \mathrm{d}(x,d') + \mathrm{d}(d',d) = i$. □

Lemma 2.12. *Let x, a, b and c points of $\mathcal{S}$ such that $d(a,b) = d(b,c) = 1$, $d(a,c) = 2$, $d(x,a) = d(x,c) = i$ and $d(x,b) = i-1$ for a certain $i \in \mathbb{N} \setminus \{0\}$. If x is ovoidal with respect to the quad $\mathcal{C}(a,c)$, then there exists a path of length at most 3 between a and c which is completely contained in $\Gamma_i(x)$.*

Proof. The quad $Q := \mathcal{C}(a,c)$ only contains points at distance $i-1$ and i from x. If there exists a point in $\Gamma_1(a) \cap \Gamma_1(c)$ at distance i from x, then the lemma holds. So, we may suppose that any common neighbour of a and c is contained in $\Gamma_{i-1}(x)$. Let d_1 denote a point of ab different from a and b. Then $\mathrm{d}(d_1,c) = 2$. Let d_2 denote any common neighbour of d_1 and c different from b. Then $\mathrm{d}(a,d_2) = 2$. By our assumption, any point of $\Gamma_1(c) \cap Q$ at distance $i-1$ from x is collinear with a. As a consequence, d_2 lies at distance i from x. The path (a,d_1,d_2,c) now satisfies all required conditions. □

Lemma 2.13. *Let x, a, b and c be points of $\mathcal{S}$ such that $d(a,b) = d(b,c) = 1$, $d(a,c) = 2$, $d(x,a) = d(x,c) = i$ and $d(x,b) = i-1$ for a certain $i \in \mathbb{N} \setminus \{0,1\}$. If x is classical with respect to $Q := \mathcal{C}(a,c)$ and $d(x,Q) = i-2$, then there exists a path of length at most 3 between a and c which is completely contained in $\Gamma_i(x)$.*

Proof. The quad Q only contains points at distance $i-2$, $i-1$ and i from x. Suppose that any common neighbour of a and c lies at distance $i-1$ from x. By Theorem 1.21, there exists a path of length 3 in $\Gamma_2(\pi_Q(x)) \cap Q$ connecting a and c. This path is completely contained in $\Gamma_i(x)$. □

Definitions. If $\gamma = (y_0, \ldots, y_k)$ denotes a path of $\mathcal{S}$, then we define $b(\gamma) := y_0$ and $e(\gamma) := y_k$. For every point x of $\mathcal{S}$, let Ω_x denote the set of all paths $(y_0, \ldots, y_k)$ in $\mathcal{S}$ for which $(S(x,y_i) \setminus S(x,y_{i-1})) \cap S(x,y_0) \neq \emptyset$ for every $i \in \{1, \ldots, k\}$ such that $\mathrm{d}(x,y_i) > \mathrm{d}(x,y_{i-1})$. For each such path γ, we define $i(\gamma) := \sum_{i=0}^{k} 3^{\mathrm{d}(x,y_0)-\mathrm{d}(x,y_i)}$. For every two different points x and y of $\mathcal{S}$ at distance $i \geq 1$ from each other, we define the following sets:

- $H_1(x,y) = \{u \in \mathcal{P} \,|\, S(x,u) \subseteq S(x,y)\}$;
- $H_2(x,y)$ contains all points u which are contained on a shortest path between x and a point of the component C_y of (the subgraph induced by) $\Gamma_i(x)$ to which y belongs.
- $H_3(x,y) = \{e(\gamma) \,|\, \gamma \in \Omega_x \text{ and } b(\gamma) = y\}$.

For every point x of $\mathcal{S}$, we define $H_1(x,x) = H_2(x,x) = H_3(x,x) = \{x\}$.

Theorem 2.14. *For all points x and y of $\mathcal{S}$, $H_1(x,y) = H_2(x,y) = H_3(x,y)$.*

Proof. We may suppose that $x \neq y$. If y' is a point of C_y, then by Lemma 2.9, $S(x,y') = S(x,y)$. By Lemma 2.10, it then follows that $H_2(x,y) \subseteq H_1(x,y)$. If u is a point of $H_1(x,y)$, then the path of $\mathcal{S}$ which consists of a shortest path between y and x followed by a shortest path between x and u belongs to Ω_x by Lemma 2.10. As a consequence, $H_2(x,y) \subseteq H_1(x,y) \subseteq H_3(x,y)$. We will now prove that $H_3(x,y) \subseteq H_2(x,y)$. So, let z denote an arbitrary point of $H_3(x,y)$. Let γ denote a path of Ω_x with $b(\gamma) = y$, $e(\gamma) = z$ and $i(\gamma)$ as small as possible.

- Suppose that there exist successive points u, v and w in γ such that $\mathrm{d}(x,v) = \mathrm{d}(x,w) = \mathrm{d}(x,u) - 1$. By Lemma 2.11, any common neighbour $\tilde{v}$ of u and w different from v lies at distance $\mathrm{d}(x,u)$ from x. Now, if we change the subpath (u,v,w) of γ by $(u,\tilde{v},w)$, then we obtain a path $\tilde{\gamma} \in \Omega_x$ with $b(\tilde{\gamma}) = y$, $e(\tilde{\gamma}) = z$ and $i(\tilde{\gamma}) < i(\gamma)$, a contradiction.

- Suppose that there exist successive points u, v and w in γ such that $\mathrm{d}(x,u) = \mathrm{d}(x,w) = \mathrm{d}(x,v) + 1$. If u, v and w are on a line and $u = w$, then omitting the points v and w in the path γ, we obtain a path $\tilde{\gamma}$ with $b(\tilde{\gamma}) = y$, $e(\tilde{\gamma}) = z$ and $i(\tilde{\gamma}) < i(\gamma)$, a contradiction. If u, v and w are on a line and $u \neq w$, then omitting the point v in the path γ, we obtain a path $\tilde{\gamma}$ with $b(\tilde{\gamma}) = y$, $e(\tilde{\gamma}) = z$ and $i(\tilde{\gamma}) < i(\gamma)$, a contradiction. So, we may suppose that $\mathrm{d}(u,w) = 2$. Let Q denote the unique quad through u and w. There are three possibilities.

  (i) x is ovoidal with respect to Q. By Lemma 2.12, we can replace the subpath (u,v,w) of γ by a path of length at most 3 between u and w which only contains points at distance $\mathrm{d}(x,u)$ from x. In this way, we obtain a path $\tilde{\gamma}$ with $b(\tilde{\gamma}) = y$, $e(\tilde{\gamma}) = z$ and $i(\tilde{\gamma}) < i(\gamma)$, a contradiction.

  (ii) x is classical with respect to Q and $\mathrm{d}(x,Q) = \mathrm{d}(x,u) - 2$. By Lemma 2.13, we can replace the subpath (u,v,w) of γ by a path of length at most 3 between u and w containing only points at distance $\mathrm{d}(x,u)$ from x. In this way, we obtain a path $\tilde{\gamma}$ with $b(\tilde{\gamma}) = y$, $e(\tilde{\gamma}) = z$ and $i(\tilde{\gamma}) < i(\gamma)$, a contradiction.

  (iii) x is classical with respect to Q and $i := \mathrm{d}(x,Q) = \mathrm{d}(x,u) - 1$. Obviously, $S(x,v) \subseteq S(x,u) \cap S(x,w)$. Suppose that there exists a line L which is contained in $S(x,u) \cap S(x,w)$ but not in $S(x,v)$. This line cannot contain a point at distance $i-1$ from Q, otherwise L would be contained in $S(x,v)$. The line L contains two different points of $\Gamma_i(Q)$ and hence is completely contained in $X_{i,C}(Q)$ by Theorem 1.23. Every point of Q at distance i from L lies on the line $\pi_Q(L)$. So, $u,v,w \in \pi_Q(L)$, a contradiction. As a consequence, $S(x,v) = S(x,u) \cap S(x,w)$. Now, let $\tilde{v}$ be a common neighbour of u and w different from v. Then

$(S(x,\tilde{v}) \setminus S(x,u)) \cap S(x,y) \supseteq (S(x,w) \setminus S(x,u)) \cap S(x,y) = (S(x,w) \setminus S(x,v)) \cap S(x,y) \neq \emptyset$. So, replacing the subpath (u,v,w) of γ by $(u,\tilde{v},w)$ we find a path $\tilde{\gamma}$ with $b(\tilde{\gamma}) = y$, $e(\tilde{\gamma}) = z$ and $i(\tilde{\gamma}) < i(\gamma)$, a contradiction.

- We will show that every point of γ lies at distance at most $\mathrm{d}(x,y)$ from x. Suppose the contrary and let u denote the first point on γ for which this does not hold and let v denote the point just before u on γ. Then $j := \mathrm{d}(x,v) = \mathrm{d}(x,y)$ and $\mathrm{d}(x,u) = \mathrm{d}(x,y)+1$. By the previous reasoning, we know that the part of the path γ before u is completely contained in $\Gamma_j(x)$. This would imply that $S(x,v) = S(x,y)$ and $(S(x,u) \setminus S(x,v)) \cap S(x,y) = \emptyset$, a contradiction.

Let y^* denote the last point on the path γ for which $\mathrm{d}(x,y^*) = \mathrm{d}(x,y)$. Then $y^* \in C_y$ and by the previous reasoning we know that z is contained on a shortest path between y^* and x. This proves that $H_3(x,y) \subseteq H_2(x,y)$. As a consequence, $H_1(x,y) = H_2(x,y) = H_3(x,y)$. □

Definition. For every two points x and y of $\mathcal{S}$, we define $H(x,y) := H_1(x,y) = H_2(x,y) = H_3(x,y)$.

Lemma 2.15. *For all points x and y of $\mathcal{S}$, $H(x,y)$ contains all points of a geodesic path between x and any of its points.*

Proof. Suppose that u is contained on a geodesic path between x and a point $v \in H(x,y)$. Then $S(x,u) \subseteq S(x,v) \subseteq S(x,y)$. This proves that $u \in H(x,y)$. □

Lemma 2.16. *For all points x and y of $\mathcal{S}$, $H(x,y)$ is a subspace.*

Proof. Let u and v denote two different collinear points of $H(x,y)$ and let w denote a third point on the line uv. Without loss of generality, we may suppose that $\mathrm{d}(x,u) \geq \mathrm{d}(x,v)$. If $\mathrm{d}(x,w) = \mathrm{d}(x,u) - 1$, then $w \in H(x,y)$ by Lemma 2.15. If $\mathrm{d}(x,w) = \mathrm{d}(x,u)$, then $S(x,w) = S(x,u) \subseteq S(x,y)$ by Lemma 2.9. Hence, also $w \in H(x,y)$ in that case. This proves that $H(x,y)$ is a subspace. □

Lemma 2.17. *Let x and y be two different points of $\mathcal{S}$ and let x_1 be a point of $H(x,y)$ collinear with x such that $d(x_1,y) = d(x,y)$. Then $H(x_1,y) = H(x,y)$.*

Proof. The set $H(x_1,y)$ is a subspace. By symmetry, it suffices to show that $H(x_1,y) \subseteq H(x,y)$.

Step 1. *If L is a line of $H(x_1,y)$ intersecting $H(x,y)$ in a unique point, then $L \cap H(x,y) = \{\pi_L(x)\} = \{\pi_L(x_1)\}$.*

Proof. Put $L \cap H(x,y) = \{u\}$. If $u \neq \pi_L(x)$, then by Lemma 2.15, $H(x,y)$ contains the point $\pi_L(x)$ since it is on a shortest path between the points x and u of $H(x,y)$. This is however impossible since u is the only point of $H(x,y)$ on L. So, $\pi_L(x) = u$. Suppose $\pi_L(x_1) \neq \pi_L(x)$. Then the line xx_1 is parallel with L. Consider now the path γ which consists of a shortest path between y and x followed by a shortest

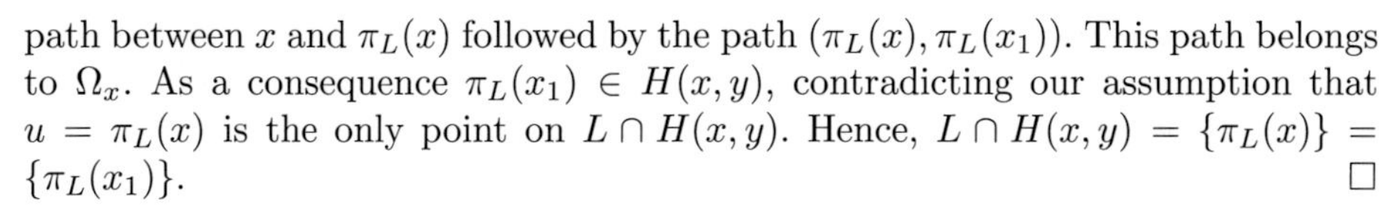

path between x and $\pi_L(x)$ followed by the path $(\pi_L(x), \pi_L(x_1))$. This path belongs to Ω_x. As a consequence $\pi_L(x_1) \in H(x,y)$, contradicting our assumption that $u = \pi_L(x)$ is the only point on $L \cap H(x,y)$. Hence, $L \cap H(x,y) = \{\pi_L(x)\} = \{\pi_L(x_1)\}$. $\square$

Step 2. Let C'_y denote the component of y in $\Gamma_i(x_1)$. Then $C'_y \subseteq H(x,y)$.

Proof. Let $v \in C'_y$ and consider a path $y = u_0, u_1, \ldots, u_k = v$ in $\mathcal{C}'_y$ connecting y and v. Using induction on i and using Step 1, one easily sees that $u_i \in H(x,y)$ for every $i \in \{0, \ldots, k\}$. In particular, $v \in H(x,y)$. $\square$

Step 3. $H(x_1,y) \subseteq H(x,y)$.

Proof. Let z be an arbitrary point of $H(x_1,y)$. Then z is on a geodesic path $u_0, u_1, \ldots, u_k$ of length $k = \mathrm{d}(x_1,y) - \mathrm{d}(x_1,z)$ between a point u_0 of C'_y and the point $u_k = z$. By Step 2, we have $u_0 \in H(x,y)$. Using induction on i and using Step 1, one easily sees that $u_i \in H(x,y)$ for every $i \in \{0, \ldots, k\}$. In particular, $z \in H(x,y)$. $\square$

Lemma 2.18. *If x and y are two different points of $\mathcal{S}$, then for every $u \in H(x,y)$, there exists a point $v \in H(x,y)$ at distance $d(x,y)$ from u such that $H(u,v) = H(x,y)$.*

Proof. Put $\mathrm{d}(x,y) = i$. By connectedness of $H(x,y)$ it suffices to prove this for every point u collinear with x. By Lemma 2.17, the property holds if $\mathrm{d}(u,y) = i$. So, suppose that $\mathrm{d}(u,y) = i - 1$. Since $S(y,u) \neq S(y,x)$, see Lemma 2.10, there exists a line $L \in S(y,x) \setminus S(y,u)$. Let y' denote a point of L contained in $\Gamma_i(x) \setminus \{y\}$. Then y' has distance i from x and u. As a consequence, $H(u,y') = H(x,y') = H(x,y)$. (The latter equality follows from the fact that $S(x,y') = S(x,y)$, see Lemma 2.9.) $\square$

Lemma 2.19. *For all points x and y of $\mathcal{S}$, $H(x,y)$ is convex.*

Proof. Let u_1 and u_2 denote two arbitrary points of $H(x,y)$. Let v_1 be a point of $H(x,y)$ such that $H(u_1,v_1) = H(x,y)$. By Lemma 2.15, every point on a shortest path between u_1 and u_2 belongs to $H(u_1,v_1) = H(x,y)$. As a consequence, $H(x,y)$ is convex. $\square$

Lemma 2.20. *For all points x and y of $\mathcal{S}$, $H(x,y)$ has diameter $d(x,y)$.*

Proof. Let u_1 and u_2 denote two arbitrary points of $H(x,y)$. Let v_1 be a point of $H(x,y)$ at distance $\mathrm{d}(x,y)$ from u_1 such that $H(u_1,v_1) = H(x,y)$. Since $u_2 \in H(u_1,v_1)$, u_2 is contained on a shortest path between u_1 and a point at distance $\mathrm{d}(x,y)$ from u_1. As a consequence, $\mathrm{d}(u_1,v_1) \leq \mathrm{d}(x,y)$. $\square$

Corollary 2.21. *For all points x and y of $\mathcal{S}$, $H(x,y)$ induces a convex sub-$[2 \cdot d(x,y)]$-gon.*

Theorem 2.22. *Let x and y be two points of $\mathcal{S}$. Then every convex subpolygon H through x and y contains $H(x,y)$. As a consequence, $H(x,y) = \mathcal{C}(x,y)$.*

Proof. Put $i := \mathrm{d}(x, y)$ and let C_y denote the component of $\Gamma_i(x)$ to which y belongs. Suppose that z and z' are two different collinear points of $\Gamma_i(x)$ and let z'' denote the unique point of the line zz' at distance $i - 1$ from x. If $z \in H$, then $z'' \in H$ since z'' is on a shortest path between two vertices of H. As a consequence, the line zz'' is completely contained in H. In particular, $z' \in H$. Since $y \in H$, it now follows that every point of C_y is contained in H. Now, every point u of $H(x, y)$ is contained on a shortest path between x and a point of C_y and hence also belongs to H. This proves that $H(x, y) \subseteq H$. Since $\mathcal{C}(x, y)$ is the smallest convex subpolygon through the points x and y, we necessarily have $\mathcal{C}(x, y) = H(x, y)$. □

We will now prove Theorems 2.3 and 2.4.

Theorem 2.23. *Let x and y denote points of $\mathcal{S}$ at distance i from each other. Then $H(x, y) = \mathcal{C}(x, y)$ is the unique convex sub-$2i$-gon through x and y.*

Proof. Let $H'(x, y)$ denote a convex sub-$2i$-gon through x and y. By Theorem 2.22, $H(x, y) \subseteq H'(x, y)$. Let $\mathcal{L}_x$ denote the set of lines through x contained in $H'(x, y)$. Since the maximal distance between two points of $H'(x, y)$ is equal to i, every line of $\mathcal{L}_x$ contains a point at distance $i - 1$ from y. As a consequence, $\mathcal{L}_x \subseteq S(x, y)$. Now, let u denote an arbitrary point of $H'(x, y)$. Since $H'(x, y)$ is convex, $S(x, u) \subseteq \mathcal{L}_x$. Hence, $S(x, u) \subseteq S(x, y)$ and $u \in H(x, y)$. This also proves that $H'(x, y) \subseteq H(x, y)$. □

Theorem 2.24. *Let H be a convex sub-2δ-gon of $\mathcal{S}$ and let L be a line of $\mathcal{S}$ intersecting H in a unique point x. Then $\mathcal{C}(H, L)$ is a convex sub-$(2\delta + 2)$-gon.*

Proof. Let $y \in L \setminus \{x\}$ and $z \in H$ at distance δ from x. By Theorem 1.5, y is classical with respect to H. Hence, $\mathrm{d}(y, z) = \delta + 1$ and $\mathcal{C}(y, z)$ is a convex sub-$(2\delta + 2)$-gon. Since $y, z \in \mathcal{C}(H, L)$, $\mathcal{C}(y, z) \subseteq \mathcal{C}(H, L)$. Now, since x is on a shortest path between y and z, $x \in \mathcal{C}(y, z)$, $L = \mathcal{C}(x, y) \subseteq \mathcal{C}(y, z)$ and $\mathcal{C}(x, z) \subseteq \mathcal{C}(y, z)$. By Theorem 2.23, $H = \mathcal{C}(x, z)$. So, $\mathcal{C}(H, L) \subseteq \mathcal{C}(y, z)$. It follows that $\mathcal{C}(H, L) = \mathcal{C}(y, z)$. This proves the theorem. □

Theorem 2.25. *Let H_1 be a convex sub-$2i$-gon of $\mathcal{S}$ and let H_2 and H_3 be two different convex sub-$2(i + 1)$-gons through H_1. Then*

- *$H_2 \cap H_3 = H_1$;*
- *there exists a unique convex sub-$2(i + 2)$-gon through H_2 and H_3.*

Proof. From Theorem 2.24, it follows that $H_2 \cap H_3 = H_1$. Let u denote a point of H_2 such that $\mathrm{d}(u, H_1) = 1$, let v denote the unique point of H_1 collinear with u and let w denote a point of H_3 at distance $i + 1$ from v. Since u is classical with respect to H_3 and $u \notin H_3$, $\mathrm{d}(u, w) = \mathrm{d}(u, v) + \mathrm{d}(v, w) = i + 2$. Obviously, every convex sub-$2(i + 2)$-gon through H_2 and H_3 must coincide with $\mathcal{C}(u, w)$. So, it suffices now to prove that $\mathcal{C}(u, w)$ contains H_2 and H_3. Since $\mathcal{C}(u, w)$ contains the point v, $H_3 = \mathcal{C}(v, w)$ is contained in $\mathcal{C}(u, w)$. In particular, $\mathcal{C}(u, w)$ contains H_1. Now, $\mathcal{C}(u, w)$ also contains $\mathcal{C}(H_1, uv)$. By Theorem 2.24, $\mathcal{C}(H_1, uv) = H_2$. This proves the theorem. □

The following corollary of Theorem 2.25 is precisely Theorem 2.5.

Corollary 2.26. *Let Δ be the rank d geometry whose elements of type $i \in \{1, \dots, d\}$ are the convex sub-$2(i-1)$-gons of $\mathcal{S}$ (natural incidence). Then Δ belongs to the following diagram:*

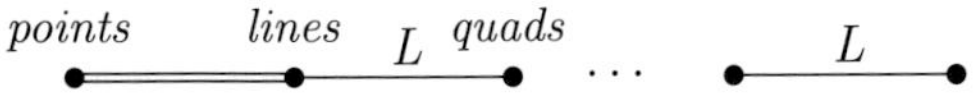

points *lines* L *quads* $\dots$ L

Theorem 2.27. *The subgraph induced by Γ on $\Gamma_d(x)$ is connected for every point x of $\mathcal{S}$.*

Proof. Let y and y' denote two points of $\Gamma_d(x)$ and let C_y denote the connected component of $\Gamma_d(x)$ to which y belongs. Since $S(x, y) = \mathcal{L}(\mathcal{S}, x) = S(x, y')$, the point y' is contained on a shortest path between x and a point y'' of C_y. Hence, $y' = y'' \in C_y$. □

In Section 2.4, we will determine an upper bound for the diameter of $\Gamma_d(x)$.

Theorem 2.28. *A point x is classical with respect to a convex sub-2δ-gon H of $\mathcal{S}$ if and only if there exists a point $y \in H$ at distance $d(x, H) + \delta$ from x.*

Proof. Suppose first that x is classical with respect to H and let y denote a point of H at distance δ from $\pi_H(x)$. Then $\mathrm{d}(x, y) = \mathrm{d}(x, \pi_H(x)) + \mathrm{d}(\pi_H(x), y) = \mathrm{d}(x, H) + \delta$.

Conversely, suppose that there exists a point $y \in H$ such that $\mathrm{d}(x, y) = \mathrm{d}(x, H) + \delta$ and let x' denote a point of H at distance $\mathrm{d}(x, H)$ from x. Every point of $\Gamma_1(y) \cap \Gamma_{\delta-1}(x')$ lies at distance at most $\mathrm{d}(x, H) + \delta - 1$ from x. As a consequence, every point of $\Gamma_1(y) \cap \Gamma_{\delta-1}(x')$ lies at distance precisely $\mathrm{d}(x, H) + \delta - 1$ from x. Since every line of H through y contains a unique point at distance $\delta - 1$ from x', every point of $\Gamma_1(y) \cap \Gamma_\delta(x')$ lies at distance $\mathrm{d}(x, H) + \delta$ from x. By the connectedness of $\Gamma_\delta(x') \cap H$ it follows that every point of $\Gamma_\delta(x') \cap H$ has distance $\mathrm{d}(x, H) + \delta$ from x. Now, consider an arbitrary point z of H. Then z is contained on a shortest path between x' and a point y' of $\Gamma_\delta(x') \cap H$. From $\mathrm{d}(x, z) \leq \mathrm{d}(x, x') + \mathrm{d}(x', z)$ and $\mathrm{d}(x, z) \geq \mathrm{d}(x, y') - \mathrm{d}(z, y') = \mathrm{d}(x, y') - \mathrm{d}(x', y') + \mathrm{d}(x', z) = \mathrm{d}(x, x') + \mathrm{d}(x', z)$, it follows that $\mathrm{d}(x, z) = \mathrm{d}(x, x') + \mathrm{d}(x', z)$. This means that x is classical with respect to H and that $\pi_H(x) = x'$. □

Theorem 2.29. *Let H be a convex sub-2δ-gon of $\mathcal{S}$ and let x be an arbitrary point of H. Then for every $i \in \{0, \dots, d - \delta\}$, there exists a point y satisfying:*

- *$d(y, H) = d(y, x) = i$;*
- *y is classical with respect to H.*

Proof. We will prove this by induction on i. Obviously, the theorem holds if $i = 0$. So, suppose $i \geq 1$ and that y' is a point of $\mathcal{S}$ such that (a) $\mathrm{d}(y', H) = \mathrm{d}(y', x) = i - 1$ and (b) y' is classical with respect to H. Let z denote a point of H at distance δ

from x. Then $\mathrm{d}(y',z) = \mathrm{d}(y',x) + \mathrm{d}(x,z) = i - 1 + \delta$. Since $i - 1 + \delta < d$, there exists a point y collinear with y' not contained in $\mathcal{C}(y',z)$. Since x is contained in a shortest path between y' and z, $H = \mathcal{C}(x,z) \subseteq \mathcal{C}(y',z)$. Now, by Theorem 1.5, y is classical with respect to $\mathcal{C}(y',z)$. Hence, $\mathrm{d}(y,x) = \mathrm{d}(y,y') + \mathrm{d}(y',x) = i$, $\mathrm{d}(y,H) = \mathrm{d}(y,y') + \mathrm{d}(y',H) = i$ and $\mathrm{d}(y,u) = \mathrm{d}(y,y') + \mathrm{d}(y',u) = \mathrm{d}(y,y') + \mathrm{d}(y',x) + \mathrm{d}(x,u) = \mathrm{d}(y,x) + \mathrm{d}(x,u)$ for every $u \in H$. This proves the theorem. □

Theorem 2.30. *A convex sub-$2(d-1)$-gon H of $\mathcal{S}$ is big in $\mathcal{S}$ if and only if every quad which meets H either is contained in H or intersects H in a line.*

Proof. If H is big, then by Theorem 1.7 it follows that every quad which meets H either is contained in H or intersects H in a line.

Conversely, suppose that every quad which meets H either is contained in H or intersects H in a line. If H were not big, then there would exist two points x and y with $x \in \Gamma_2(H)$, $y \in H$ and $\mathrm{d}(x,y) = 2$. The quad $\mathcal{C}(x,y)$ would then intersect H in a point, a contradiction. Hence, H is big. □

We can improve Theorem 2.30 as follows.

Theorem 2.31. *Let H be a convex subpolygon of $\mathcal{S}$ and let x be an arbitrary point of H. Then H is big in $\mathcal{S}$ if and only if every quad through x either is contained in H or intersects H in a line.*

Proof. If H is big in $\mathcal{S}$, then every quad through x either is contained in H or intersects H in a line, see Theorem 2.30. Conversely, suppose that every quad through x either is contained in H or intersects H in a line. Suppose that H is not big in $\mathcal{S}$. Choose points y and y' satisfying $\mathrm{d}(y,H) = 2$ and $y' \in \Gamma_2(y) \cap H$ with $\mathrm{d}(x,y')$ as small as possible. If $y' = x$, then $\mathcal{C}(x,y) \cap H = \{x\}$, a contradiction. So, $y' \neq x$ and we can take a point z' collinear with y' at distance $\mathrm{d}(x,y') - 1$ from x. In the hex $\mathcal{C}(z',y',y)$ we take a line K through y which is not contained in the quad $\mathcal{C}(y,y')$. Let z denote the unique point of K at distance 2 from z'. By the minimality of $\mathrm{d}(x,y')$, z is collinear with a point $\tilde{z} \in H$. Since $\mathrm{d}(z,z') = 2$ and $\mathrm{d}(z,y') = 3$, $\mathrm{d}(\tilde{z},z') = 1$ and $\mathrm{d}(\tilde{z},y') = 2$. Hence the hex $\mathcal{C}(z',y',y)$ intersects H in the quad $R := \mathcal{C}(\tilde{z},z',y')$. Since y has distance 2 or 3 to every point of R, $\Gamma_2(y) \cap R$ is an ovoid O of R. Now, O contains a point u not collinear with z'. If v denotes a common neighbour of u and y, then the line yv has a unique point w at distance 2 from z'. Since $\mathrm{d}(v,z') = 1 + \mathrm{d}(u,z') = 3$, $w \neq v$ and w has distance 2 from H, contradicting the minimality of $\mathrm{d}(x,y')$. As a consequence, H is big in $\mathcal{S}$. □

Theorem 2.32. *Let H be a convex sub-$2m$-gon of $\mathcal{S}$ which is classical in $\mathcal{S}$ and let x be a point of H. If H' is a convex sub-$2(d-m+\delta)$-gon through x, then* $\mathrm{diam}(H \cap H') \geq \delta$.

Proof. Let y denote a point of H' at distance $d - m + \delta$ from x. Since y is classical with respect to H, $\mathrm{d}(y,\pi_H(y)) \leq d - m$. Hence, $\mathrm{d}(x,\pi_H(y)) \geq \delta$. Now, $\pi_H(y)$ belongs to H' since it is contained in a shortest path between x and

y. Since $x, \pi_H(y) \in H' \cap H$, $\mathcal{C}(x, \pi_H(y)) \subseteq H' \cap H$. Hence, $\mathrm{diam}(H' \cap H) \geq \mathrm{diam}(\mathcal{C}(x, \pi_H(y))) \geq \delta$. □

2.3 Proof of Theorem 2.6

We take the proof of [45]. For a convex subpolygon F of a dense near polygon $\mathcal{S}'$ and an $i \in \mathbb{N}$, we define the following sets:

- $\mathcal{GC}(\mathcal{S}')$: the set of all convex subpolygons of $\mathcal{S}'$,
- $\mathcal{GC}_i(\mathcal{S}')$: the set of all convex sub-$2i$-gons of $\mathcal{S}'$,
- $\mathcal{GC}_i(\mathcal{S}', F)$: the set of all convex sub-$2i$-gons of $\mathcal{S}'$ containing F.

For every point x of $\mathcal{S}'$ and every $i \in \mathbb{N}$, we define $\mathcal{GC}_i(\mathcal{S}', x) := \mathcal{GC}_i(\mathcal{S}', \{x\})$ and $n_i(\mathcal{S}', x)$ denotes the number of points of $\mathcal{S}'$ at distance i from x. By Theorem 2.3,

$$n_i(\mathcal{S}', x) = \sum_{U \in \mathcal{GC}_i(\mathcal{S}', x)} n_i(U, x). \tag{2.1}$$

Using Theorem 1.2, we easily see that the following property holds.

Property. Let $\mathcal{S}'$ be a finite dense near $2d'$-gon, $d' \geq 2$, of order (s, t) and let x be a point of $\mathcal{S}'$. Then

$$n_{d'}(\mathcal{S}', x) = v - \sum_{i=0}^{d'-1} n_i(\mathcal{S}', x), \tag{2.2}$$

$$n_{d'-1}(\mathcal{S}', x) = \frac{1}{s+1}\left(v + \sum_{i=0}^{d'-2} [(-s)^{d'-i} - 1] \cdot n_i(\mathcal{S}', x)\right), \tag{2.3}$$

where v denotes the total number of points of $\mathcal{S}'$. As a consequence, if the numbers $n_i(\mathcal{S}', x)$, $i \in \{0, \ldots, d'-2\}$, do not depend on x, then also the numbers $n_{d'-1}(\mathcal{S}', x)$ and $n_{d'}(\mathcal{S}', x)$ do not depend on x.

We will prove Theorem 2.6 by induction on the pair (d, i). We will use the following ordering on the set of all such pairs: $(d', i') < (d, i)$ if either $d' < d$ or ($d' = d$ and $i' < i$). Obviously, the theorem is true if $(d, i) \leq (2, 0)$. Suppose therefore that $(d, i) > (2, 0)$ and that the theorem holds for any finite dense near $2d'$-gon $\mathcal{S}'$ and any $i' \in \{0, \ldots, d'\}$ satisfying $(d', i') < (d, i)$ (= Induction Hypothesis). In particular, the theorem holds for any proper convex subpolygon of $\mathcal{S}$.

Suppose first that not every line of $\mathcal{S}$ is incident with the same number of points. Then by Theorem 1.12, there exist dense near polygons $\mathcal{S}_1$ and $\mathcal{S}_2$ satisfying $\mathcal{S} \cong \mathcal{S}_1 \times \mathcal{S}_2$, $\mathrm{diam}(\mathcal{S}_1) < \mathrm{diam}(\mathcal{S})$ and $\mathrm{diam}(\mathcal{S}_2) < \mathrm{diam}(\mathcal{S})$. For every point (x, y)

of $\mathcal{S}_1 \times \mathcal{S}_2$, we have $n_i(\mathcal{S}_1 \times \mathcal{S}_2, (x, y)) = \sum_{j=0}^{i} n_j(\mathcal{S}_1, x) \cdot n_{i-j}(\mathcal{S}_2, y)$ and, by the Induction Hypothesis, this number does not depend on the chosen point (x, y).

So, we may suppose that $\mathcal{S}$ has an order (s, t), where $t = t_{\mathcal{S}}$. By connectedness of $\mathcal{S}$, it suffices to show that $n_i(\mathcal{S}, x) = n_i(\mathcal{S}, y)$ for two arbitrary collinear points x and y. By the above Property, we only need to consider the case $i \leq d - 2$. By equation (2.1), we have

$$n_i(\mathcal{S}, x) = \sum_{U \in \mathcal{GC}_i(\mathcal{S}, xy)} n_i(U, x) + \sum_{U \in \mathcal{GC}_i(\mathcal{S}, x) \setminus \mathcal{GC}_i(\mathcal{S}, xy)} n_i(U, x). \tag{2.4}$$

By Theorem 2.24,

$$\sum_{U \in \mathcal{GC}_i(\mathcal{S}, x) \setminus \mathcal{GC}_i(\mathcal{S}, xy)} n_i(U, x) = \sum_{V \in \mathcal{GC}_{i+1}(\mathcal{S}, xy)} \; \sum_{U \in \mathcal{GC}_i(V, x) \setminus \mathcal{GC}_i(V, xy)} n_i(U, x).$$

Now,

$$\sum_{U \in \mathcal{GC}_i(V, x) \setminus \mathcal{GC}_i(V, xy)} n_i(U, x) = n_i(V, x) - \sum_{U \in \mathcal{GC}_i(V, xy)} n_i(U, x)$$

by equation (2.4). So,

$$n_i(\mathcal{S}, x) = \sum_{U \in \mathcal{GC}_i(\mathcal{S}, xy)} n_i(U, x) + \sum_{V \in \mathcal{GC}_{i+1}(\mathcal{S}, xy)} \left(n_i(V, x) - \sum_{U \in \mathcal{GC}_i(V, xy)} n_i(U, x) \right). \tag{2.5}$$

All convex subpolygons occurring in the right-hand side of the latter equation contain the point y and have diameter at most $i + 1$. Since $i + 1 \leq d - 1$, we can apply the Induction Hypothesis. It follows that

$$n_i(\mathcal{S}, x) = \sum_{U \in \mathcal{GC}_i(\mathcal{S}, xy)} n_i(U, y) + \sum_{V \in \mathcal{GC}_{i+1}(\mathcal{S}, xy)} \left(n_i(V, y) - \sum_{U \in \mathcal{GC}_i(V, xy)} n_i(U, y) \right).$$

Now, applying equation (2.5) with y instead of x, we obtain that

$$n_i(\mathcal{S}, x) = n_i(\mathcal{S}, y).$$

As mentioned earlier, the theorem now follows from the connectedness of $\mathcal{S}$. □

2.4 Upper bound for the diameter of $\Gamma_d(x)$

In this section, let $\mathcal{S} = (\mathcal{P}, \mathcal{L}, \mathrm{I})$ be a dense near $2d$-gon and let x denote an arbitrary point of $\mathcal{S}$. In Theorem 2.27, we showed that the graph induced by Γ on $\Gamma_d(x)$ is connected. The aim of this section is to derive an upper bound for the diameter of $\Gamma_d(x)$. We follow the approach of [37].

Definitions. A path $\gamma = (x_0, x_1, \ldots, x_k)$ in $\mathcal{S}$ is called *saw-edged* (with respect to x) if the following three conditions are satisfied:

(1) $\mathrm{d}(x, x_0) = \mathrm{d}(x, x_k)$;

(2) $\mathrm{d}(x, x_i) \in \{\mathrm{d}(x, x_0), \mathrm{d}(x, x_0) + 1\}$ for all $i \in \{0, \ldots, k\}$;

(3) if $\mathrm{d}(x, x_i) = \mathrm{d}(x, x_0) + 1$, then $\mathrm{d}(x, x_{i-1}) = \mathrm{d}(x, x_{i+1}) = \mathrm{d}(x, x_0)$.

Let $l(\gamma) = k$ denote the length of γ. We call $\bar{l}(\gamma) = l(\gamma) + t(\gamma)$ the *modified length* of γ; here $t(\gamma)$ denotes the number of *teeth* of γ, i.e. the number of vertices x_i, $i \in \{0, \ldots, k\}$, at distance $\mathrm{d}(x, x_0) + 1$ from x.

Theorem 2.33. *Let $y, z \in \mathcal{P}$ such that $d(x, y) = d(x, z)$. If y and z are connected by a path of length δ consisting only of points at distance at most $d(x, y)$ from x, then y and z are connected by a saw-edged path γ with $\bar{l}(\gamma) \leq \frac{3}{2}\delta$.*

Proof. We use induction on δ. Clearly, the theorem holds if $\delta = 0$ or $\delta = 1$. Let $\delta = 2$ and let u and u' be two common neighbours of y and z. We may suppose that $\mathrm{d}(x, u) = \mathrm{d}(x, u') = \mathrm{d}(x, y) - 1$. Choose now collinear points $v \in uz \setminus \{u, z\}$ and $v' \in yu' \setminus \{y, u'\}$, then the path (y, v', v, z) is saw-edged and has modified length 3. Suppose therefore that $\delta \geq 3$ and consider a path $y = x_0, x_1, \ldots, x_\delta = z$ for which $\mathrm{d}(x, x_i) \leq \mathrm{d}(x, y)$ for all $i \in \{0, \ldots, \delta\}$. If $\mathrm{d}(x, x_i) = \mathrm{d}(x, y)$ for some $i \in \{1, \ldots, \delta - 1\}$ then there exists a saw-edged path of modified length at most $\frac{3}{2}i + \frac{3}{2}(\delta - i) = \frac{3}{2}\delta$ connecting y and z. Suppose therefore that $\mathrm{d}(x, x_i) < \mathrm{d}(x, y)$ for all $i \in \{1, \ldots, \delta - 1\}$. By induction, x_1 and $x_{\delta-1}$ are connected by a saw-edged path γ_1 with $\bar{l}(\gamma_1) \leq \frac{3}{2}(\delta - 2) = \frac{3}{2}\delta - 3$. The path γ_1 can be extended to a path γ_2 of length $k = l(\gamma_1) + 2$ connecting y and z. By Lemma 2.11, the path γ_2 can be replaced by a path $\gamma_3 = (a_0, \ldots, a_k)$ which satisfies the following properties:

(a) $a_0 = y$, $a_k = z$;

(b) there are exactly $t(\gamma_1) + 1$ points a_i, $i \in \{0, \ldots, k\}$, satisfying $\mathrm{d}(x, a_i) = \mathrm{d}(x, y) - 1$; all the other points of the path γ_3 lie at distance $\mathrm{d}(x, y)$ from x.

(c) If $\mathrm{d}(x, a_i) = \mathrm{d}(x, y) - 1$ for some $i \in \{0, \ldots, k\}$, then $\mathrm{d}(x, a_{i-1}) = \mathrm{d}(x, a_{i+1}) = \mathrm{d}(x, y)$.

If $\mathrm{d}(x, a_i) = \mathrm{d}(x, y) - 1$ for some $i \in \{0, \ldots, k\}$, then the path (a_{i-1}, a_i, a_{i+1}) can be replaced by a saw-edged path of modified length at most 3. Hence y and z are connected by a saw-edged path of length at most

$$l(\gamma_3) - 2(t(\gamma_1) + 1) + 3(t(\gamma_1) + 1) = l(\gamma_1) + t(\gamma_1) + 3 = \bar{l}(\gamma_1) + 3 \leq \frac{3}{2}\delta. \qquad \square$$

Corollary 2.34. (a) *If y and z are points of $\mathcal{S}$ such that $d(x, y) = d(x, z)$, then they are connected by a saw-edged path γ with $\bar{l}(\gamma) \leq 3\, d(x, y)$.*

(b) *For every point x of $\mathcal{S}$, $\Gamma_d(x)$ has diameter at most $\lfloor \frac{3}{2}d \rfloor$.*

2.5 Upper bounds for $t+1$ in the case of slim dense near polygons

Recall that a near polygon is called *slim* if every line is incident with precisely three points. In this section, we will prove the following result.

Theorem 2.35. *There exist constants M_i, $i \in \mathbb{N} \setminus \{0\}$, such that $t_{\mathcal{S}} \leq M_i$ for every slim dense near $2i$-gon $\mathcal{S}$. In particular, $\mathcal{S}$ is finite.*

This theorem has been proved in [20] for generalized quadrangles (see also Theorem 1.32), in [12] for near hexagons and in [37] for general near polygons. We follow the approach of [37].

We will prove Theorem 2.35 by induction on the diameter of the near polygon. Obviously, the theorem holds if the diameter is equal to 1 ($M_1 = 0$). By Theorem 1.32, the result is also true if the diameter is equal to 2 ($M_2 = 4$). So, suppose that $\mathcal{S}$ is a dense slim near $2d$-gon with $d \geq 3$ and that Theorem 2.35 holds for any dense slim near polygon of diameter at most $d-1$. In particular, we suppose that the theorem holds for any convex subpolygon of $\mathcal{S}$. Let M be a positive integer such that every convex sub-$2(d-1)$-gon H of $\mathcal{S}$ has order $(2, t_H)$ with $t_H \leq M$. Let x denote an arbitrary point of $\mathcal{S}$ and put $t := t_{\mathcal{S}}$.

Lemma 2.36. *If there is a cycle of length $2n+1$, $n > 1$, in $\Gamma_d(x)$, then $t+1 \leq (2n+1)(M+1)$.*

Proof. Let $y_0, y_1, \ldots, y_{2n+1} = y_0$ be a cycle of length $2n+1$ in $\Gamma_d(x)$. Let z_i, $i \in \{0, \ldots, 2n\}$, denote the unique third point on the line $y_i y_{i+1}$. Suppose that $t+1 > (2n+1)(M+1)$. Then there exists a line L through x which is not contained in one of the subpolygons $H(x, z_i)$. Now, let u be the point of L at distance $d-1$ from y_0. If $\mathrm{d}(u, z_i) \leq d-1$ for a certain $i \in \{0, \ldots, 2n\}$, then $\mathrm{d}(x, z_i) = d-1$ implies that $\mathrm{d}(w, z_i) = d-2$ for a certain point w on xu or that L is contained in $H(x, z_i)$, a contradiction. Hence $\mathrm{d}(u, z_i) = d$ for every $i \in \{0, \ldots, 2n\}$. Since $\mathrm{d}(u, y_0) = d-1$ and $\mathrm{d}(u, z_0) = d$, we necessarily have that $\mathrm{d}(u, y_1) = d$. Since $\mathrm{d}(u, y_1) = d$ and $\mathrm{d}(u, z_1) = d$, $\mathrm{d}(u, y_2) = d-1$. Repeating this argument several times, one finds that $\mathrm{d}(u, y_{2n+1}) = d$, contradicting $d-1 = \mathrm{d}(u, y_0) = \mathrm{d}(u, y_{2n+1})$. Hence $t+1 \leq (2n+1)(M+1)$. □

Corollary 2.37. *At least one of the following possibilities occurs:*

(1) $t+1 \leq (2\lfloor \frac{3d}{2} \rfloor + 1)(M+1)$,

(2) $\Gamma_d(x)$ *is a bipartite graph.*

Proof. Let $\mathrm{d}'(\cdot, \cdot)$ denote the distance in $\Gamma_d(x)$. Suppose that $\Gamma_d(x)$ is not bipartite. Let y denote an arbitrary point of $\Gamma_d(x)$. Define $V_+ := \{z \in \Gamma_d(x) \mid \mathrm{d}(y, z) \text{ is even}\}$ and $V_- := \{z \in \Gamma_d(x) \mid \mathrm{d}(y, z) \text{ is odd}\}$. Since $\Gamma_d(x)$ is not bipartite, there exists an edge connecting two vertices of V_ϵ for a certain $\epsilon \in \{+, -\}$. This edge gives rise to a closed path which contains y and whose length l is odd and at most equal to $2\lfloor \frac{3d}{2} \rfloor + 1$. The result then follows from Lemma 2.36. □

Theorem 2.38. *If $\Gamma_d(x)$ is bipartite, then* $t+1 \leq \lfloor \frac{3d}{2} \rfloor (M+1)^2$.

Proof. Let $\mathrm{d}'(\cdot,\cdot)$ denote the distance in $\Gamma_d(x)$. Let y be a fixed vertex of $\Gamma_d(x)$. For every $z \in \Gamma_d(x)$, let C_z be the set of lines through x containing a point of $\Gamma_{d-1}(y) \cap \Gamma_{d-1}(z)$, and let D_z denote the set of all the other lines through x. If $\mathrm{d}'(y,z)$ is even, then we put $A_z := D_z$ and $B_z := C_z$; otherwise $A_z := C_z$ and $B_z := D_z$. Clearly $A_y = \emptyset$. For two collinear points z and z' in $\Gamma_d(x)$, let $E_{z,z'}$ denote the set of all lines through x contained in $H(x,z'')$ with z'' the unique third point of the line zz'. Since $\Gamma_1(x) \cap (\Gamma_{d-1}(z) \cap \Gamma_{d-1}(z')) = \Gamma_1(x) \cap \Gamma_{d-2}(z'')$, we have that $A_{z'} = A_z \, \Delta \, E_{z,z'}$, the symmetrical difference of A_z and $E_{z,z'}$. Hence for every point z of $\Gamma_d(x)$, $|A_z| \leq \lfloor \frac{3d}{2} \rfloor (M+1)$. Now, let z be a point of $\Gamma_d(x)$ for which $|A_z|$ is maximal. Let $A_z = \{L_1, \ldots, L_{|A_z|}\}$. For every $i \in \{1, \ldots, |A_z|\}$, let H_i denote the unique convex subpolygon through z and the unique point of L_i at distance $d-1$ from z. If there would be a line zz', $z' \in \Gamma_d(x)$, through z not contained in any of these subpolygons, then $|A_{z'}| = |A_z| + |E_{z,z'}| > |A_z|$, a contradiction. Hence $t+1 \leq \lfloor \frac{3d}{2} \rfloor (M+1)^2$. □

Theorem 2.35 now follows from Corollary 2.37 and Theorem 2.38. In certain cases, we are able to improve the upper bound given in Lemma 2.36.

Theorem 2.39. *Let H be a convex sub-2δ-gon of $\mathcal{S}$. If there exists a $(2n+1)$-cycle in $\Gamma_\delta(x) \cap H$ for some point $x \in H$, then* $t+1 \leq (2n+1)(M+1) - 2n(d-\delta)$.

Proof. By Theorem 2.29, there exists a point y at distance $\mathrm{d}(y,x) = \mathrm{d}(y,H) = d-\delta$ from x such that y is classical with respect to H. If $\mathcal{C}(x,y)$ has order $(2,N)$, then a similar reasoning as in the proof of Lemma 2.36 yields $(t+1)-(N+1) \leq (2n+1)((M+1)-(N+1))$. Hence $t+1 \leq (2n+1)(M+1) - 2n(N+1)$. The theorem now follows since $N+1 \geq d-\delta$. [In a dense near $2\tilde{d}$-gon, the number of lines through a given point is at least $\tilde{d}$, as one can easily show by induction on $\tilde{d}$, making use of Theorem 2.3.] □

By Theorem 1.40, the previous theorem applies if $H \cong Q(5,2)$. In Chapter 6, we will give other examples of near polygons H for which the theorem applies (Theorems 6.4, 6.54, 6.68 and 6.80).

2.6 Slim dense near polygons with a big convex subpolygon

In this section, we will prove a result which will be very useful for classifying slim dense near polygons containing a big convex subpolygon.

2.6.1 Statement of the result

Let $\mathcal{S}$ be a dense slim near $2d$-gon and let F denote a big convex subpolygon of $\mathcal{S}$. Define the following geometry $\Omega(\mathcal{S},F)$.

- The POINTS of $\Omega(\mathcal{S}, F)$ are the lines of $\mathcal{S}$ which intersect F in a unique point.
- The LINES of $\Omega(\mathcal{S}, F)$ are the (not necessarily convex) subgrids of $\mathcal{S}$ which intersect F in a line.
- Incidence is containment.

We will prove the following result.

Theorem 2.40. *For every $i \in \{1,2\}$, let $\mathcal{S}_i$ denote a dense slim near $2d$-gon and let F_i denote a big convex subpolygon of $\mathcal{S}_i$. Then $\Omega(\mathcal{S}_1, F_1) \cong \Omega(\mathcal{S}_2, F_2)$ if and only if there exists an automorphism from $\mathcal{S}_1$ to $\mathcal{S}_2$ mapping F_1 to F_2.*

Obviously, if there exists an automorphism from $\mathcal{S}_1$ to $\mathcal{S}_2$ mapping F_1 to F_2, then $\Omega(\mathcal{S}_1, F_1) \cong \Omega(\mathcal{S}_2, F_2)$. We will prove the other direction in the following section.

2.6.2 Proof of Theorem 2.40

We will take the proof from [34]. Let θ denote an isomorphism from $\Omega(\mathcal{S}_1, F_1)$ to $\Omega(\mathcal{S}_2, F_2)$. Let $\mathrm{d}_i(\cdot,\cdot)$ denote the distance in $\mathcal{S}_i$, $i \in \{1,2\}$. For every $j \in \mathbb{N}$ and every POINT K of $\Omega(\mathcal{S}_i, F_i)$, $i \in \{1,2\}$, $\Gamma_j^*(K)$ denotes the set of POINTS of $\Omega(\mathcal{S}_i, F_i)$ at distance j from K. We prove Theorem 2.40 in a sequence of lemmas.

Definition. If x is a point of $\mathcal{S}_i$, $i \in \{1,2\}$, then we denote the point $\pi_{F_i}(x)$ also by $\pi(x)$.

Lemma 2.41. *Two POINTS K and L of $\Omega(\mathcal{S}_i, F_i)$, $i \in \{1,2\}$, are parallel regarded as lines of $\mathcal{S}_i$ if and only if the distance between K and L in $\Omega(\mathcal{S}_i, F_i)$ is equal to $d_i(K \cap F_i, L \cap F_i)$.*

Proof. We first notice that if K and L are two POINTS of $\Omega(\mathcal{S}_i, F_i)$, $i \in \{1,2\}$, then $\mathrm{d}_i(K, L) = \mathrm{d}_i(K \cap F_i, L \cap F_i)$.

Suppose that K and L are parallel and let $x_0, \dots, x_k$ be a path of length $k := \mathrm{d}_i(K, L)$ between the points $x_0 \in K \setminus F_i$ and $x_k \in L \setminus F_i$. Then $\gamma = x_0\,\pi(x_0)), \dots, x_k\,\pi(x_k))$ is a path of length k in $\Omega(\mathcal{S}_i, F_i)$ connecting K and L, proving one direction of the lemma.

Conversely, suppose that the distance between K and L in $\Omega(\mathcal{S}_i, F_i)$ is equal to $k = \mathrm{d}_i(K \cap F_i, L \cap F_i)$. Then $\mathrm{d}_i(p, L) \leq k$ for every point p on K. Since $\mathrm{d}_i(K, L) = k$, $\mathrm{d}_i(p, L) = k$ for every point p on K. This shows that K and L are parallel lines. □

Lemma 2.42. *There exists an isomorphism μ from F_1 to F_2 such that $\mu(K \cap F_1) = \theta(K) \cap F_2$ for every POINT K of $\Omega(\mathcal{S}_1, F_1)$.*

Proof. Let $i \in \{1,2\}$. Call two POINTS K and L of $\Omega(\mathcal{S}_i, F_i)$ equivalent if they intersect F_i in the same point. We show that two POINTS K and L of $\Omega(\mathcal{S}_i, F_i)$ are equivalent if and only if $\Gamma_1^*(K) \cap \Gamma_1^*(L)$ contains two collinear POINTS of $\Omega(\mathcal{S}_i, F_i)$.

By Theorem 1.7, the quad $Q := \mathcal{C}(K, L)$ intersects F_i in a line $\{a_1, a_2, a_3\}$ with $\{a_1\} = K \cap L$. If M_i, $i \in \{2, 3\}$, denotes any line of Q through a_i different from $Q \cap F_i$, then M_2 and M_3 are two collinear POINTS in $\Gamma_1^*(K) \cap \Gamma_1^*(L)$. Conversely, suppose that M and N are two collinear POINTS in $\Gamma_1^*(K) \cap \Gamma_1^*(L)$. Since $K \cap F_i$, $M \cap F_i$ and $N \cap F_i$ are three mutually collinear points, $K \cap F_i$ is the unique third point on the line through $M \cap F_i$ and $N \cap F_i$. The same holds for $L \cap F_i$. Hence, K and L are equivalent POINTS in $\Omega(\mathcal{S}_i, F_i)$.

If K denotes an arbitrary POINT of $\Omega(\mathcal{S}_1, F_1)$ intersecting F_1 in a point x, then we put $\mu(x) := \theta(K) \cap F_2$. We show that μ is well-defined. If K' denotes another POINT of $\Omega(\mathcal{S}_1, F_1)$ through x, then $\Gamma_1^*(K) \cap \Gamma_1^*(K')$ contains two collinear POINTS of $\Omega(\mathcal{S}_1, F_1)$. Since θ is an isomorphism, $\Gamma_1^*(\theta(K)) \cap \Gamma_1^*(\theta(K'))$ contains two collinear POINTS of $\Omega(\mathcal{S}_2, F_2)$. As a consequence $\theta(K) \cap F_2 = \theta(K') \cap F_2$. This proves that μ is well-defined. Obviously, μ is a bijection between F_1 and F_2. If K and L are collinear POINTS of $\Omega(\mathcal{S}_1, F_1)$, then $\theta(K)$ and $\theta(L)$ are collinear POINTS of $\Omega(\mathcal{S}_2, F_2)$. From this, it follows that μ maps collinear points of F_1 to collinear points of F_2. Hence, μ is an isomorphism. □

Lemma 2.43. *Two POINTS K and L of $\Omega(\mathcal{S}_1, F_1)$ are parallel regarded as lines of $\mathcal{S}_1$ if and only if $\theta(K)$ and $\theta(L)$ are parallel.*

Proof. Suppose that K and L are parallel. By Lemma 2.41, the distance between K and L in $\Omega(\mathcal{S}_1, F_1)$ is equal to $\mathrm{d}_1(K \cap F_1, L \cap F_1)$. Since θ is an isomorphism, the distance between $\theta(K)$ and $\theta(L)$ in $\Omega(\mathcal{S}_2, F_2)$ is equal to $\mathrm{d}_1(K \cap F_1, L \cap F_1) = \mathrm{d}_2(\mu(K \cap F_1), \mu(L \cap F_1)) = \mathrm{d}_2(\theta(K) \cap F_2, \theta(L) \cap F_2)$. By Lemma 2.41, $\theta(K)$ and $\theta(L)$ are parallel. This proves one direction of the lemma. The other direction follows by symmetry. □

Let (x, y) denote a pair of opposite points in F_1 (i.e. $\mathrm{d}_1(x, y) = d - 1$), let A denote an arbitrary POINT of $\Omega(\mathcal{S}_1, F_1)$ through y, let $B_1, \ldots, B_k$ denote all the POINTS of $\Omega(\mathcal{S}_1, F_1)$ through x, and let a_1 and a_2 denote arbitrary points of $A \setminus F_1$ and $\theta(A) \setminus F_2$ respectively. Since $d_1(x, y) = d - 1$ and F_1 is convex, the lines $B_1, \ldots, B_k$ are parallel with A. For every permutation ϕ of $\{1, \ldots, k\}$, let μ_ϕ denote the following bijection between the point sets of $\mathcal{S}_1$ and $\mathcal{S}_2$.

- For every point z of F_1, let $\mu_\phi(z) := \mu(z)$;
- For every point z of $\mathcal{S}_1 \setminus F_1$, take the smallest $i \in \{1, \ldots, k\}$ such that $z\,\pi(z) \parallel B_{\phi(i)}$, and let z'_ϕ denote the unique point of $B_{\phi(i)}$ nearest to z. Since each line of $\mathcal{S}_2$ has exactly three points, there exists a unique point z''_ϕ on $\theta(B_{\phi(i)}) \setminus F_2$ for which $\mathrm{d}_1(z'_\phi, a_1) = \mathrm{d}_2(z''_\phi, a_2)$. By Lemma 2.43, the lines $\theta(B_{\phi(i)})$ and $\theta(z\,\pi(z))$ are parallel. The point $\mu_\phi(z)$ is now defined as the unique point of $\theta(z\,\pi(z))$ nearest to z''_ϕ.

Lemma 2.44. *All the maps μ_ϕ are equal.*

Proof. Let u and v be two collinear points in $\Gamma_d(x)$ and suppose that all the points $\mu_\phi(u)$ are equal. Let w denote the third point on the line uv. There are three possibilities.

(a) If $w \in F_1$, then $\mu_\phi(v)$ is the unique point of $\theta(uv)$ different from $\mu(w)$ and $\mu_\phi(u)$.

(b) If $w \notin F_1$ and $w\,\pi(w) \parallel B_{\phi(1)}$, then $u'_\phi = v'_\phi$ is the unique point of $B_{\phi(1)}$ nearest to w. Hence $u''_\phi = v''_\phi$. Since $\theta(w\,\pi(w)) \parallel \theta(B_{\phi(1)})$, $\mu_\phi(u)$ and $\mu_\phi(v)$ are the unique points of $\theta(u\,\pi(u))$ and $\theta(v\,\pi(v))$, respectively, collinear with $\mu_\phi(w)$. Since the lines $\theta(w\,\pi(w))$, $\theta(u\,\pi(u))$ and $\theta(v\,\pi(v))$ are contained in a grid, $\mu_\phi(v)$ is the unique point of $\theta(v\,\pi(v))$ collinear with $\mu_\phi(u)$.

(c) If $w \notin F_1$ and $w\,\pi(w)$ is not parallel with $B_{\phi(1)}$, then $uv \parallel B_{\phi(1)}$. Hence $u'_\phi \neq v'_\phi$ and $u''_\phi \neq v''_\phi$. Since $\theta(w\,\pi(w))$ and $\theta(B_{\phi(1)})$ are not parallel, the unique line through $\mu_\phi(u)$ intersecting $\theta(w\,\pi(w))$ and $\theta(v\,\pi(v))$ is parallel with $\theta(B_{\phi(1)})$. The point $\mu_\phi(v)$ necessarily lies on the intersection of this line with $\theta(v\,\pi(v))$. Hence $\mu_\phi(v)$ is the unique point of $\theta(v\,\pi(v))$ collinear with $\mu_\phi(u)$.

In each of the three considered cases, $\mu_\phi(v)$ does not depend on ϕ. Since $\Gamma_d(x)$ is connected and $\mu_\phi(a_1) = a_2$ for every permutation ϕ, $\mu_\phi(z)$ does not depend on ϕ for every $z \in \Gamma_d(x)$.

Suppose now that u is a point of $\mathcal{S}_1$ with the property that $\mu_\phi(u)$ is independent of ϕ, and let v be an arbitrary point collinear with u on a shortest path from u to x. We will prove that also $\mu_\phi(v)$ is independent of ϕ. This trivially holds if $v \in F_1$. Suppose therefore that $v \notin F_1$, then the line uv is disjoint from F_1. For every permutation ϕ of $\{1,\ldots,k\}$, another permutation $\bar{\phi}$ of $\{1,\ldots,k\}$ can be defined in the following way: take the smallest $i \in \{1,\ldots,k\}$ such that $B_{\phi(i)} \parallel v\,\pi(v)$ and choose $\bar{\phi}$ such that $\bar{\phi}(1) = \phi(i)$. Since $v'_\phi = u'_{\bar{\phi}}$ and $v''_\phi = u''_{\bar{\phi}}$, $\mu_\phi(v)$ is the unique point of $\theta(v\,\pi(v))$ collinear with $\mu_{\bar{\phi}}(u)$. Hence $\mu_\phi(v)$ is independent of ϕ.

Now, let w denote an arbitrary point of $\mathcal{S}_1$. By Section 2.2, there exists a path γ of length $d - \mathrm{d}_1(x,w)$ between w and a point w' of $\Gamma_d(x)$. We already know that $\mu_\phi(w')$ is independent of ϕ. It follows by induction that $\mu_\phi(r)$ is independent of ϕ for every point r on γ. In particular, $\mu_\phi(w)$ is independent of ϕ. This proves our lemma. $\square$

Put $\mu := \mu_\phi$ with ϕ any permutation of $\{1,\ldots,k\}$. The following lemma completes the proof of Theorem 2.40.

Lemma 2.45. *Two points u and v of $\mathcal{S}_1$ are collinear if and only if $\mu(u)$ and $\mu(v)$ are collinear points of $\mathcal{S}_2$.*

Proof. Suppose that u and v are collinear points of $\mathcal{S}_1$. Obviously, $\mu(u)$ and $\mu(v)$ are collinear if the line uv meets F_1. So, suppose that the line uv is completely contained in $\mathcal{S}_1 \setminus F_1$. Let w be the unique point of uv nearest to x, and let ϕ be a permutation of $\{1,\ldots,k\}$ such that $B_{\phi(1)} \parallel w\,\pi(w)$. Then $u'_\phi = v'_\phi$ is the unique point of $B_{\phi(1)}$ nearest to w. Hence $u''_\phi = v''_\phi$. Following a similar reasoning as in the proof of Lemma 2.44, case (b), we immediately see that $\mu_\phi(u)$ and $\mu_\phi(v)$ are collinear. This proves one direction of the lemma. The other direction follows from symmetry. □

Chapter 3

Regular near polygons

3.1 Introduction

In this chapter $\mathcal{S} = (\mathcal{P}, \mathcal{L}, \mathrm{I})$ denotes a finite regular near $2d$-gon, $d \geq 2$, with parameters s, t, t_i ($i \in \{0, \ldots, d\}$), i.e. $\mathcal{S}$ has order (s, t) and for every two points x and y at distance i there are $t_i + 1$ lines through y containing a (necessarily unique) point of $\Gamma_{i-1}(x)$. The remaining $t - t_i$ lines through y contain besides the point y only points of $\Gamma_{i+1}(x)$. Notice that $t_0 = -1$, $t_1 = 0$ and $t_d = t$. The number $|\Gamma_i(x)|$, $i \in \{0, \ldots, d\}$, is independent from the chosen point x and equal to

$$k_i = \frac{s^i \prod_{j=0}^{i-1}(t - t_j)}{\prod_{j=1}^{i}(t_j + 1)}.$$

The total number of points is equal to

$$v = k_0 + k_1 + \cdots + k_d.$$

If x, y and z are points such that $\mathrm{d}(x, y) = i$, $\mathrm{d}(y, z) = 1$ and $\mathrm{d}(x, z) = i + 1$, then $\Gamma_1(x) \cap \Gamma_{i-1}(y) \subseteq \Gamma_1(x) \cap \Gamma_i(z)$. As a consequence, $t_i \leq t_{i+1}$ for every $i \in \{0, \ldots, d-1\}$. Since $\Gamma_d(x) \neq \emptyset$ for at least one (and hence all) point(s) of $\mathcal{S}$, we must have that $t_{d-1} \neq t$.

3.2 Some restrictions on the parameters

In this section $\mathcal{S}$ is supposed to be dense; so, $s \geq 2$ and $t_2 \geq 1$.

Theorem 3.1. *For all $i, j, k \in \{0, \ldots, d\}$ with $i < j < k$,* $\frac{\prod_{l=i}^{j-1}(t_k - t_l)}{\prod_{l=i}^{j-1}(t_j - t_l)} \in \mathbb{N}$.

Proof. Let F and F' denote two convex subpolygons of $\mathcal{S}$ such that $\mathrm{diam}(F) = i$, $\mathrm{diam}(F') = k$ and $F \subset F'$. Then the number of convex sub-$2j$-gons G satisfying $F \subset G \subset F'$ is equal to $\frac{\prod_{l=i}^{j-1}(t_k - t_l)}{\prod_{l=i}^{j-1}(t_j - t_l)}$. (Let x be a given point of F. Count in two different ways the number of tuples $(L_{i+1}, L_{i+2}, \ldots, L_j)$, where $L_{i+1}, \ldots, L_j$ are lines through x such that $\mathcal{C}(F, L_{i+1}, \ldots, L_l)$, $l \in \{i+1, \ldots, j\}$, is a convex sub-$2l$-gon contained in F'.) □

Let Q denote an arbitrary quad of $\mathcal{S}$. As in Section 1.8, let $X_{i,C}$, respectively $X_{i,O}$, denote the set of points of $\mathcal{S}$ at distance i from Q which are classical, respectively ovoidal, with respect to Q. We will calculate the numbers $|X_{i,C}|$ and $|X_{i,O}|$ and derive some restrictions on the parameters s, t and t_i from that.

Lemma 3.2. *Let L denote a line of $\mathcal{S}$ and x a point at distance $i \in \{0, \ldots, d-1\}$ from L. Then*

(a) *x is contained in $t_i + 1$ lines which contain a (necessarily unique) point at distance $i - 1$ from L;*

(b) *x is contained in $t_{i+1} - t_i$ lines which contain only points at distance i from L;*

(c) *x is contained in $t - t_{i+1}$ lines which contain points at distance $i+1$ from L.*

Proof. (a) If x is collinear with a point y at distance $i-1$ from L, then $\pi_L(y) = \pi_L(x)$ and y is one of the $t_i + 1$ points collinear with x at distance $i-1$ from $\pi_L(x)$.

(b) If a line M through x contains only points at distance i from L, then M is contained in the convex sub-$2(i+1)$-gon $\mathcal{C}(x, L)$. Conversely, if M is one of the $t_{i+1} - t_i$ lines of $\mathcal{C}(x, L)$ through x not containing points at distance $i-1$ from $\pi_L(x)$, then every point of M has distance i from L.

(c) This follows from (a), (b) and the fact that there are precisely $t+1$ lines through x. □

Corollary 3.3. *The total number of points at distance $i \in \{0, \ldots, d\}$ from a given line is equal to $(s+1)s^i \frac{\prod_{j=1}^{i}(t-t_j)}{\prod_{j=1}^{i}(t_j+1)}$.*

Lemma 3.4. *If $x \in X_{i,C}$, then*

(a) *x is contained in $t_i + 1$ lines meeting $X_{i-1,C}$ (in a unique point);*

(b) *x is contained in $(t_2+1)(t_{i+1} - t_i)$ lines contained within $X_{i,C}$;*

(c) *x is contained in $t - t_{i+2}$ lines meeting $X_{i+1,C}$.*

(d) *x is contained in $t_{i+2} - t_i - (t_2+1)(t_{i+1} - t_i)$ lines meeting $X_{i+1,O}$.*

Proof. (a) If x is collinear with a point y of $X_{i-1,C} \cup X_{i-1,O}$, then by Theorem 1.23, y is classical with respect to Q and $\pi_Q(y) = \pi_Q(x)$. So, y is one of the $t_i + 1$ points collinear with x at distance $i-1$ from $\pi_Q(x)$.

(b) If L is a line through x contained in X_i, then by Theorem 1.23, every point of L is classical with respect to Q and $\pi_Q(L)$ is a line of Q through $\pi_Q(x)$. Moreover, L is contained in $\mathcal{C}(x, \pi_Q(L))$. Now, there are $t_2 + 1$ lines in Q through $\pi_Q(x)$. Fix such a line L' and consider the convex sub-$2(i+1)$-gon $\mathcal{C}(L', x)$. In this convex subpolygon, there are $t_{i+1} - t_i$ lines through x which do not contain a point at distance $i-1$ from Q and all these lines are contained in $X_{i,C}$.

(c) Let z denote a point of Q at distance 2 from $\pi_Q(x)$. Then $\mathrm{d}(x, z) = i + 2$.

If y is a point of $X_{i+1,C}$ collinear with x, then $\pi_Q(y) = \pi_Q(x)$, $\mathrm{d}(y, z) = i + 3$ and y is not contained in the convex subpolygon $\mathcal{C}(x, z)$.

Conversely, if y is a point collinear with x not contained in the convex subpolygon $\mathcal{C}(x, z)$, then by Theorem 1.5, $\mathrm{d}(y, u) = 1 + \mathrm{d}(x, u)$ for every point u of $\mathcal{C}(x, z)$. Since x is classical with respect to Q, also y is classical with respect to Q. Since x has distance i from Q, y has distance $i + 1$ from Q.

It now easily follows that there are $t - t_{i+2}$ lines through x meeting $X_{i+1,C}$.

(d) This follows from Theorem 1.23, from (a), (b), (c) and from the fact that there are $t + 1$ lines through x. □

Corollary 3.5. *Let $i \in \{0, \ldots, d-1\}$. Then there are $s^i \frac{\prod_{j=2}^{i+1}(t-t_j)}{\prod_{j=1}^{i}(t_j+1)}$ points x in $X_{i,C}$ for which $\pi_Q(x)$ is a given point of Q. As a consequence,*

$$|X_{i,C}| = (s+1)(st_2+1)s^i \frac{\prod_{j=2}^{i+1}(t-t_j)}{\prod_{j=1}^{i}(t_j+1)}.$$

Lemma 3.6. *For every $i \in \{1, \ldots, d\}$,*

$$(st_2+1) \cdot |X_{i,O}| + st_2 \cdot |X_{i-1,C}| + |X_{i,C}| = (s+1)(st_2+1)s^i \frac{\prod_{j=1}^{i}(t-t_j)}{\prod_{j=1}^{i}(t_j+1)}.$$

Proof. Let L denote a fixed line of Q. Every point x of $\mathcal{S}$ at distance i from L, either belongs to $X_{i-1,C}$, $X_{i,C}$ or $X_{i,O}$. If $x \in X_{i-1,C}$, then $\pi_Q(x)$ does not belong to the line L. If $x \in X_{i,C}$, then $\pi_Q(x)$ belongs to the line L. As a consequence, $|\Gamma_i(L)| = |X_{i,O}| + |X_{i-1,C}| \cdot \frac{(s+1)st_2}{(s+1)(st_2+1)} + |X_{i,C}| \cdot \frac{s+1}{(s+1)(st_2+1)}$. The lemma now follows from Corollary 3.3. □

Using Corollary 3.5 and Lemma 3.6, we are now able to calculate $|X_{i,O}|$.

Corollary 3.7. (a) *For every $i \in \{0, \ldots, d-1\}$,*

$$|X_{i,O}| = (s+1)s^i[t_{i+1} - t_2(t_i+1)] \frac{\prod_{j=2}^{i}(t-t_j)}{\prod_{j=1}^{i}(t_j+1)}.$$

(b) *For every* $i \in \{0, \ldots, d-1\}$, $t_{i+1} \geq t_2(t_i + 1)$.

(c) $\mathcal{S}$ *is classical if and only if* $t_{i+1} = t_2(t_i + 1)$ *for every* $i \in \{0, \ldots, d-1\}$.

Remark. If $t_2 \geq 2$ and $t_3 = t_2(t_2 + 1)$, then $\mathcal{S}$ is classical by [11].

Theorem 3.8. *If* $d \geq 3$ *and* $t_3 \neq t_2(t_2 + 1)$, *then* $t_3 + 1 \geq (t_2 + 1)(st_2 + 1)$.

Proof. Since $t_3 \neq t_2(t_2 + 1)$, there exists a point $x \in X_{2,O}$ by Corollary 3.7. Consider the st_2+1 quads $\mathcal{C}(x, y)$ where y is one of the points of the ovoid $\Gamma_2(x) \cap Q$ of Q. These quads define $(t_2 + 1)(st_2 + 1)$ lines through x which are all contained in the hex $\mathcal{C}(x, Q)$. All these lines are different. If two of them were equal, then there would exist a point in $\Gamma_1(Q)$ which is collinear with two different points of Q. This is impossible since Q is convex. Hence, $(t_2 + 1)(st_2 + 1) \leq t_3 + 1$. □

3.3 Eigenvalues of the collinearity matrix

Let $\mathcal{M}$ denote the set of all $(v \times v)$-matrices over $\mathbb{C}$ whose rows and columns are indexed by the points of $\mathcal{S}$. For every $i \in \{0, \ldots, d\}$, let A_i be the following matrix of $\mathcal{M}$ $(x, y \in \mathcal{P})$:

$$\begin{aligned} (A_i)_{xy} &= 1 \qquad \text{if } \mathrm{d}(x, y) = i; \\ (A_i)_{xy} &= 0 \qquad \text{otherwise.} \end{aligned}$$

The matrix A_0 is equal to the identity matrix I and $A := A_1$ is called the *collinearity matrix* of $\mathcal{S}$. If J denotes the $(v \times v)$-matrix with each entry equal to 1, then

$$A_0 + A_1 + \cdots + A_d = J,$$

$$AJ = s(t + 1)J.$$

Since A is symmetric, it is diagonizable. Let $m(X) \in \mathbb{C}[X]$ denote the minimal (monic) polynomial of A.

Theorem 3.9. *For every* $i \in \{0, \ldots, d\}$,

$$AA_i = s(t - t_{i-1})A_{i-1} + (s - 1)(t_i + 1)A_i + (t_{i+1} + 1)A_{i+1}.$$

Here t_{d+1} *and* t_{-1} *are arbitrary elements of* $\mathbb{Z}$ *and* $A_{-1} = A_{d+1} = 0$.

Proof. We have $(AA_i)_{xy} = \sum_z A_{xz}(A_i)_{zy} = |\Gamma_1(x) \cap \Gamma_i(y)|$. As a consequence, $(AA_i)_{xy}$ is equal to 0 if $\mathrm{d}(x, y) \notin \{i-1, i, i+1\}$, equal to $s(t-t_{i-1})$ if $\mathrm{d}(x, y) = i-1$, equal to $(s - 1)(t_i + 1)$ if $\mathrm{d}(x, y) = i$ and equal to $t_{i+1} + 1$ if $\mathrm{d}(x, y) = i + 1$. □

Corollary 3.10. *For every* $i \in \{0, \ldots, d\}$, *there exists a polynomial* $p_i(X) \in \mathbb{Q}[X]$ *of degree* i *such that* $A_i = p_i(A)$.

Corollary 3.11. *If* $p(X) \neq 0$ *is a polynomial of* $\mathbb{C}[X]$ *with degree less than* $d + 1$, *then* $p(A) \neq 0$.

Proof. This follows directly from Corollary 3.10 and the fact that $A_0, A_1, \ldots, A_d$ are linearly independent elements of $\mathcal{M}$ (regarded as a v^2-dimensional vector space over $\mathbb{C}$). $\square$

Corollary 3.12. *$m(X) = a[X - s(t+1)] \cdot [p_0(X) + p_1(X) + \cdots + p_d(X)]$ for a certain element a in $\mathbb{Q} \setminus \{0\}$.*

Proof. This follows from Corollary 3.11 and the fact that $[A - s(t+1)I][p_0(A) + p_1(A) + \cdots + p_d(A)] = [A - s(t+1)I]J = AJ - s(t+1)J = 0$. $\square$

Since A is symmetric, the number of distinct eigenvalues of A is equal to the degree of the polynomial $m(X)$, i.e. equal to $d+1$. We will denote the eigenvalues of A by $\lambda_0, \lambda_1, \ldots, \lambda_d$. Without loss of generality, we may suppose that $\lambda_0 > \lambda_1 > \cdots > \lambda_d$. Since A is symmetric, $\lambda_0, \ldots, \lambda_d \in \mathbb{R}$.

Theorem 3.13. $\lambda_0 = s(t+1)$.

Proof. The sum of the entries on each row is equal to $s(t+1)$. Hence, $|\lambda_i| \leq s(t+1)$ for every $i \in \{0, \ldots, d\}$. Since $AJ = s(t+1)J$, $s(t+1)$ is an eigenvalue of A. $\square$

Theorem 3.14. $\lambda_d = -(t+1)$.

Proof. Let N be the incidence matrix of A (the rows of N are indexed by the points of $\mathcal{S}$ and the columns of N are indexed by the lines of $\mathcal{S}$). Then $N \cdot N^T = (t+1)I + A$. Since $N \cdot N^T$ is positive semi-definite, every eigenvalue of A is at least $-(t+1)$. Now, put

$$M := \sum_{i=0}^{d} (-\frac{1}{s})^i A_i.$$

Obviously, $M \neq 0$. We have

$$\begin{aligned}
AM &= \sum_{i=0}^{d} (-\frac{1}{s})^i AA_i \\
&= \sum_{i=0}^{d} (-\frac{1}{s})^i [s(t - t_{i-1})A_{i-1} + (s-1)(t_i+1)A_i + (t_{i+1}+1)A_{i+1}] \\
&= \sum_{i=0}^{d} -\frac{t - t_i}{(-s)^i} A_i + \sum_{i=0}^{d} \frac{(s-1)(t_i+1)}{(-s)^i} A_i + \sum_{i=0}^{d} \frac{(-s)(t_i+1)}{(-s)^i} A_i \\
&= \sum_{i=0}^{d} -(t+1)(-\frac{1}{s})^i A_i \\
&= -(t+1)M.
\end{aligned}$$

So, every nonzero column of M is an eigenvector of A with eigenvalue $-(t+1)$. $\square$

For every $i \in \{0, \dots, d\}$, let f_j denote the multiplicity of the eigenvalue λ_j. We have

$$f_0 + f_1 + \cdots + f_d = v.$$

Theorem 3.15. $f_0 = 1$.

Proof. If $\bar{x} = [x_1 \cdots x_v]^T$ denotes an eigenvector of A corresponding with the eigenvalue $s(t+1)$, then we may suppose that $x_0, x_1, \dots, x_v \in \mathbb{R}$. If $i \in \{1, \dots, v\}$ such that $|x_i|$ is maximal, then by considering the i-th entry of $A\bar{x}$, we see that $x_j = x_i$ if p_j is collinear with p_i. By the connectedness of $\mathcal{S}$ it then follows that all entries of $\bar{x}$ are equal to each other. This proves the theorem. □

Now, let $\mathcal{U}$ be the $(d+1)$-dimensional subspace of $\mathcal{M}$ generated by all matrices $A_0, A_1, \dots, A_d$. Then not only

$$\mathcal{U} = \langle A_0, \dots, A_d \rangle$$

but also

$$\mathcal{U} = \langle A^0, A^1, \dots, A^d \rangle.$$

Theorem 3.16. *For all $M_1, M_2 \in \mathcal{U}$, $M_1 \cdot M_2 \in \mathcal{U}$.*

Proof. For every $i \in \{1, 2\}$, there exists a polynomial $q_i(X) \in \mathbb{C}[x]$ of degree at most d such that $q_i(A) = M_i$. Now, let $r(X)$ be the remainder of the division of $q_1(X) \cdot q_2(X)$ by $m(X)$. Then $M_1 \cdot M_2 = r(A) \in \mathcal{U}$. □

Theorem 3.16 implies that $\mathcal{U}$ is an algebra, if we take the ordinary addition and multiplication of matrices as operations. This algebra is the so-called *Bose–Mesner algebra* [6] of the association scheme associated with the regular near polygon $\mathcal{S}$, see also Chapter 20 of [5].

Theorem 3.17. *There exist constants $p_{ij}^l \in \mathbb{N}$ $(i, j, l \in \{0, \dots, d\})$ such that $A_i A_j = \sum_{l=0}^{d} p_{ij}^l A_l$ for all $i, j \in \{0, \dots, d\}$.*

Proof. This follows from Theorem 3.16 and the fact that all entries of $A_i A_j$ are nonnegative integers. □

Corollary 3.18. *Let $i, j, l \in \{0, \dots, d\}$. For all points x and y at distance l, the number of points z satisfying $d(z, x) = i$ and $d(z, y) = j$ is equal to p_{ij}^l.*

Calculation of the multiplicities

Since $\langle A^0, A^1, \dots, A^d \rangle = \langle A_0, \dots, A_d \rangle$, there exist constants a_{ij}, $i, j \in \{0, \dots, d\}$, such that $A^i = \sum_{j=1}^{d} a_{ij} A_j$ for every $i \in \{0, \dots, d\}$. Obviously,

$$\mathrm{Tr}(A^i) = v \cdot a_{i0}.$$

On the other hand, we have

$$\mathrm{Tr}(A^i) = \sum_{j=0}^{d} f_j \lambda_j^i,$$

for every $i \in \{0, \ldots, d\}$. Now, the eigenvalues $\lambda_1, \ldots, \lambda_d$ and the numbers $v \cdot a_{i0} - s^i(t+1)^i$ are functions of the parameters of $\mathcal{S}$. Since the determinant

$$\begin{vmatrix} 1 & \lambda_1 & \cdots & \lambda_1^{d-1} \\ \vdots & \vdots & \ddots & \vdots \\ 1 & \lambda_d & \cdots & \lambda_d^{d-1} \end{vmatrix}$$

is nonzero, the multiplicities $f_1, \ldots, f_d$ can be calculated from the nonsingular system

$$\sum_{j=1}^{d} f_j \lambda_j^i = v \cdot a_{i0} - s^i(t+1)^i \qquad (0 \le i \le d-1).$$

As a consequence, we can determine the multiplicities as functions of the parameters s, t, t_i $(2 \le i \le d-1)$. Since the multiplicities have to be strictly positive integers, we obtain additional restrictions on the parameters. We will now list the eigenvalues and multiplicities for the regular near hexagons and near octagons. Calculations can be found in [87] and [88].

Example 1: The case of regular near hexagons

Put

$$m = 1 + s(t+1) + \frac{s^2(t+1)t}{t_2+1} + \frac{s^3t(t-t_2)}{t_2+1}.$$

The eigenvalues of the collinearity matrix A are $\lambda_0 := s(t+1)$, λ_1, λ_2 and $\lambda_3 := -(t+1)$, where λ_1 and λ_2 are the roots of the polynomial $x^2 - (s-1)(t_2+2)x + (s^2-s+1)t_2 - st + (s-1)^2$. The multiplicity of the eigenvalue $s(t+1)$ is equal to 1, the multiplicity of the eigenvalue $-(t+1)$ is equal to

$$f_3 = s^3 \frac{(t_2+1) + s(t_2+1)t + s^2t(t-t_2)}{s^2(t_2+1) + st(t_2+1) + t(t-t_2)},$$

and if $\{i, j\} = \{1, 2\}$, then the multiplicity of λ_i is given by

$$\frac{\lambda_j(m-1) + s(t+1) - (\lambda_j + t + 1)f_3}{\lambda_j - \lambda_i}.$$

Example 2: The case of regular near octagons

Put

$$
\begin{aligned}
b &= -(s-1)(t_2+t_3+3),\\
c &= (s^2-s+1)t_2t_3+(2s^2-3s+2)(t_2+t_3)-st(t_2+2)+3(s-1)^2,\\
d &= -(s-1)[(t_2+1)(t_3+1)(s^2+1)-s(t+1)(t_2+t_3+2)],\\
m &= 1+s(t+1)+\frac{s^2(t+1)t}{t_2+1}+\frac{s^3(t+1)t(t-t_2)}{(t_2+1)(t_3+1)}+\frac{s^4t(t-t_2)(t-t_3)}{(t_2+1)(t_3+1)},\\
e &= s(t+1)(m-s(t+1)),\\
f &= -s(t+1)[(s-1)m-s^2(t+1)^2].
\end{aligned}
$$

Then the eigenvalues of the collinearity matrix are $\lambda_0 = s(t+1)$, λ_1, λ_2, λ_3 and $\lambda_4 = -(t+1)$, where λ_1, λ_2 and λ_3 are the roots of the polynomial x^3+bx^2+cx+d. The multiplicity of the eigenvalue $s(t+1)$ is equal to 1. If $\{i,j,k,l\} = \{1,2,3,4\}$, then the multiplicity of the eigenvalue λ_i is equal to

$$\frac{e(\lambda_j+\lambda_k+\lambda_l)-s\lambda_i(\lambda_j\lambda_k+\lambda_j\lambda_l+\lambda_k\lambda_l)+(m-1)\lambda_j\lambda_k\lambda_l+f}{(\lambda_j-\lambda_i)(\lambda_k-\lambda_i)(\lambda_l-\lambda_i)}.$$

3.4 Upper bounds for t

The multiplicity of $-(t+1)$ is of the form $\frac{p_1(t)}{p_2(t)}$, where $p_1(t)$ and $p_2(t)$ are polynomials in t whose coefficients are again polynomial functions in s, t_2, ..., t_{d-1}. In general, for given values of s, t_2, ..., t_{d-1}, we can use the fact that the multiplicity must be integral to derive an upper bound for t. For instance, in the case of regular near hexagons, the multiplicity of the eigenvalue $-(t+1)$ is equal to

$$s^3\frac{(t_2+1)+s(t_2+1)t+s^2t(t-t_2)}{s^2(t_2+1)+st(t_2+1)+t(t-t_2)}.$$

If $s \neq 1$, then the fact that this number is integral can be used to derive an upper bound for t as a function of the parameters s and t_2.

There is a well-known result in the literature which gives an upper bound for t in terms of s and t_2. It is called the *Mathon bound.*

Theorem 3.19 ([16],[76],[77]). *If $\mathcal{S}$ is a regular near hexagon with parameters $s \neq 1$, t_2 and t, then $t \leq s^3 + t_2(s^2-s+1)$.*

In the case $s = 2$, we have the following possibilities:

- If $s = 2$ and $t_2 = 0$, then $t \leq 8$.
- If $s = 2$ and $t_2 = 1$, then $t \leq 11$.
- If $s = 2$ and $t_2 = 2$, then $t \leq 14$.

- If $s = 2$ and $t_2 = 4$, then $t \leq 20$.

These bounds are better than the ones we have derived in Section 2.5.

Remark. The *Krein conditions*, due to Scott [85], give additional restrictions on the parameters, see also [13], [16] and [77]. The Mathon bound is in fact one of these Krein conditions.

3.5 Slim dense regular near hexagons

In this section, we suppose that $\mathcal{S}$ is a regular near hexagon with parameters $s = 2$, t_2 and t. If $t_2 \neq 0$, then $\mathcal{S}$ is a dense near hexagon and every two points at distance 2 are contained in a quad, which is necessarily isomorphic to $\mathbb{L}_3 \times \mathbb{L}_3$, $W(2)$ or $Q(5,2)$. So, either $t_2 \in \{0,1,2,4\}$. The following parameters survive the above-mentioned restrictions on the parameters (Theorem 3.1, Corollary 3.7, Theorem 3.8, Theorem 3.19, Conditions on the parameters for k_i and f_i, $i \in \{0,\dots,d\}$, to be integers).

t_2	t	#
0	1	1
0	2	2
0	8	1
1	2	1
1	11	1
2	6	1
2	14	1
4	20	1

The first three examples in the table are generalized hexagons. The unique generalized hexagon of order $(2,1)$ is the flag-geometry of $\mathrm{PG}(2,2)$. So, the points, respectively lines, of the generalized hexagon are the flags, respectively points and lines, of $\mathrm{PG}(2,2)$ (natural incidence). In [25], it was shown that there are two generalized hexagons of order (2,2), namely the split Cayley hexagon $H(2)$ and its point-line dual. Also in [25] it has been shown that there exists a unique generalized hexagon of order $(2,8)$ (the point-line dual of a twisted triality hexagon). For more details, we also refer to [100]. In [7], it was shown that there exists a unique regular near hexagon with parameters $(s,t_2,t) = (2,1,11)$. We will define it in Section 6.5. In [8], it was shown that there exists a unique near hexagon with parameters $(2,2,14)$. We will define it in Section 6.6. The near hexagons with parameters (s,t_2,t) equal to $(2,1,2)$, $(2,2,6)$ and $(2,4,20)$ are classical near hexagons because $t = t_2(t_2+1)$. If $(s,t_2,t) = (2,1,2)$, then the classical near hexagon is the direct product of three lines; if $(s,t_2,t) = (2,2,6)$, then the the classical near hexagon is isomorphic to $DQ(6,2)$; if $(s,t_2,t) = (2,4,20)$, then the classical near hexagon is isomorphic to $DH(5,4)$.

3.6 Slim dense regular near octagons

In this section, we suppose that $\mathcal{S}$ is a regular near octagon with parameters $s = 2$, t_2, t_3 and t. If $t_2 \neq 0$, then $\mathcal{S}$ is a dense near octagon and every two points at distance 2, respectively 3, are contained in a unique quad, respectively hex. Hence, if $t_2 \geq 1$, then (t_2, t_3) is equal to either $(1, 2)$, $(1, 11)$, $(2, 6)$, $(2, 14)$ or $(4, 20)$. The following parameters survive the above-mentioned restrictions on the parameters (an upper bound for t follows from the fact that the multiplicity of the eigenvalue $-(t + 1)$ is an integer):

t_2	t_3	t	#
0	0	1	1
0	0	4	≥ 1
0	3	4	1
0	8	24	?
1	2	3	1
2	6	14	1
4	20	84	1

The first two examples in the table are generalized octagons. There is a unique generalized octagon of order $(2, 1)$. It is the point-line dual of the double of the generalized quadrangle $W(2)$. There is a unique example known of a generalized octagon of order $(2, 4)$ (a so-called Ree–Tits octagon). It is not known whether there are more such generalized octagons. In [25], it has been shown that there exists a unique regular near octagon with parameters $(s, t_2, t_3, t) = (2, 0, 3, 4)$. Its point graph is the graph with vertices the 315 involutions of the Hall–Janko group ([27]) whose centralizers contain Sylow-2-subgroups, and two involutions are adjacent whenever they commute. It is possible, see [24] or Section 13.6 of [13], to give an alternative description of this near octagon which exploits the quaternionic root system given in [23], see also [98] and [101]. It is not known whether there exists a regular near octagon with parameters $(s, t_2, t_3, t) = (2, 0, 8, 24)$. The regular near octagons with parameters (s, t_2, t_3, t) equal to $(2, 1, 2, 3)$, $(2, 2, 6, 14)$, $(2, 4, 20, 84)$ are classical near octagons since $t_3 = t_2(t_2 + 1)$ and $t = t_2(t_3 + 1)$. If $(s, t_2, t_3, t) = (2, 1, 2, 3)$, then $\mathcal{S}$ is the direct product of four lines of size 3; if $(s, t_2, t_3, t) = (2, 2, 6, 14)$, then $\mathcal{S}$ is isomorphic to $DQ(8, 2)$; if $(s, t_2, t_3, t) = (2, 4, 20, 84)$, then $\mathcal{S}$ is isomorphic to $DH(7, 4)$.

Chapter 4

Glued near polygons

Let $\mathcal{S}_1$ and $\mathcal{S}_2$ be two near polygons and let δ be a nonnegative integer. We will show that if $\mathcal{S}_1$ and $\mathcal{S}_2$ satisfy certain nice properties, then (a) new near polygon(s) of diameter $\mathrm{diam}(\mathcal{S}_1)+\mathrm{diam}(\mathcal{S}_1)-\delta$ can be derived from $\mathcal{S}_1$ and $\mathcal{S}_2$. We call these new near polygons glued near polygons of type δ. The contents of this chapter are based on the papers [29], [30], [33], [35], [38], [41], [46], [58] and [59].

The glued near polygons of type 0 are precisely the product near polygons which we have defined in Section 1.6. We start this chapter with some characterizations of product near polygons.

4.1 Characterizations of product near polygons

We refer to Section 1.6 for the definition of direct product of two near polygons. The aim of this section is to give a number of characterizations of product near polygons.

Theorem 4.1. *Let F be a big convex sub-$2(n-1)$-gon of a dense near $2n$-gon $\mathcal{S}$, $n \geq 2$, then the following are equivalent:*

(a) *there exists a line L and a point x_L on L such that $\mathcal{S} \cong F \times L$ and $\phi(F \times \{x_L\}) = F$ for a certain isomorphism ϕ from $F \times L$ to $\mathcal{S}$;*

(b) *through every point of F there is a unique line which is not contained in F;*

(c) *every quad intersecting F in a line is a grid.*

Proof. (a) $\Rightarrow$ (b): This is trivial.

(b) $\Rightarrow$ (c): If Q is a quad which intersects F in a line, then $t_Q = 1$ since through every point of F there is a unique line which is not contained in F. Hence Q is a grid.

(c) $\Rightarrow$ (b): Suppose the contrary. Then there exists a point $x \in F$ and two lines L_1 and L_2 through x which are not contained in F. By Theorem 1.7, the quad

$Q := \mathcal{C}(L_1, L_2)$ intersects F in a line, proving that $t_Q \geq 2$. This contradicts the fact that Q is a grid.

(b) $\Rightarrow$ (a): Let x_1 and x_2 denote two points of F at distance $n-1$ from each other. Let L_i, $i \in \{1,2\}$, denote the unique line through x_i which is not contained in F. The point $\pi_{L_2}(x_1)$ is contained in F since it is on a shortest path between x_1 and x_2. Hence, $\pi_{L_2}(x_1) = x_2$. In a similar way, one shows that $x_1 = \pi_{L_1}(x_2)$. If L_1 and L_2 were not parallel, then $\mathrm{d}(y,z) = \mathrm{d}(y,x_1) + \mathrm{d}(x_1,x_2) + \mathrm{d}(x_2,z) = n+1$ for every point $y \in L_1 \setminus \{x_1\}$ and every point $z \in L_2 \setminus \{x_2\}$, a contradiction. Hence, $L_1 \parallel L_2$. For every point y of L_1, let F_y be the convex sub-$(2n-2)$-gon $\mathcal{C}(y, \pi_{L_2}(y))$. Obviously, $F_{x_1} = F$, $F_y \cap L_1 = \{y\}$ and $F_y \cap L_2 = \{\pi_{L_2}(y)\}$.

Property. If y_1 and y_2 are two different points of L_1, then F_{y_1} and F_{y_2} are disjoint.

Proof. Suppose that u is a common point of F_{y_1} and F_{y_2}. By Theorem 1.5, y_2 is on a shortest path between y_1 and u. Hence $y_2 \in F_{y_1}$, contradicting $L_1 \cap F_{y_1} = \{y_1\}$. □

Property. Let $y \in L_1 \setminus \{x_1\}$ and $z \in F_y$. Then $z\pi_F(z)$ is the unique line through z not contained in F_y.

Proof. Let z' denote a point of F_y at distance $n-1$ from z. Let M denote an arbitrary line through z different from $z\pi_F(z)$. We must show that $M \subseteq F_y$, or equivalently, that M contains a point at distance $n-2$ from z'. The grid-quad $\mathcal{C}(M, z\pi_F(z))$ intersects F in a line M'. Let u denote the unique point of M' at distance $n-2$ from $\pi_F(z')$ and let L' denote the unique line through u not contained in F. The line L' meets M. The point z' has distance $n-1$ from u and the convex sub-$2(n-1)$-gon $\mathcal{C}(u, z')$ is not completely contained in F. As a consequence, L' contains a point u' at distance $n-2$ from z'. Since $\mathrm{d}(\pi_F(z), z') = n$, the point z' is necessarily classical with respect to $\mathcal{C}(M, z\pi_F(z))$. Moreover, u' is the unique point of $\mathcal{C}(M, z\pi_F(z))$ nearest to z'. Since $\mathrm{d}(z', z) = n-1$, u' is collinear with z and hence $u' \in M$. This was precisely what we needed to show. □

Property. For every $y \in L_1$, F_y is big in $\mathcal{S}$.

Proof. Suppose the contrary and let x be a point of $\mathcal{S}$ at distance 2 from F_y. If $x' \in \Gamma_2(x) \cap F_y$, then the quad $\mathcal{C}(x, x')$ intersects F_y only in the point x'. The quad $\mathcal{C}(x, x')$ then has at least two lines through x' not contained in F_y, a contradiction. So, F_y is big in $\mathcal{S}$. □

Let A denote the set of lines of $\mathcal{S}$ which meet F in a unique point. By the previous property, every F_y, $y \in L_1$, intersects each line of A in a unique point. By Theorem 1.10, the projection from F_y to F is an isomorphism for every $y \in L_1$. Obviously, the map $x \mapsto (\pi_F(x), \pi_{L_1}(x))$ defines an isomorphism between the near polygons $\mathcal{S}$ and $F \times L_1$. □

Theorem 4.2. *Let $\mathcal{S}$ be a dense near polygon and let T_1 and T_2 be two partitions of $\mathcal{S}$ in convex subpolygons satisfying the following properties:*

- *every element of T_1 intersects every element of T_2 in a point,*
- *every line of $\mathcal{S}$ is contained in precisely one element of $T_1 \cup T_2$.*

Then

(a) *all elements of T_1 are isomorphic;*

(b) *all elements of T_2 are isomorphic;*

(c) *$\mathcal{S} \cong F_1 \times F_2$ for any $F_1 \in T_1$ and any $F_2 \in T_2$.*

Proof. For every point x of $\mathcal{S}$, let A_x, respectively B_x, denote the unique element of T_1, respectively T_2, through x.

Step 1. *If x and y are two collinear points such that $xy \subseteq B_x$, then $A_x \cap B_z$ and $A_y \cap B_z$ are collinear for every point z of $\mathcal{S}$.*

Proof. Without loss of generality we may suppose that z is a point of A_x. We will prove Step 1 by induction on the distance $\mathrm{d}(x, z)$. Obviously, the property holds if $z = x$. Suppose therefore that $\mathrm{d}(x, z) \geq 1$ and that the property holds for a point $z' \in \Gamma_1(z)$ at distance $\mathrm{d}(x, z) - 1$ from x. Let u' denote the unique point in $A_y \cap B_{z'}$, then z' and u' are collinear. Since the quad $\mathcal{C}(z, z', u')$ is a grid, the points z and u' have a unique common neighbour u different from z'. The line zu is not contained in $A_x = A_z$; so, it is contained in B_z. Similarly, since the line $u'u$ is not contained in $B_{z'} = B_{u'}$, it is contained in $A_{u'} = A_y$. The property now follows from the fact that $A_y \cap B_z = \{u\}$ and $A_x \cap B_z = \{z\}$. □

Step 2. *All elements of T_1 are isomorphic. All elements of T_2 are isomorphic.*

Proof. Take two elements A_{x_1} and A_{x_2} in T_1. For every point y of A_{x_1}, let $\theta(y)$ be the unique point in $A_{x_2} \cap B_y$. Then θ is a bijection between A_{x_1} and A_{x_2}. By Step 1, θ is an isomorphism. So, A_{x_1} and A_{x_2} are isomorphic. In a similar way one proves that all elements of T_2 are isomorphic. □

Step 3. *$\mathcal{S} \cong A \times B$ for any $A \in T_1$ and any $B \in T_2$.*

Proof. For every point x in $\mathcal{S}$, put $\theta(x) := (B_x \cap A, A_x \cap B)$. Obviously, θ is a bijection between the point sets of $\mathcal{S}$ and $A \times B$. We show that θ maps collinear points of $\mathcal{S}$ to collinear points of $A \times B$. If x and y are collinear points of $\mathcal{S}$ such that the line xy is contained in B_x, then $B_x \cap A = B_y \cap A$ and $\mathrm{d}(A_x \cap B, A_y \cap B) = 1$ by Step 1. If x and y are collinear points of $\mathcal{S}$ such that the line xy is contained in A_x, then $A_x \cap B = A_y \cap B$ and $\mathrm{d}(B_x \cap A, B_y \cap A) = 1$ by Step 1. Hence, θ maps collinear points of $\mathcal{S}$ to collinear points of $A \times B$. In a similar way, one shows that θ^{-1} maps collinear points of $A \times B$ to collinear points of $\mathcal{S}$. So, θ is an isomorphism between $\mathcal{S}$ and $A \times B$. □

Theorem 4.3. *Let $\mathcal{S}$ be a dense near polygon and let T_1 and T_2 be two partitions of $\mathcal{S}$ in convex subpolygons such that the following holds.*

- *$A_x \cap B_x = \{x\}$ for every point x of $\mathcal{S}$. Here A_x, respectively B_x, denotes the unique element of T_1, respectively T_2, through x.*
- *Every line of $\mathcal{S}$ is contained in a unique element of $T_1 \cup T_2$.*

Then

(a) *all elements of T_1 are isomorphic;*

(b) *all elements of T_2 are isomorphic;*

(c) *$\mathcal{S} \cong A \times B$ for any $A \in T_1$ and any $B \in T_2$.*

Proof. In view of Theorem 4.2, we must show that every element of T_1 intersects every element of T_2 in a point. Let x and y be arbitrary points of $\mathcal{S}$. We will show that $|A_x \cap B_y| = 1$. Let y' denote a point of A_x at minimal distance from y. Obviously, $A_{y'} = A_x$. If L is a line of $A_{y'} \cap \mathcal{C}(y, y')$ through y', then the unique point on L nearest to y belongs to A_x and has distance $\mathrm{d}(y, y') - 1$ from y, contradicting our assumption that y' is a point of A_x at minimal distance from y. So, every line of $\mathcal{C}(y, y')$ through y' is contained in $B_{y'}$. By Theorem 2.14, $\mathcal{C}(y, y') \subseteq B_{y'}$. Hence, $y \in B_{y'}$ and $B_y = B_{y'}$. It now follows that $A_x \cap B_y = A_{y'} \cap B_{y'} = \{y'\}$. □

Theorem 4.4. *Let $\mathcal{S}$ be a dense near $2(n_1 + n_2)$-gon and let F_1 and F_2 be two convex subpolygons for which the following holds:*

- $\mathrm{diam}(F_i) = n_i \geq 1$ *($i \in \{1, 2\}$);*
- *F_1 intersects F_2 in a point x;*
- *every line through x is contained in either F_1 or F_2.*

Then $\mathcal{S} \cong F_1 \times F_2$.

Proof. We prove the theorem in several steps.

Step 1. If L is a line of F_2 through x, then $\mathcal{C}(F_1, L) \cong F_1 \times L$.

Proof. Let A_i, $i \in \{n_1, \ldots, n_1 + n_2\}$, be convex subpolygons through x satisfying: (i) $\mathrm{diam}(A_i) = i$, (ii) $A_{n_1} = F_1$, (iii) $A_{n_1+1} = \mathcal{C}(F_1, L)$ and (iv) $A_i \subset A_{i+1}$ for every $i \in \{n_1, \ldots, n_1 + n_2 - 1\}$. Let $B_i := A_i \cap F_2$, $i \in \{n_1, \ldots, n_1 + n_2\}$. Since $A_i \neq A_{i+1}$, the set of lines of A_i through x is different from the set of lines of A_{i+1} through x. Since every line through x is contained in $F_1 \cup F_2$ and since $F_1 \subseteq A_i$ and $F_1 \subseteq A_{i+1}$, B_i and B_{i+1} are different. It follows that B_i, $i \in \{n_1, \ldots, n_1 + n_2\}$, are convex subpolygons of F_2 through x satisfying (i) $B_{n_1} = \{x\}$, (ii) $B_i \neq B_{i+1}$ for every $i \in \{n_1, \ldots, n_1 + n_2 - 1\}$, and (iii) $B_{n_1+n_2} = F_2$. Hence, $\mathrm{diam}(B_i) = i - n_1$ for every $i \in \{n_1, \ldots, n_1 + n_2\}$. It follows that $B_{n_1+1} = \mathcal{C}(F_1, L) \cap F_2 = L$.

We now show that through every point y of F_1, there is a unique line of $\mathcal{C}(F_1, L)$ not contained in F_1. Suppose this property holds for a certain $y_1 \in F_1$ and let y_2 be a point of $\Gamma_1(y_1) \cap F_1$. The set of quads of $\mathcal{C}(F_1, L)$ through $y_1 y_2$ not contained in F_1 partition the set of lines of $\mathcal{C}(F_1, L)$ through y_i, $i \in \{1, 2\}$, not contained in F_1. It follows that the property also holds for y_2. Since the property holds for $x \in F_1$, the property holds for every $y \in F_1$ by the connectedness of F_1. So, through every point y of F_1, there is a unique line of $\mathcal{C}(F_1, L)$ not contained in F_1. This implies that F_1 is big in $\mathcal{C}(F_1, L)$ (see the proof of the last property in Theorem 4.1). From Theorem 4.1, it then follows that $\mathcal{C}(F_1, L) \cong F_1 \times L$. □

For every point y of F_i, $i \in \{1, 2\}$, let $F_{3-i}(y)$ denote the convex subpolygon generated by all lines through y not contained in F_i. Clearly $F_1(x) = F_1$ and $F_2(x) = F_2$.

Step 2. *For every point y of $\mathcal{S}$, there exists a point $y_1 \in F_1$ and a point $y_2 \in F_2$ such that $y \in F_2(y_1) \cap F_1(y_2)$.*

Proof. Let y_i, $i \in \{1, 2\}$, denote a point of F_i nearest to y. If $\mathcal{C}(y, y_1) \cap F_1$ contains a line through y_1, then this line would contain a point at distance $\mathrm{d}(y, y_1) - 1$ from y, a contradiction. Hence $\mathcal{C}(y, y_1) \subseteq F_2(y_1)$ and $y \in F_2(y_1)$. Similarly, one proves that $y \in F_1(y_2)$. □

Step 3. *For every point y of F_1,* $\mathrm{diam}(F_2(y)) = n_2$ *and* $F_2(y) \cap F_1 = \{y\}$. *Similarly, for every point y of F_2,* $\mathrm{diam}(F_1(y)) = n_1$ *and* $F_1(y) \cap F_2 = \{y\}$.

Proof. Suppose that y_1 and y_2 are collinear points in F_1 such that $F_2(y_1) \cap F_1 = \{y_1\}$ and $\mathrm{diam}(F_2(y_1)) = n_2$. If L is a line through y_2 not contained in F_1, then $\mathcal{C}(L, y_1 y_2)$ intersects $F_2(y_1)$ in a line and hence $L \subseteq \mathcal{C}(y_1 y_2, F_2(y_1))$. Now, by Step 1 applied to y_1 instead of x, $\mathcal{C}(y_1 y_2, F_2(y_1)) \cong y_1 y_2 \times F_2(y_1)$ and hence $\mathrm{diam}(F_2(y_2)) = n_2$ and $F_2(y_2) \cap F_1 = \{y_2\}$. Since $\mathrm{diam}(F_2(x)) = n_2$ and $F_2(x) \cap F_1 = \{x\}$, the result follows from the connectedness of F_1. □

Step 4. *For all points $y \in F_1$ and $z \in F_2$, $F_2(y)$ intersects $F_1(z)$ in a unique point. Moreover, every line through this point is contained in either $F_2(y)$ or $F_1(z)$.*

Proof. By Step 3, we may suppose that $y \neq x \neq z$. Suppose y_1 and y_2 are two distinct collinear points of F_1 such that the claim holds for the pair $(y_1, z) \in F_1 \times F_2$. By Step 1 applied to y_1 instead of x, $\mathcal{C}(y_1 y_2, F_2(y_1)) \cong y_1 y_2 \times F_2(y_1)$. It follows that $|F_2(y_2) \cap F_1(z)| = 1$. Suppose L were a line through $F_2(y_2) \cap F_1(z)$ not contained in $F_2(y_2) \cup F_1(z)$ and hence also not in $\mathcal{C}(y_1 y_2, F_2(y_1))$. Let $L' \subseteq \mathcal{C}(y_1 y_2, F_2(y_1))$ denote the unique line through $F_2(y_2) \cap F_1(z)$ meeting $F_2(y_1)$. The quad $\mathcal{C}(L', L)$ is not contained in $F_1(z)$ and hence intersects $F_2(y_1)$ in a line L''. (Recall that every line through $F_1(z) \cap F_2(y_1)$ is contained in $F_1(z) \cup F_2(y_1)$.) It follows that $\mathcal{C}(L, L') = \mathcal{C}(L', L'') \subseteq \mathcal{C}(y_1 y_2, F_2(y_1))$, a contradiction. It follows that the claim also holds for (y_2, z).

Now, since the claim holds for the pair (x, z), the result follows from the connectedness of F_1. □

Step 5. *If y_1 and y_2 are different points of F_1, then $F_2(y_1)$ is disjoint from $F_2(y_2)$. Similarly, if y_1 and y_2 are different points of F_2, then $F_1(y_1)$ and $F_1(y_2)$ are disjoint.*

Proof. Let y_1 and y_2 be two different elements of F_1 and suppose that z is a common point of $F_2(y_1)$ and $F_2(y_2)$. Let z' be an element of F_2 such that $z \in F_1(z')$. Since $F_1(z') \cap F_2(y_1) = \{z\}$, the lines of $F_2(y_1)$ through z are precisely those lines through z which are not contained in $F_1(z')$ (recall Step 4). Since this property also holds for $F_2(y_2)$, we necessarily have $F_2(y_1) = F_2(y_2)$. As a consequence $\{y_1\} = F_1 \cap F_2(y_1) = F_1 \cap F_2(y_2) = \{y_2\}$, contradicting $y_1 \neq y_2$. Hence, $F_2(y_1)$ and $F_2(y_2)$ are disjoint. □

Step 6. $\mathcal{S} \cong F_1 \times F_2$.

Proof. Put $T_1 := \{F_1(x) \,|\, x \in F_2\}$ and $T_2 := \{F_2(x) \,|\, x \in F_1\}$. By Steps 2 and 5, T_1 and T_2 define partitions of $\mathcal{S}$. By Step 4 and Theorem 4.2, $\mathcal{S} \cong F_1 \times F_2$. □

4.2 Admissible δ-spreads

Definition. Let $\mathcal{S}$ be a near polygon. A *δ-spread* of $\mathcal{S}$ ($\delta \in \{0, \dots, d\}$) is a partition of $\mathcal{S}$ in convex sub-2δ-gons. A δ-spread is called *admissible* if every two elements of it are parallel. By Theorem 1.10, any two elements of an admissible δ-spread are isomorphic.

Theorem 4.5. *Let T be an admissible δ-spread of $\mathcal{S}$ and let F denote a convex subpolygon of $\mathcal{S}$. Let T' denote the set of all elements of T meeting F and put $T_F := \{G \cap F | G \in T'\}$. Then T_F is an admissible δ'-spread of F for some $\delta' \in \{0, \dots, \delta\}$. Hence, if an element of T is contained in F, then every element of T which meets F is contained in F.*

Proof. Clearly T_F determines a partition of F in convex subpolygons. Take now two elements F_1 and F_2 in T_F. Let G_i, $i \in \{1, 2\}$, denote the unique element of T' through F_i and let $\pi_{i,3-i}$ denote the projection from G_i to G_{3-i}. If x_1 and x_2 are points in F_1 and F_2, respectively, then $\pi_{1,2}(x_1)$ is contained in a shortest path between x_1 and x_2 and so belongs to F (since F is convex). We even can say that $\pi_{1,2}(x_1) \in F_2$. This proves that $\pi_{1,2}(F_1) \subseteq F_2$. By symmetry, $\pi_{1,2}^{-1}(F_2) \subseteq F_1$ and so $\pi_{1,2}(F_1) = F_2$ and $\pi_{2,1}(F_2) = F_1$. It is now easily seen that F_1 and F_2 are parallel. □

Definition. If $\mathcal{S}$ is a near polygon and if T_1 and T_2 are two partitions of $\mathcal{S}$ in convex subpolygons of diameter at least $\delta + 1$, then we say that $\{T_1, T_2\} \in \Delta_\delta(\mathcal{S})$ if the following conditions are satisfied:

(1) every element of T_1 intersects every element of T_2 in a convex sub-2δ-gon;

(2) every line of $\mathcal{S}$ is contained in an element of $T_1 \cup T_2$;

(3) for every element F of T_i ($i \in \{1, 2\}$), the δ-spread S_F of F obtained by intersecting F with all elements of T_{3-i} is an admissible spread of F.

4.3 Construction and elementary properties of glued near polygons

Let $\mathcal{B}$ denote a near $2\delta_{\mathcal{B}}$-gon. Let $\mathcal{A}_i$, $i \in \{1,2\}$, denote a near polygon of diameter at least $\delta_{\mathcal{B}}+1$ admitting an admissible $\delta_{\mathcal{B}}$-spread S_i consisting of convex subpolygons isomorphic to $\mathcal{B}$. Put $\delta_i = \mathrm{diam}(\mathcal{A}_i)$ and let $S_i = \{F_1^{(i)}, \ldots, F_{n_i}^{(i)}\}$. In S_i, we choose a special element $F_1^{(i)}$ which we call the *base element* of S_i. For all $i \in \{1,2\}$ and all $j,k \in \{1,\ldots,n_i\}$, let $\pi_{j,k}^{(i)}$ denote the projection from $F_j^{(i)}$ to $F_k^{(i)}$. We denote the distances in $\mathcal{A}_1$, $\mathcal{A}_2$ and $\mathcal{B}$ respectively by $\mathrm{d}_1(\cdot,\cdot)$, $\mathrm{d}_2(\cdot,\cdot)$ and $\mathrm{d}_{\mathcal{B}}(\cdot,\cdot)$.

Lemma 4.6. *The maximal distance between two elements of S_i, $i \in \{1,2\}$, is equal to $\delta_i - \delta_{\mathcal{B}}$.*

Proof. The lemma follows from (a) and (b) below.

(a) Let F and F' denote two arbitrary elements of S_i. Let $x \in F$ and $x' \in F'$ such that $\mathrm{d}_i(x,x') = \mathrm{d}_i(F,F')$. Let y be a point of F at distance $\delta_{\mathcal{B}}$ from x. Then $\mathrm{d}_i(x',y) = \mathrm{d}_i(x',x) + \mathrm{d}_i(x,y) = \mathrm{d}_i(F,F') + \delta_{\mathcal{B}} \leq \delta_i$, proving that $\mathrm{d}_i(F,F') \leq \delta_i - \delta_{\mathcal{B}}$.

(b) Let x and y be two points of $\mathcal{A}_i$ at distance δ_i from each other and let F_x and F_y denote the respective elements of S_i through x. Then $\delta_i = \mathrm{d}_i(x,y) = \mathrm{d}_i(x, \pi_{F_y}(x)) + \mathrm{d}_i(\pi_{F_y}(x), y) \leq \mathrm{d}_i(F_x,F_y) + \delta_{\mathcal{B}}$, proving that $\delta_i - \delta_{\mathcal{B}} \leq \mathrm{d}_i(F_x,F_y)$. □

For every $i \in \{1,2\}$, consider an isomorphism $\theta_i : \mathcal{B} \to F_1^{(i)}$. We put $\Phi_{j,k}^{(i)} := \theta_i^{-1} \circ \pi_{k,1}^{(i)} \circ \pi_{j,k}^{(i)} \circ \pi_{1,j}^{(i)} \circ \theta_i$ $(1 \leq j,k \leq n_i)$ and $\Pi_i := \langle \Phi_{j,k}^{(i)} | 1 \leq j,k \leq n_i \rangle$. Since $\Phi_{j,k}^{(i)}$ is a composition of isomorphisms, it is itself an isomorphism and so $\Pi_i \leq \mathrm{Aut}(\mathcal{B})$.

Lemma 4.7. $\Pi_i = \langle \Phi_{j,k}^{(i)} | 1 \leq j,k \leq n_i$ *and* $d_i(F_j^{(i)}, F_k^{(i)}) = 1 \rangle$.

Proof. Let $F_j^{(i)}$ and $F_k^{(i)}$ denote two arbitrary elements of S_i and put $n := \mathrm{d}_i(F_j^{(i)}, F_k^{(i)})$. Choose elements $F_{l_0}^{(i)}, F_{l_1}^{(i)}, \ldots, F_{l_n}^{(i)}$ of S_i such that $F_j^{(i)} = F_{l_0}^{(i)}$, $F_k^{(i)} = F_{l_n}^{(i)}$ and $\mathrm{d}_i(F_{l_m}^{(i)}, F_{l_{m+1}}^{(i)}) = 1$ for every $m \in \{0,\ldots,n-1\}$. The map $\mu := \pi_{l_{n-1},l_n}^{(i)} \circ \cdots \circ \pi_{l_1,l_2}^{(i)} \circ \pi_{l_0,l_1}^{(i)}$ is, as composition of isomorphisms, again an isomorphism. Now, μ maps every point x of $F_j^{(i)}$ to a point of $F_k^{(i)}$ at distance at most n from x. So, we must have that

$$\pi_{j,k}^{(i)} = \mu = \pi_{l_{n-1},l_n}^{(i)} \circ \cdots \circ \pi_{l_1,l_2}^{(i)} \circ \pi_{l_0,l_1}^{(i)}.$$

This equation is equivalent with

$$\Phi_{j,k}^{(i)} = \Phi_{l_{n-1},l_n}^{(i)} \circ \cdots \circ \Phi_{l_1,l_2}^{(i)} \circ \Phi_{l_0,l_1}^{(i)}.$$

The lemma now readily follows. □

Consider now the following graph Γ with vertex set $\mathcal{B}\times S_1\times S_2$. Two different vertices $\alpha = (x, F^{(1)}_{i_1}, F^{(2)}_{j_1})$ and $\beta = (y, F^{(1)}_{i_2}, F^{(2)}_{j_2})$ are adjacent if and only if exactly one of the following three conditions is satisfied:

(A) $F^{(1)}_{i_1} = F^{(1)}_{i_2}$, $F^{(2)}_{j_1} = F^{(2)}_{j_2}$, and $\mathrm{d}_{\mathcal{B}}(x,y) = 1$,

(B) $F^{(2)}_{j_1} = F^{(2)}_{j_2}$, $\mathrm{d}_1(F^{(1)}_{i_1}, F^{(1)}_{i_2}) = 1$ and $y = \Phi^{(1)}_{i_1,i_2}(x)$,

(C) $F^{(1)}_{i_1} = F^{(1)}_{i_2}$, $\mathrm{d}_2(F^{(2)}_{j_1}, F^{(2)}_{j_2}) = 1$ and $y = \Phi^{(2)}_{j_1,j_2}(x)$.

An edge $\{\alpha,\beta\}$ of Γ is called of *type* (A), (B) or (C) depending on which one of the above possibilities occurs. Distances in Γ will be denoted by $\mathrm{d}(\cdot,\cdot)$. The following property is obvious.

Lemma 4.8. *If α, β and γ are three different mutually adjacent vertices, then $\{\alpha,\beta\}$, $\{\beta,\gamma\}$ and $\{\alpha,\gamma\}$ have the same type.*

For every element $F_1 \in S_1$ and $F_2 \in S_2$, let $\mathcal{A}_1(F_2)$, respectively $\mathcal{A}_2(F_1)$, denote the set of all vertices of Γ whose third, respectively middle, coordinate is equal to F_2, respectively F_1. We put $\mathcal{B}(F_1,F_2) := \mathcal{A}_1(F_2)\cap\mathcal{A}_2(F_1)$, $T_i := \{\mathcal{A}_i(F) | F \in S_{3-i}\}$ and $S := \{\mathcal{B}(F_1,F_2) | F_1 \in S_1 \text{ and } F_2 \in S_2\}$.

Lemma 4.9. *Let $F_1 \in S_1$ and $F_2 \in S_2$. Then*

(i) *the graph induced by Γ on $\mathcal{A}_1(F_2)$ is isomorphic to the point graph of $\mathcal{A}_1$;*

(ii) *the graph induced by Γ on $\mathcal{A}_2(F_1)$ is isomorphic to the point graph of $\mathcal{A}_2$;*

(iii) *the graph induced by Γ on $\mathcal{B}(F_1,F_2)$ is isomorphic to the point graph of $\mathcal{B}$.*

Proof. For a vertex $\alpha = (x, F^{(1)}_i, F_2)$ of $\mathcal{A}_1(F_2)$, let $\mu(\alpha)$ be the following point of $\mathcal{A}_1$:

$$\mu(\alpha) := \pi^{(1)}_{1,i} \circ \theta_1(x).$$

If α and β are two vertices of $\mathcal{A}_1(F_2)$, then one easily verifies that $\mu(\alpha)\sim\mu(\beta)$ if and only if $\{\alpha,\beta\}$ is of type (A) or (B). Hence, μ defines an isomorphism between the graph induced by Γ on $\mathcal{A}_1(F_2)$ and the point graph of $\mathcal{A}_1$. This proves (i). In a similar way, one proves (ii). Claim (iii) is obvious. □

Lemma 4.10. *Every two adjacent vertices of Γ are contained in a unique maximal clique.*

Proof. Let α and β denote two adjacent vertices of Γ. If $\{\alpha,\beta\}$ is of type (A), then $\alpha,\beta \in \mathcal{B}(F_1,F_2)$ for certain $F_1 \in S_1$ and $F_2 \in S_2$ and, by Lemma 4.8, every maximal clique through α and β belongs to $\mathcal{B}(F_1,F_2)$. If $\{\alpha,\beta\}$ is of type (B), then $\alpha,\beta \in \mathcal{A}_1(F_2)$ for a certain $F_2 \in S_2$ and, by Lemma 4.8, every maximal clique through α and β belongs to $\mathcal{A}_1(F_2)$. If $\{\alpha,\beta\}$ is of type (C), then $\alpha,\beta \in \mathcal{A}_2(F_1)$ for a certain $F_1 \in S_1$ and, by Lemma 4.8, every maximal clique through α and β belongs to $\mathcal{A}_2(F_1)$. Now, $\mathcal{B}(F_1,F_2)$, $\mathcal{A}_1(F_2)$ and $\mathcal{A}_2(F_1)$ are near polygons by Lemma 4.9 and hence every two adjacent vertices in one of these near polygons are contained in a unique maximal clique, proving the lemma. □

Let $\mathcal{S}_\Gamma$ denote the partial linear space whose points are the vertices of Γ and whose lines are the maximal cliques of Γ (natural incidence). A set X of vertices in Γ is called *classical* if for every vertex x in Γ there exists a (necessarily unique) vertex $x' \in X$ such that $\mathrm{d}(x,y) = \mathrm{d}(x,x') + d(x',y)$ for every vertex $y \in X$. One easily verifies that every classical set is convex.

Theorem 4.11. *The following statements are equivalent:*

(i) *Every element of $T_1 \cup T_2$ is classical.*

(ii) *The groups Π_1 and Π_2 commute.*

(iii) *$\mathcal{S}_\Gamma$ is a near polygon and every element of $T_1 \cup T_2$ is convex.*

Proof. (a) Suppose that every element of $T_1 \cup T_2$ is classical (and hence also convex). Let (x, L) denote an arbitrary point-line pair of $\mathcal{S}_\Gamma$. By Lemma 4.8, there exists an element F of $T_1 \cup T_2$ through L. Since F is classical, there exists a point $x' \in F$ such that $\mathrm{d}(x,y) = \mathrm{d}(x,x') + \mathrm{d}(x',y)$ for every point y of L. Since F is convex, distances in the near polygon F are inherited from the corresponding distances in Γ. As a consequence, L contains a unique point nearest to x, namely the unique point of L nearest to x' in the near polygon F. This proves (i) $\Rightarrow$ (iii).

Take now an arbitrary point x of $\mathcal{B}$ and arbitrary elements $i_1, i_2 \in \{1, \ldots, n_1\}$ and $j_1, j_2 \in \{1, \ldots, n_2\}$ such that $\mathrm{d}_1(F^{(1)}_{i_1}, F^{(1)}_{i_2}) = 1$ and $\mathrm{d}_2(F^{(2)}_{j_1}, F^{(2)}_{j_2}) = 1$. Consider now the points $\alpha = (x, F^{(1)}_{i_1}, F^{(2)}_{j_1})$ and $\beta = (\Phi^{(1)}_{i_1,i_2} \circ \Phi^{(2)}_{j_1,j_2}(x), F^{(1)}_{i_2}, F^{(2)}_{j_2})$. Clearly, $\mathrm{d}(\alpha, \beta) = 2$. The point β is collinear with the point $\gamma := (\Phi^{(2)}_{j_2,j_1} \circ \Phi^{(1)}_{i_1,i_2} \circ \Phi^{(2)}_{j_1,j_2}(x), F^{(1)}_{i_2}, F^{(2)}_{j_1})$ of $\mathcal{A}_1(F^{(2)}_{j_1})$. Since $\mathcal{A}_1(F^{(2)}_{j_1})$ is classical and $\alpha \in \mathcal{A}_1(F^{(2)}_{j_1})$, we necessarily have $2 = \mathrm{d}(\alpha,\beta) = \mathrm{d}(\beta,\gamma) + \mathrm{d}(\gamma,\alpha) = 1 + \mathrm{d}(\alpha,\gamma)$ or $\alpha \sim \gamma$. As a consequence $\Phi^{(2)}_{j_2,j_1} \circ \Phi^{(1)}_{i_1,i_2} \circ \Phi^{(2)}_{j_1,j_2}(x) = \Phi^{(1)}_{i_1,i_2}(x)$. Since this holds for every point x of $\mathcal{B}$, we have that $\Phi^{(1)}_{i_1,i_2}$ and $\Phi^{(2)}_{j_1,j_2}$ commute for all $i_1, i_2 \in \{1, \ldots, n_1\}$ and all $j_1, j_2 \in \{1, \ldots, n_2\}$ with $\mathrm{d}_1(F^{(1)}_{i_1}, F^{(1)}_{i_2}) = 1$ and $\mathrm{d}_2(F^{(2)}_{j_1}, F^{(2)}_{j_2}) = 1$. By Lemma 4.7 it now follows that Π_1 and Π_2 commute. Hence (i) $\Rightarrow$ (ii).

(b) Suppose that Π_1 and Π_2 commute. If α, β and γ are points of $\mathcal{S}_\Gamma$ such that $\mathrm{d}(\alpha,\gamma) = \mathrm{d}(\gamma,\beta) = 1$ and $\mathrm{d}(\alpha,\beta) = 2$, then we have the following (since Π_1 and Π_2 commute):

- if $\{\alpha,\gamma\}$ has type (C) and $\{\gamma,\beta\}$ has type (A), then there exists a (unique) common neighbour γ' of α and β such that $\{\alpha,\gamma'\}$ has type (A) and $\{\gamma',\beta\}$ has type (C);

- if $\{\alpha,\gamma\}$ has type (C) and $\{\gamma,\beta\}$ has type (B), then there exists a (unique) common neighbour γ' of α and β such that $\{\alpha,\gamma'\}$ has type (B) and $\{\gamma',\beta\}$ has type (C).

We will now prove that every element F of $T_1 \cup T_2$ is classical. Without loss of generality, we may suppose that $F \in T_2$. Take an arbitrary point $\alpha \in \mathcal{S}_\Gamma$ and let F' denote the unique element of T_1 through α. Since every element of S_1 is classical in

$\mathcal{A}_1$, $F \cap F'$ is classical in F'. Let α' denote the unique element of $F \cap F'$ nearest to α. Let β denote an arbitrary point of F. By the remark above we know that there exists a shortest path between α and β containing a point γ such that no step of type (C) occurs between α and γ and only steps of type (C) occur between γ and β. So, $\gamma \in F \cap F'$ and $\mathrm{d}(\alpha,\beta) = \mathrm{d}(\alpha,\gamma) + \mathrm{d}(\gamma,\beta) = \mathrm{d}(\alpha,\alpha') + \mathrm{d}(\alpha',\gamma) + \mathrm{d}(\gamma,\beta) = \mathrm{d}(\alpha,\alpha') + \mathrm{d}(\alpha',\beta)$. Hence α is classical with respect to F. Since α is arbitrary, F is classical. This proves (ii) $\Rightarrow$ (i).

(c) Suppose now that $\mathcal{S}_\Gamma$ is a near polygon and that every element of $T_1 \cup T_2$ is convex. If F is an arbitrary element of $T_1 \cup T_2$, then since F is convex, every point at distance 1 from F is classical with respect to F. As in (a) one can then prove that Π_1 and Π_2 commute. Hence (iii) $\Rightarrow$ (ii). □

Definitions. Every near polygon $\mathcal{S}$ which can be derived in the above described way from a near 2δ-gon $\mathcal{B}$, near polygons $\mathcal{A}_1$ and $\mathcal{A}_2$, spreads S_1 and S_2, base elements $F_1^{(1)}$ and $F_1^{(2)}$ and isomorphisms θ_1 and θ_2, such that the groups Π_1 and Π_2 commute, will be called a *glued near polygon of type* δ. We will then say that $\mathcal{S}$ is of type $\mathcal{A}_1 \otimes_{\mathcal{B}} \mathcal{A}_2$, of type $\mathcal{A}_1 \otimes_\delta \mathcal{A}_2$ or shortly of type $\mathcal{A}_1 \otimes \mathcal{A}_2$.

Remarks. (a) Suppose that $\mathcal{A}_1$ is isomorphic to the direct product of $\mathcal{B}$ and $\mathcal{A}_1'$ ($\mathrm{diam}(\mathcal{A}_1') \geq 1$) and that $S_1 = \{F_x \,|\, x \in \mathcal{A}_1'\}$ where $F_x := \{(y,x) \,|\, y \in \mathcal{B}\}$. Then S_1 is an admissible $\delta_{\mathcal{B}}$-spread of $\mathcal{A}_1$. We call a δ-spread of a near polygon *trivial* if it is obtained in this way. Suppose also that S_2 is an admissible $\delta_{\mathcal{B}}$-spread of $\mathcal{A}_2$ and that every element of S_2 is isomorphic to $\mathcal{B}$. One easily sees that the group Π_1 is always trivial for any isomorphism θ_1 between $\mathcal{B}$ and a base element in S_1. Hence, by Theorem 4.11 a near polygon $\mathcal{S}_{\theta_1,\theta_2}$ arises for any isomorphism θ_1 between $\mathcal{B}$ and an element of S_1 and any isomorphism θ_2 between $\mathcal{B}$ and an element of S_2. All these near polygons are isomorphic to $\mathcal{A}_1' \times \mathcal{A}_2$ since the map $(y, F_x, F_i^{(2)}) \mapsto (x, \pi_{1,i}^{(2)} \circ \theta_2(y))$ is an isomorphism between $\mathcal{S}_{\theta_1,\theta_2}$ and $\mathcal{A}_1' \times \mathcal{A}_2$ (see the proof of Lemma 4.9).

(b) By (a), glued near polygons exist for any type $\delta \geq 2$. Let $\mathcal{B}$ be a glued near polygon of diameter δ and put $\mathcal{A}_1 = \mathcal{B} \times \mathcal{A}_1'$, $\mathcal{A}_2 = \mathcal{B} \times \mathcal{A}_2'$ for certain near polygons $\mathcal{A}_1'$ and $\mathcal{A}_2'$ of diameter at least 1. By (a), $\mathcal{A}_i$ has an admissible δ-spread S_i all whose elements are isomorphic to $\mathcal{B}$. For any two isomorphisms θ_1 and θ_2 between $\mathcal{B}$ and base elements of S_1 and S_2, respectively, we obtain a glued near polygon of type $\delta_{\mathcal{B}}$ which is isomorphic to $\mathcal{A}_1' \times \mathcal{B} \times \mathcal{A}_2'$. In this case, the glued near polygon of type $\delta_{\mathcal{B}}$ is also of type 0. In general, we can say that any known glued near polygon of type $\delta \geq 2$ is also of type 0 or 1.

Above we remarked that the group Π_1 is trivial if S_1 is a trivial spread of $\mathcal{A}_1$. Also the converse holds as we will show now.

Theorem 4.12. *The group* Π_1 *is trivial if and only if* S_1 *is a trivial spread of* $\mathcal{A}_1$*.*

Proof. Suppose Π_1 is trivial. For every point x of $F_1^{(1)}$, the set $\Delta_x := \{y \in \mathcal{A}_1 \,|\, \mathrm{d}_1(y,x) = \mathrm{d}_1(y, F_1^{(1)})\}$ is a subspace of $\mathcal{A}_1$ by Theorem 1.9. Let W_2 denote the set of all these subspaces and put $W_1 := S_1$. It is easily seen that every element of W_1 intersects every element of W_2 in a unique point and since Π_1 is trivial, every line must be contained in an element of $W_1 \cup W_2$. Now, put $\overline{F}_1^{(1)} := F_1^{(1)}$ and let $\overline{F}_1^{(2)}$ denote an arbitrary element of W_2. For every point x of $\mathcal{A}_1$, put $x_1 := \pi_{F_1^{(1)}}(x)$ and let x_2 be the unique point in the intersection of $\overline{F}_1^{(2)}$ and the unique element of W_1 through x. The map $x \mapsto (x_1, x_2)$ defines a bijection between the point set of $\mathcal{A}_1$ and the vertex set of $\Gamma_1 \times \Gamma_2$, where Γ_i denotes the point graph of $\overline{F}_1^{(i)}$. We will show that x and x' are collinear points of $\mathcal{A}_1$ if and only if (x_1, x_2) and (x_1', x_2') are adjacent vertices of $\Gamma_1 \times \Gamma_2$. Obviously, this holds if x and x' belong to the same element of W_1. Suppose that x and x' belong to different elements of W_1, say G_x and $G_{x'}$. Then $x_1 = x_1'$ and $\mathrm{d}_1(G_x, G_{x'}) = 1$. Let x_2'' denote the unique point of $G_{x'}$ collinear with x_2. Then $x_2 x_2''$ is not contained in an element of W_1 and hence is contained in $\overline{F}_1^{(2)}$. Hence, $x_2'' \in F_1^{(2)} \cap G_{x'}$. It follows that $x_2'' = x_2'$. Hence, $\mathrm{d}_1(x_2, x_2') = 1$. With a similar reasoning one shows that $\mathrm{d}_1(x, x') = 1$ if $x_1 = x_1'$ and $\mathrm{d}_1(x_2, x_2') = 1$. This proves that the point graph of $\mathcal{A}_1$ is isomorphic to $\Gamma_1 \times \Gamma_2$. Since Γ_1 is the point graph of a near polygon, it now follows that also Γ_2 is the point graph of a near polygon. The theorem now readily follows. □

In the sequel we will suppose that Π_1 and Π_2 commute. Then $\mathcal{S}_\Gamma$ is a near polygon.

Theorem 4.13. $\{T_1, T_2\} \in \Delta_{\delta_\mathcal{B}}(\mathcal{S})$.

Proof. Since the spread S_1 of $\mathcal{A}_1$ is admissible, the $\delta_\mathcal{B}$-spread $\{\mathcal{B}(F_1, F_2) \mid F_1 \in S_1\}$ of $\mathcal{A}_1(F_2)$ is admissible, see the proof of Lemma 4.9. Similarly, the $\delta_\mathcal{B}$-spread $\{\mathcal{B}(F_1, F_2) \mid F_2 \in S_2\}$ of $\mathcal{A}_2(F_1)$ is admissible. □

Theorem 4.14. *If $\alpha = (x, F_{i_1}^{(1)}, F_{j_1}^{(2)})$ and $\beta = (y, F_{i_2}^{(1)}, F_{j_2}^{(2)})$ are two points of $\mathcal{S}_\Gamma$, then $d(\alpha, \beta) = d_1(F_{i_1}^{(1)}, F_{i_2}^{(1)}) + d_2(F_{j_1}^{(2)}, F_{j_2}^{(2)}) + d_\mathcal{B}(\Phi_{j_1,j_2}^{(2)} \circ \Phi_{i_1,i_2}^{(1)}(x), y)$. As a consequence,* $\mathrm{diam}(\mathcal{S}_\Gamma) = \delta_1 + \delta_2 - \delta_\mathcal{B}$.

Proof. The element $\mathcal{A}_1(F_{j_2}^{(2)})$ of T_1 is classical and hence it contains a unique point γ_1 nearest to α. The point γ_1 is also the unique point of $\mathcal{B}(F_{i_1}^{(1)}, F_{j_2}^{(2)})$ nearest to α in the subpolygon $\mathcal{A}_2(F_{i_1}^{(1)})$ which is isomorphic to $\mathcal{A}_2$. Hence $\mathrm{d}(\alpha, \gamma_1) = \mathrm{d}_2(F_{j_1}^{(2)}, F_{j_2}^{(2)})$. The element $\mathcal{B}(F_{i_2}^{(1)}, F_{j_2}^{(2)})$ is classical in $\mathcal{A}_1(F_{j_2}^{(2)})$ and hence it contains a unique point γ_2 nearest to γ_1. Since $\mathcal{A}_1(F_{j_2}^{(2)})$ is isomorphic to $\mathcal{A}_1$, we have $\mathrm{d}(\gamma_1, \gamma_2) = \mathrm{d}_1(F_{i_1}^{(1)}, F_{i_2}^{(1)})$. Now, $\mathrm{d}(\alpha, \beta) = \mathrm{d}(\alpha, \gamma_1) + \mathrm{d}(\gamma_1, \beta) = \mathrm{d}(\alpha, \gamma_1) + \mathrm{d}(\gamma_1, \gamma_2) + \mathrm{d}(\gamma_2, \beta) = \mathrm{d}_1(F_{i_1}^{(1)}, F_{i_2}^{(1)}) + \mathrm{d}_2(F_{j_1}^{(2)}, F_{j_2}^{(2)}) + \mathrm{d}_\mathcal{B}(\Phi_{j_1,j_2}^{(2)} \circ \Phi_{i_1,i_2}^{(1)}(x), y)$. This proves the first part of the lemma. Now we choose i_1 ,i_2, j_1 and j_2 such that $\mathrm{d}_1(F_{i_1}^{(1)}, F_{i_2}^{(1)})$ and $\mathrm{d}_2(F_{j_1}^{(2)}, F_{j_2}^{(2)})$ reach their maximal values $\delta_1 - \delta_\mathcal{B}$ and $\delta_2 - \delta_\mathcal{B}$, respectively, see Lemma 4.6. Now, we can choose x and y such that

$d_{\mathcal{B}}(\Phi^{(2)}_{j_1,j_2} \circ \Phi^{(1)}_{i_1,i_2}(x), y)$ reaches the maximal value $\delta_{\mathcal{B}}$. For these choices, $d(\alpha, \beta) = (\delta_1 - \delta_{\mathcal{B}}) + (\delta_2 - \delta_{\mathcal{B}}) + \delta_{\mathcal{B}} = \delta_1 + \delta_2 - \delta_{\mathcal{B}}$ and this is precisely the diameter of $\mathcal{S}_\Gamma$. □

Theorem 4.15. *The spreads T_1, T_2 and S of $\mathcal{S}_\Gamma$ are admissible.*

Proof. By Theorem 4.11, each element of T_1 is classical. Now, choose two arbitrary elements $F_1 = \mathcal{A}_1(F_i^{(2)})$ and $F_2 = \mathcal{A}_1(F_j^{(2)})$ in T_1. By Theorem 4.14, every point of F_1 has distance $d_2(F_i^{(2)}, F_j^{(2)})$ from F_2. Hence T_1 is admissible. By symmetry also T_2 is admissible. Each element of S is, as intersection of two classical elements, again classical, see Lemma 1.6. As before, one derives that S is admissible by relying on Theorem 4.14. □

Theorem 4.16. *The base elements in S_1 and S_2 do not play a special role, i.e. $\mathcal{S}_\Gamma$ can be obtained starting from any base element $F^{(1)}_\lambda$ in S_1 and any base element $F^{(2)}_\mu$ in S_2.*

Proof. Using the base element $F^{(1)}_\lambda$ and $F^{(2)}_\mu$ and the isomorphisms $\theta_1' := \pi^{(1)}_{1,\lambda} \circ \theta_1$ and $\theta_2' := \pi^{(2)}_{1,\mu} \circ \theta_2$, we can define a graph Γ' and an incidence structure $\mathcal{S}_{\Gamma'}$. We will construct an isomorphism between $\mathcal{S}_\Gamma$ and $\mathcal{S}_{\Gamma'}$. Consider the following bijection ϕ between the set of vertices of Γ and the set of vertices of Γ':

$$\phi[(x, F^{(1)}_{i_1}, F^{(2)}_{i_2})] = (\Phi^{(1)}_{i_1,\lambda} \circ \Phi^{(2)}_{i_2,\mu}(x), F^{(1)}_{i_1}, F^{(2)}_{i_2}).$$

Consider now two vertices $\alpha = (x, F^{(1)}_{i_1}, F^{(2)}_{i_2})$ and $\beta = (y, F^{(1)}_{i_1'}, F^{(2)}_{i_2'})$ of Γ. Clearly, α and β are adjacent vertices of type (A) in Γ if and only if $\phi(\alpha)$ and $\phi(\beta)$ are adjacent vertices of type (A) in Γ'. Now, α and β are adjacent vertices of type (B) in Γ if and only if $i_2 = i_2'$, $d_1(F^{(1)}_{i_1}, F^{(1)}_{i_2}) = 1$ and $y = \Phi^{(1)}_{i_1,i_1'}(x)$. The condition $y = \Phi^{(1)}_{i_1,i_1'}(x)$ is equivalent with the following condition (use the definition of the Φ's and recall that Π_1 and Π_2 commute):

$$\Phi^{(1)}_{i_1',\lambda} \circ \Phi^{(2)}_{i_2,\mu}(y) = [\theta_1'^{-1} \circ \pi^{(1)}_{i_1',\lambda} \circ \pi^{(1)}_{i_1,i_1'} \circ \pi^{(1)}_{\lambda,i_1} \circ \theta_1'] \circ [\Phi^{(1)}_{i_1,\lambda} \circ \Phi^{(2)}_{i_2,\mu}(x)].$$

Hence, α and β are adjacent vertices of type (B) in Γ if and only if $\phi(\alpha)$ and $\phi(\beta)$ are adjacent vertices of type (B) in Γ'. By symmetry, this property also holds for adjacent vertices of type (C). This proves the theorem. □

4.4 Basic characterization result for glued near polygons

In Theorem 4.13, we have noticed that $\Delta_\delta(\mathcal{S}) \neq \emptyset$ for every glued near polygon $\mathcal{S}$ of type δ. In the following theorem, we will show that the converse holds in the case of dense near polygons. Theorem 4.17 is an improvement of Theorem 4.2 and is called the *basic characterization theorem for glued near polygons.*

Theorem 4.17. *Let $\mathcal{S}$ be a dense near polygon. If $\{T_1, T_2\} \in \Delta_\delta(\mathcal{S})$ for a certain $\delta \geq 0$, then*

- *all elements of T_1 are isomorphic;*
- *all elements of T_2 are isomorphic;*
- *$\mathcal{S}$ is of type $F_1 \otimes_\delta F_2$ for any $F_1 \in T_1$ and any $F_2 \in T_2$.*

We prove Theorem 4.17 in a series of lemmas.

For every point x of $\mathcal{S} = (\mathcal{P}, \mathcal{L}, \mathrm{I})$ and every $i \in \{1, 2\}$, let $F_i(x)$ denote the unique element of T_i through x. Put $I(x) = F_1(x) \cap F_2(x)$. By our assumptions, $S := \{I(x) \,|\, x \in \mathcal{P}\}$ is a δ-spread of $\mathcal{S}$ and any two elements of S are isomorphic.

Lemma 4.18. *For every two points x and y of $\mathcal{S}$, there exists a shortest path between x and y containing a point of $F_2(x) \cap F_1(y)$.*

Proof. Put $\mathrm{d}(x, y) = k$. We will define points z_i, $i \in \{0, \ldots, k\}$, in the following way.

- Put $z_0 := y$.
- If for a certain $i \in \{0, \ldots, k-1\}$, $z_i \in F_2(x)$, then z_{i+1} denotes a neighbour of z_i at distance $\mathrm{d}(x, z_i) - 1$ from x. Obviously, $z_{i+1} \in F_2(x)$.
- If for a certain $i \in \{0, \ldots, k-1\}$, $z_i \notin F_2(x)$, then z_{i+1} denotes a neighbour of z_i at distance $\mathrm{d}(x, z_i) - 1$ from x and contained in $F_1(z_i)$. If such a point did not exist, then $\mathcal{C}(x, z_i)$ would intersect $F_1(z_i)$ in a point and hence would be contained in $F_2(z_i)$. Then we would have $x \in F_2(z_i)$ or $z_i \in F_2(x)$, a contradiction.

The path $z_0, \ldots, z_k$ satisfies the conditions of the lemma. □

Lemma 4.19. *Every element $T_1 \cup T_2$ is classical in $\mathcal{S}$.*

Proof. Let x denote an arbitrary point of $\mathcal{S}$ and let F denote an arbitrary element of T_1. Since $F_2(x) \cap F$ is classical in $F_2(x)$, it contains a unique point x' nearest to x. Now, let y denote an arbitrary point of F and let y' denote a point of $F_2(x) \cap F$ on a shortest path between x and y, see Lemma 4.18. Since $F_2(x) \cap F$ is classical in $F_2(x)$, there exists a shortest path between x and y' containing the point x'. Hence, there exists also a shortest path between x and y containing x', proving that $\mathrm{d}(x, y) = \mathrm{d}(x, x') + \mathrm{d}(x', y)$. Hence, F is classical in $\mathcal{S}$. In a similar way one shows that also every element of T_2 is classical in $\mathcal{S}$. □

Lemma 4.20. *Every element of S is classical in $\mathcal{S}$. The spread S is admissible.*

Proof. Each element of S is, as intersection of two classical convex subpolygon, itself classical, see Theorem 1.6. Let K_1 and K_2 denote two arbitrary elements of S. Let F_1 denote the unique element of T_1 through K_1 and let F_2 denote the unique element of T_2 through K_2. Put $K_3 = F_1 \cap F_2$. For every point u_1 of K_1 and every point u_2 of K_2, there exists a shortest path between u_1 and u_2 containing a point u_3 of K_3. So, $\mathrm{d}(K_1, K_2) \geq \mathrm{d}(K_1, K_3) + \mathrm{d}(K_3, K_2)$. Now, for every point v_1 of K_1, there exists a point $v_3 \in K_3$ at distance $\mathrm{d}(K_1, K_3)$ from v_1 and a point $v_2 \in K_2$ at distance $\mathrm{d}(K_3, K_2)$ from v_3. As a consequence, $\mathrm{d}(K_1, K_2) = \mathrm{d}(K_1, K_3) + \mathrm{d}(K_3, K_2)$ and for every point v_1 of K_1, there exists a point $v_2 \in K_2$ at distance $\mathrm{d}(K_1, K_2)$ from v_1. This proves that K_1 and K_2 are parallel. □

Lemma 4.21. *The spreads T_1 and T_2 are admissible spreads of $\mathcal{S}$. As a consequence, all elements of T_i, $i \in \{1, 2\}$, are isomorphic.*

Proof. We will show by induction that any two elements of T_i, $i \in \{1, 2\}$, are parallel. Without loss of generality, we may suppose that $i = 1$. Let F_1 and G_1 denote two arbitrary elements of T_1.

(a) Suppose that $\mathrm{d}(F_1, G_1) = 1$. Let $x \in F_1$ and $x' \in G_1$ be points such that $\mathrm{d}(x, x') = 1$. By Theorem 4.5, the convex subpolygon $\mathcal{C}(I(x), x')$ contains $I(x')$. We show that $G_1 \subset \mathcal{C}(F_1, x')$. It suffices to show that every line L through x' not contained in $F_2(x)$ is contained in $\mathcal{C}(F_1, x')$. The quad $\mathcal{C}(L, xx')$ is not contained in $F_2(x)$ and hence intersects F_1 in a line L'. So, $\mathcal{C}(L, xx') = \mathcal{C}(xx', L') \subseteq \mathcal{C}(F_1, x')$. Hence, $L \subseteq \mathcal{C}(F_1, x')$ as claimed. Now, F_1 and G_1 are classical and hence also big in $\mathcal{C}(F_1, x')$. It follows that F_1 and G_1 are parallel.

(b) Suppose that $\mathrm{d}(F_1, G_1) \geq 2$. Let $k \in F_1$ and $l \in G_1$ be points such that $\mathrm{d}(k, l) = \mathrm{d}(F_1, G_1)$. Let m denote a point in $\Gamma_1(l)$ at distance $\mathrm{d}(k, l) - 1$ from k and let H_1 denote the unique element of T_1 through m. Then $\mathrm{d}(F_1, H_1) \leq \mathrm{d}(F_1, G_1) - 1$ and $\mathrm{d}(H_1, G_1) = 1$. Hence $G_1 \| H_1$ by the previous step. By the induction hypothesis, $F_1 \| H_1$. Hence, for every point k' of F_1, there exists a point m' in H_1 at distance $d(F_1, H_1)$ from k' and a point l' in G_1 collinear with m'. So, for every point k' in F_1, there exists a point l' in G_1 at distance at most $\mathrm{d}(F_1, H_1) + 1 \leq \mathrm{d}(F_1, G_1)$ (and hence exactly $\mathrm{d}(F_1, G_1)$) from k'. It now follows that F_1 and G_1 are parallel. □

Lemma 4.22. *$\mathcal{S}$ is a glued near polygon of type $F_1 \otimes_\delta F_2$ for every $F_1 \in T_1$ and every $F_2 \in T_2$.*

Proof. (i) If we intersect F_i, $i \in \{1, 2\}$, with all elements of T_{3-i}, then we obtain an admissible δ-spread S_i of F_i. If we consider $F_1 \cap F_2$ as base element in both spreads S_1 and S_2 and if we take θ_1 and θ_2 equal to the trivial permutation of this base element, then by Section 4.3, we have all ingredients to construct an incidence structure which we will denote by $F_1 \otimes F_2$. We will construct an isomorphism between $\mathcal{S}$ and $F_1 \otimes F_2$ and from the existence of such an isomorphism it will follow that $\mathcal{S}$ and $F_1 \otimes F_2$ are glued near polygons.

(ii) For every point x of $\mathcal{S}$, we define $\theta(x) := (\pi(x), F_2(x) \cap F_1, F_1(x) \cap F_2)$ where $\pi(x)$ denotes the unique point of $F_1 \cap F_2$ nearest to x. Obviously, $\theta(x)$ is a point of $F_1 \otimes F_2$. Now, consider the equation $\theta(x) = (x', K_1, K_2)$ $(*)$ with

(x', K_1, K_2) a given point of $F_1 \otimes F_2$. Let G_i, $i \in \{1,2\}$, be the unique element of T_i through K_{3-i}, and let y denote the unique point of $G_1 \cap G_2$ nearest to x'. Since $F_1 \cap F_2 \parallel G_1 \cap G_2$, the point y is the unique solution of the equation $(*)$. It follows that θ is a bijection between the point sets of $\mathcal{S}$ and $F_1 \otimes F_2$.

(iii) For every point x of $\mathcal{S}$, let $\pi_i(x)$, $i \in \{1,2\}$, denote the unique point of F_i nearest to x. By Lemma 4.18, $\pi_i(x) \in F_{3-i}(x) \cap F_i$. Since F_i is classical in $\mathcal{S}$, $\pi(x)$ is the unique point of $F_1 \cap F_2$ nearest to $\pi_i(x)$. Since $F_1 \cap F_2 \parallel F_{3-i}(x) \cap F_i$, $\pi_i(x)$ is the unique point of $F_{3-i}(x) \cap F_i$ nearest to $\pi(x)$.

(iv) We will now show that θ determines an isomorphism between $\mathcal{S}$ and $F_1 \otimes F_2$. By Lemma 4.21, two different points x and y of $\mathcal{S}$ are collinear if and only if $F_i(x) = F_i(y)$ and $\mathrm{d}(\pi_i(x), \pi_i(y)) = 1$ for at least one $i \in \{1,2\}$. By (iii) and Section 4.3 this is precisely the condition for $\theta(x) = (\pi(x), F_2(x) \cap F_1, F_1(x) \cap F_2)$ and $\theta(y) = (\pi(y), F_2(y) \cap F_1, F_1(y) \cap F_2)$ to be collinear points of $F_1 \otimes F_2$. Hence θ is an isomorphism between the collinearity graphs of $\mathcal{S}$ and $F_1 \otimes F_2$. As a consequence also the near polygons $\mathcal{S}$ and $F_1 \otimes F_2$ are isomorphic. By Theorem 4.11 it follows that $\mathcal{S}$ is glued. □

4.5 Other characterizations of glued near polygons

4.5.1 Characterization of finite glued near hexagons

An (h,k)-*cross* is the unique linear space which has a point that is incident with precisely two lines, one of length h and one of length k. If Q_i, $i \in \{1,2\}$, is a generalized quadrangle of order (s, t_i) and if $\mathcal{S}$ is a glued near hexagon of type $Q_1 \otimes Q_2$, then every local space of $\mathcal{S}$ is a (t_1+1, t_2+1)-cross. We will use this property to characterize finite glued near hexagons. We will make use of the following easy lemma.

Lemma 4.23. *Let $\mathcal{A}$ be a finite linear space that has the following properties:*

(a) *$\mathcal{A}$ has $t_1 + t_2 + 1$ points, $t_1, t_2 \in \mathbb{N} \setminus \{0,1\}$,*

(b) *$\mathcal{A}$ has a line L of length $t_1 + 1$ which meets every other line of $\mathcal{A}$;*

then the number N of lines of size 2 is at most $t_1 t_2$ with equality if and only if $\mathcal{A}$ is a (t_1+1, t_2+1)-cross.

Proof. Let V denote the set of all pairs (x,y) with x and y two different points not contained in L. For every $v = (x,y)$ of V, let $\alpha_v + 1$ denote the number of points on the line xy. Counting incident point-line pairs (p, M) with $p \notin L$, we find

$$t_2(t_1+1) = N + \sum_{v \in V} \frac{1}{\alpha_v - 1}.$$

Now, $\alpha_v \leq t_2$ for every $v \in V$. Hence $N \leq (t_1+1)t_2 - \frac{|V|}{t_2 - 1} = t_1 t_2$. If equality holds, then $\alpha_v = t_2$ for every $v \in V$ and $\mathcal{A}$ is a (t_1+1, t_2+1)-cross. □

Theorem 4.24. *Let $\mathcal{S}$ be a finite near hexagon that has the following properties:*

(a) *there exists a line which is incident with at least three points,*

(b) *every two points at distance 2 have at least two common neighbours,*

(c) *there exists a point x for which $\mathcal{L}(\mathcal{S},x)$ is a (t_1+1,t_2+1)-cross, $t_1,t_2 \in \mathbb{N}\setminus\{0\}$;*

then $\mathcal{S}$ is a glued near hexagon or is isomorphic to the direct product of a line with a generalized quadrangle.

Proof. If $\mathcal{S}$ is a product near hexagon, then it is a direct product of a line and a generalized quadrangle and we are done. In the sequel, we will suppose that $\mathcal{S}$ is not a product near polygon. By Corollary 1.13, it then follows that every line of $\mathcal{S}$ is incident with the same number of points. We denote this constant number by $s+1$. By assumption (a), $s \geq 2$. So, $\mathcal{S}$ is a dense near polygon and every two points at distance 2 from each other are contained in a unique quad. Put $t := t_{\mathcal{S}}$. By assumption (c), $t = t_1 + t_2$. If $t_1 = 1$ or $t_2 = 1$, then by Theorem 4.4, $\mathcal{S}$ is a product near polygon, a contradiction. So, $t_1, t_2 \neq 1$. Let Q_1 and Q_2 be the two quads through x with respective orders (s,t_1) and (s,t_2). By Theorem 2.31, Q_1 and Q_2 are big. We will prove that every local space $\mathcal{L}(\mathcal{S},y)$ is a (t_1+1,t_2+1)-cross. Since $\mathcal{S}$ is connected, it suffices to prove this for points y collinear with x. Without loss of generality, we may suppose that $y \in Q_1 \setminus Q_2$. Let x_i, $i \in \{1,\dots,t_2\}$, denote t_2 pairwise noncollinear points of $(Q_2 \cap \Gamma_1(x)) \setminus Q_1$, and let z_i denote the common neighbour of x_i and y different from x. Through each x_i there are $t_1 - 1$ lines $L_{i,j}$, $1 \leq j \leq t_1 - 1$, not contained in $Q_2 \cup x_i z_i$. We can define $t_1 - 1$ distinct lines $M_{i,j} \neq x_i z_i$ through z_i such that $L_{i,j}$ and $M_{i,j}$ are contained in a common quad. Let $R_{i,j}$, respectively $\tilde{R}_{i,j}$, denote the quad through $z_i y$ and $M_{i,j}$, respectively $x_i x$ and $L_{i,j}$. Since $\mathcal{L}(\mathcal{S},x)$ is a cross, $\tilde{R}_{i,j}$ is a grid. Since the lines $x_i x$, $L_{i,j}$ and $\tilde{R}_{i,j} \cap Q_1$ of the grid $\tilde{R}_{i,j}$ are contained in $\Gamma_1(R_{i,j})$, $\tilde{R}_{i,j}$ is contained in $\Gamma_1(R_{i,j})$ and projects to a subgrid of $R_{i,j}$. Since $R_{i,j}$ has a subgrid, $t_{R_{i,j}} = 1$ or $t_{R_{i,j}} \geq s$ by Theorem 1.30. Suppose that a quad $R := R_{i,j}$ through y satisfies $t_R \geq s$. If $t_R = t_2$, then clearly $\mathcal{L}(\mathcal{S},y)$ is a (t_1+1,t_2+1)-cross. Suppose therefore that $s \leq t_R < t_2$, and let $R' \neq Q_1$ denote a third quad through $R \cap Q_1$. Since Q_2 is big, R and R' project to subGQ's S and S' of Q_2. Through an arbitrary point z of $S' \setminus S$ there are $1 + st_R$ lines intersecting S and $t_{R'}$ lines completely contained in $S' \setminus S$. Now $t_2 + 1 \geq 1 + st_R + t_{R'} > s^2 + 1$, contradicting Higman's inequality (Theorem 1.28). As a consequence all quads $R_{i,j}$ are grids; hence they are all different. Since there are also t_2 grid-quads through the line xy, there are at least $t_2(t_1 - 1) + t_2 = t_1 t_2$ grid-quads through y. By Lemma 4.23, it then follows that $\mathcal{L}(\mathcal{S},y)$ is a (t_1+1,t_2+1)-cross. As mentioned earlier the connectedness of $\mathcal{S}$ now implies that every local space is a (t_1+1,t_2+1)-cross. It is now easily seen that there are two partitions T_1 and T_2 in quads satisfying the conditions of Theorem 4.17. It follows that $\mathcal{S}$ is a glued near hexagon. □

4.5.2 Characterization of general glued near polygons

We will give a characterization result for glued near polygons similar to the one given in Theorem 4.17.

Let $\mathcal{A}$ be a dense near polygon with diameter at least 2 and let T_i, $i \in \{1,2\}$, be a partition of $\mathcal{A}$ in convex subpolygons. For every point x of $\mathcal{A}$, let $F_i(x)$, $i \in \{1,2\}$, denote the unique element of T_i through x and we put $I(x) = F_1(x) \cap F_2(x)$. We also suppose that the following condition is satisfied for every point x of $\mathcal{A}$:

$$\Gamma_1(x) \subseteq F_1(x) \cup F_2(x).$$

Theorem 4.25. (i) *If x and y are two arbitrary points of $\mathcal{A}$, then there exists a shortest path between x and y containing a point of $F_2(x) \cap F_1(y)$. Hence $F_1 \cap F_2 \neq \emptyset$ for every $F_1 \in T_1$ and every $F_2 \in T_2$.*

(ii) *For every $F_1 \in T_1$ and every $F_2 \in T_2$, $\mathrm{diam}(\mathcal{A}) = \mathrm{diam}(F_1) + \mathrm{diam}(F_2) - \mathrm{diam}(F_1 \cap F_2)$.*

Proof. The proof of (i) is completely similar to the proof of Lemma 4.18. We will now show (ii).

- $\mathrm{diam}(\mathcal{A}) \leq \mathrm{diam}(F_1) + \mathrm{diam}(F_2) - \mathrm{diam}(F_1 \cap F_2)$.
  Let x denote an arbitrary point of $F_1 \cap F_2$. Since every line through x is contained in $F_1 \cup F_2$ and since every convex subspace through x intersects F_i, $i \in \{1,2\}$, in a convex subpolygon, every chain $G_0 \subset G_1 \subset \cdots \subset G_k$ of convex subpolygons through x such that $G_0 = \{x\}$ and $G_i = F_1 \cap F_2$ for a certain $i \in \{0,\ldots,k\}$ has length at most $\mathrm{diam}(F_1 \cap F_2) + (\mathrm{diam}(F_1) - \mathrm{diam}(F_1 \cap F_2)) + (\mathrm{diam}(F_2) - \mathrm{diam}(F_1 \cap F_2))$. This proves that $\mathrm{diam}(\mathcal{A}) \leq \mathrm{diam}(F_1) + \mathrm{diam}(F_2) - \mathrm{diam}(F_1 \cap F_2)$.

- $\mathrm{diam}(\mathcal{A}) \geq \mathrm{diam}(F_1) + \mathrm{diam}(F_2) - \mathrm{diam}(F_1 \cap F_2)$.
  By Theorem 2.29 there exists a point x in F_1 which is classical with respect to $F_1 \cap F_2$ and has distance $\mathrm{diam}(F_1) - \mathrm{diam}(F_1 \cap F_2)$ from a point $x' \in F_1 \cap F_2$. Now, take a point y in F_2 at distance $\mathrm{diam}(F_2)$ from x'. By (i) there exists a shortest path between x and y containing a point z of $F_1 \cap F_2$. Now, $\mathrm{d}(x,y) = \mathrm{d}(x,z) + \mathrm{d}(z,y) = \mathrm{d}(x,x') + \mathrm{d}(x',z) + \mathrm{d}(z,y) = \mathrm{d}(x,x') + \mathrm{d}(x',y) = \mathrm{diam}(F_1) + \mathrm{diam}(F_2) - \mathrm{diam}(F_1 \cap F_2)$. As a consequence, $\mathrm{diam}(\mathcal{A}) \geq \mathrm{diam}(F_1) + \mathrm{diam}(F_2) - \mathrm{diam}(F_1 \cap F_2)$. □

Lemma 4.26. *The following are equivalent for every point x^* of $\mathcal{A}$:*

(i) $|T_1| = 1$,

(ii) $I(x^*) = F_2(x^*)$,

(iii) $\Gamma_1(x^*) \setminus F_1(x^*) = \emptyset$.

Proof. We have $|T_1| = 1 \Leftrightarrow F_1(x^*) = \mathcal{A} \Leftrightarrow \Gamma_1(x^*) \setminus F_1(x^*) = \emptyset \Leftrightarrow \Gamma_1(x^*) \subseteq F_1(x^*) \Leftrightarrow F_2(x^*) \subseteq F_1(x^*) \Leftrightarrow I(x^*) = F_2(x^*)$. □

Theorem 4.27. *Suppose that*

- $I(x^*)$ *is classical in* $F_1(x^*)$,
- $\mathcal{C}(\Gamma_1(x^*) \setminus F_2(x^*)) \cap I(x^*) = \{x^*\}$,

for a certain point x^* *of* $\mathcal{A}$. *Then* $F_1(x^*) \cong I(x^*) \times F_3(x^*)$ *and* $\mathcal{A} \cong F_2(x^*) \times F_3(x^*)$, *where* $F_3(x^*) := \mathcal{C}(\Gamma_1(x^*) \setminus F_2(x^*))$.

Proof. Let y be a point of $F_3(x^*)$ at distance $\mathrm{diam}(F_3(x^*))$ from x^*. Since y is classical with respect to $I(x^*)$, we have $\mathrm{diam}(F_3(x^*)) + \mathrm{diam}(I(x^*)) \leq \mathrm{diam}(F_1(x^*))$. Since $F_1(x^*) = \mathcal{C}(F_3(x^*), I(x^*))$, we also have $\mathrm{diam}(F_1(x^*)) \leq \mathrm{diam}(F_3(x^*)) + \mathrm{diam}(I(x^*))$. So, $\mathrm{diam}(F_1(x^*)) = \mathrm{diam}(F_3(x^*)) + \mathrm{diam}(I(x^*))$ and $\mathrm{diam}(\mathcal{A}) = \mathrm{diam}(F_1(x^*)) + \mathrm{diam}(F_2(x^*)) - \mathrm{diam}(I(x^*)) = \mathrm{diam}(F_2(x^*)) + \mathrm{diam}(F_3(x^*))$. Now, the pairs $(I(x^*), F_3(x^*))$ and $(F_2(x^*), F_3(x^*))$ satisfy the conditions of Theorem 4.4. Hence $F_1(x^*) \cong I(x^*) \times F_3(x^*)$ and $\mathcal{A} \cong F_2(x^*) \times F_3(x^*)$. □

Theorem 4.28 (Section 4.5.3). *Suppose that* $\mathcal{A}$ *satisfies the following properties for every point* x *of* $\mathcal{A}$ *and every* $i \in \{1, 2\}$:

- $I(x)$ *is classical in* $F_i(x)$,
- $I(x) \subseteq \mathcal{C}(\Gamma_1(x) \setminus F_i(x))$.

Then there exist near polygons $\mathcal{A}_1$, $\mathcal{A}_2$ *and* $\mathcal{B}$ *such that*

(i) $F_1(x) \cong \mathcal{A}_1$, $F_2(x) \cong \mathcal{A}_2$ *and* $I(x) \cong \mathcal{B}$ *for every point* x *of* $\mathcal{A}$;

(ii) $\mathcal{A}$ *is a glued near polygon of type* $\mathcal{A}_1 \otimes_{\mathcal{B}} \mathcal{A}_2$.

Corollary 4.29. *If* $|T_1|, |T_2| \geq 2$ *and* $\mathrm{diam}(I(x)) \leq 1$ *for every point* $x \in \mathcal{A}$, *then there exist subpolygons* $\mathcal{A}_1$ *and* $\mathcal{A}_2$ *such that* $\mathcal{A}$ *is glued of type* $\mathcal{A}_1 \otimes_k \mathcal{A}_2$ *for a certain* $k \in \{0, 1\}$.

Proof. For every point x of $\mathcal{A}$, $\mathrm{diam}(I(x)) \leq 1$ and hence $I(x)$ is classical in $F_1(x)$ and $F_2(x)$. If also the second condition of Theorem 4.28 is satisfied, then we are done. So, we may suppose that there exist a point x^* and an $i \in \{1, 2\}$ such that $I(x^*) \not\subseteq \mathcal{C}(\Gamma_1(x^*) \setminus F_i(x^*))$. Then $X := I(x^*) \cap \mathcal{C}(\Gamma_1(x^*) \setminus F_i(x^*))$ is either equal to $\{x^*\}$ or to $\emptyset$. If $X = \emptyset$, then $\Gamma_1(x^*) \setminus F_i(x^*) = \emptyset$, contradicting Lemma 4.26. So, $X = \{x^*\}$. The corollary now easily follows from Theorem 4.27. □

4.5.3 Proof of Theorem 4.28

We prove Theorem 4.28 in a series of lemmas.

Lemma 4.30. *Every element of* $T_1 \cup T_2$ *is classical. Every element of* $S := \{I(x) | x \in \mathcal{A}\}$ *is classical.*

Proof. The proof is completely similar to the proof of Lemma 4.19. □

Lemma 4.31. *If* x *and* y *are collinear points of* $\mathcal{A}$, *then* $I(x) \parallel I(y)$, $F_1(x) \parallel F_1(y)$ *and* $F_2(x) \parallel F_2(y)$.

Proof. If $y \in I(x)$, then $I(y) = I(x)$, $F_1(x) = F_1(y)$ and $F_2(x) = F_2(y)$. Suppose therefore that $y \notin I(x)$. Without loss of generality we may suppose that $y \in F_1(x) \setminus F_2(x)$. Then $F_1(x) = F_1(y)$. We will now prove that the convex subpolygons $G := \mathcal{C}(F_2(x), y)$ and $G' := \mathcal{C}(F_2(y), x)$ coincide. If $z \in \Gamma_1(y) \setminus F_1(y)$, then the quad $\mathcal{C}(x, z)$ intersects $F_2(x)$ in a line and so z is collinear with a point z' of $F_2(x)$. Hence $z \in \mathcal{C}(z', y) \subseteq G$, $F_2(y) = \mathcal{C}(\Gamma_1(y) \setminus F_1(y)) \subseteq G$ and $G' = \mathcal{C}(F_2(y), x) \subseteq G$. In a similar way one proves that $G \subseteq G'$. Hence $G = G'$ and $d_2 := \mathrm{diam}(F_2(x)) = \mathrm{diam}(G) - 1 = \mathrm{diam}(G') - 1 = \mathrm{diam}(F_2(y))$. By Lemma 4.30, the convex sub-$2d_2$-gons $F_2(x)$ and $F_2(y)$ are classical in the convex sub-$(2d_2+2)$-gon G. Hence, every point of $F_2(x)$ (respectively $F_2(y)$) has distance 1 to $F_2(y)$ (respectively $F_2(x)$). This proves that $F_2(x) \| F_2(y)$. Clearly $\pi_{F_2(x) \to F_2(y)}(I(x)) = I(y)$ and hence also $I(x)$ and $I(y)$ are parallel. □

Lemma 4.32. *The spreads T_1, T_2 and S are admissible spreads of $\mathcal{A}$.*

Proof. Let X be one of the sets T_1, T_2 or S. Each element of X is classical; hence for all elements $K, L \in X$, $K \| L$ if and only if for every $k' \in K$ there exists a point $l' \in L$ such that $\mathrm{d}(k', l') = \mathrm{d}(K, L)$. We will prove by induction on $\mathrm{d}(K, L)$ that $K \| L$. By Lemma 4.31, $K \| L$ if $\mathrm{d}(K, L) \leq 1$. Suppose therefore that $\mathrm{d}(K, L) \geq 2$ and let $k \in K$ and $l \in L$ be points such that $\mathrm{d}(k, l) = \mathrm{d}(K, L)$. Let m denote a point in $\Gamma_1(l)$ at distance $\mathrm{d}(k, l) - 1$ from k and let M denote the unique element of X through m. Then $\mathrm{d}(K, M) \leq \mathrm{d}(K, L) - 1$ and $\mathrm{d}(M, L) = 1$. By the induction hypothesis, $K \| M$ and $M \| L$. Hence, for every point k' of K, there exists a point m' in M at distance $d(K, M)$ from k and a point l' in L collinear with m'. So, for every point k' in K, there exists a point l' in L such that $\mathrm{d}(K, L) \leq \mathrm{d}(k', l') \leq \mathrm{d}(K, M) + 1 \leq \mathrm{d}(K, L)$. Hence $K \| L$ and X is admissible. □

It is now clear that $\{T_1, T_2\} \in \Delta_\delta(A)$, where δ is the diameter of an arbitrary element of S. By Theorem 4.17, we then have:

Corollary 4.33. (a) *There exist near polygons $\mathcal{A}_1$, $\mathcal{A}_2$ and $\mathcal{B}$ such that $F_1(x) \cong \mathcal{A}_1$, $F_2(x) \cong \mathcal{A}_2$ and $I(x) \cong \mathcal{B}$ for every point x of $\mathcal{A}$.*

(b) *$\mathcal{A}$ is a glued near polygon of type $\mathcal{A}_1 \otimes_{\mathcal{B}} \mathcal{A}_2$.*

4.6 Subpolygons

We will use the same notation as in Section 4.3. Let $\mathcal{S}_\Gamma$ be a glued near polygon; so, Π_1 and Π_2 commute. Let F denote a convex subpolygon of $\mathcal{S}_\Gamma$. Without loss of generality, see Theorem 4.16, we may suppose that $F \cap \mathcal{B}(F_1^{(1)}, F_1^{(2)}) \neq \emptyset$. Put $\mathcal{B}' := \{x \in \mathcal{B} | (x, F_1^{(1)}, F_1^{(2)}) \in F\}$. Since $F \cap \mathcal{B}(F_1^{(1)}, F_1^{(2)})$ is convex, $\mathcal{B}'$ is a convex subpolygon of $\mathcal{B}$.

Lemma 4.34. *F meets $\mathcal{A}_1(F_j^{(2)})$ and $\mathcal{A}_2(F_i^{(1)}) \Leftrightarrow F$ meets $\mathcal{B}(F_i^{(1)}, F_j^{(2)})$.*

Proof. If α is a point of $\mathcal{A}_1(F_j^{(2)})$ and β is a point of $\mathcal{A}_2(F_i^{(1)})$, then there exists a shortest path between α and β containing a point of $\mathcal{B}(F_i^{(1)}, F_j^{(2)})$. Since F is convex the $\Rightarrow$-part immediately follows. The $\Leftarrow$-part also is clear since $\mathcal{B}(F_i^{(1)}, F_j^{(2)}) = \mathcal{A}_1(F_j^{(2)}) \cap \mathcal{A}_2(F_i^{(1)})$. □

Put $I := \{i \in \{1, \ldots, n_1\} \mid F \cap \mathcal{A}_2(F_i^{(1)}) \neq \emptyset\}$ and $J := \{j \in \{1, \ldots, n_2\} \mid F \cap \mathcal{A}_1(F_j^{(2)}) \neq \emptyset\}$. By Lemma 4.34, F meets $\mathcal{B}(F_i^{(1)}, F_j^{(2)})$ if and only if $i \in I$ and $j \in J$.

Lemma 4.35. *The points of F are precisely the points $(x, F_i^{(1)}, F_j^{(2)})$ where $x \in \mathcal{B}'$, $i \in I$ and $j \in J$.*

Proof. The spread S induces an admissible spread S^* in F, see Theorem 4.5. Moreover, the projection maps between two elements of S^* are inherited from the projection maps between the corresponding elements of S. So, a point $\alpha = (x, F_i^{(1)}, F_j^{(2)})$ belongs to F, if and only if the unique point α' in $\mathcal{B}(F_1^{(1)}, F_1^{(2)})$ nearest to α belongs to F. Since $\alpha' = (x, F_1^{(1)}, F_1^{(2)})$, the lemma easily follows. □

Theorem 4.36. *One of the following occurs.*

(a) *F is contained in an element of $T_1 \cup T_2$ and hence is isomorphic to a subpolygon of $\mathcal{A}_1$ or $\mathcal{A}_2$.*

(b) *F is glued and of type $F_1 \otimes F_2$, where F_i, $i \in \{1, 2\}$, is isomorphic to a convex subpolygon of $\mathcal{A}_i$.*

Proof. Again we suppose that $F \cap \mathcal{B}(F_1^{(1)}, F_1^{(2)}) \neq \emptyset$ and we use the notation as before. If $|I| = 1$ or $|J| = 1$, then we clearly have case (a). Suppose therefore that $|I|, |J| \geq 2$. The subpolygon $F \cap \mathcal{A}_i(F_1^{(3-i)})$, $i \in \{1, 2\}$, is a convex subpolygon of $\mathcal{A}_i(F_1^{(3-i)})$. The isomorphism μ defined in the proof of Lemma 4.9 (or the similar one if $i = 2$) maps $F \cap \mathcal{A}_i(F_1^{(3-i)})$ to a convex subpolygon $\mathcal{A}_i'$ of $\mathcal{A}_i$ and $F \cap \mathcal{B}(F_1^{(1)}, F_1^{(2)})$ to a convex subpolygon $\mathcal{B}'$ of $\mathcal{A}_i'$. Since $|I|, |J| \geq 2$, $\mathrm{diam}(\mathcal{A}_i') > \mathrm{diam}(\mathcal{B}')$. The elements of S^* which are completely contained in $\mathcal{A}_i(F_1^{(3-i)})$ define an admissible spread of $\mathcal{A}_i(F_1^{(3-i)})$ which, by using μ, can be transformed to an admissible spread S_i' of $\mathcal{A}_i'$. Now, define $\theta_i' := \theta_{i|\mathcal{B}'}$. We leave it as a straightforward exercise to the reader to verify that the objects $\mathcal{B}'$, $\mathcal{A}_1'$, $\mathcal{A}_2'$, S_1', S_2', θ_1' and θ_2' give rise to a glued near polygon isomorphic to F. □

The following theorem gives the convex subpolygons of $\mathcal{S}_\Gamma$ in terms of the convex subpolygons of $\mathcal{A}_1$ and $\mathcal{A}_2$.

Theorem 4.37. *Let $\mathcal{B}'$ be a nonempty set of points of $\mathcal{B}$, let $I \subseteq \{1, \ldots, n_1\}$ and let $J \subseteq \{1, \ldots, n_2\}$ such that $1 \in I$ and $1 \in J$. Then $G := \{(x, F_i^{(1)}, F_j^{(2)}) | x \in \mathcal{B}', i \in I, j \in J\}$ is convex if and only if $G_1 := \{(x, F_i^{(1)}, F_1^{(2)}) \mid x \in \mathcal{B}', i \in I\}$ and $G_2 := \{(x, F_1^{(1)}, F_j^{(2)}) | x \in \mathcal{B}', j \in J\}$ are convex.*

Proof. If G is convex, then also $G_1 = G \cap \mathcal{A}_1(F_1^{(2)})$ and $G_2 = G \cap \mathcal{A}_2(F_1^{(1)})$ are convex. Suppose now that G_1 and G_2 are convex. In order to prove that G is convex, it suffices to show that $\Gamma_1(\alpha) \cap \Gamma_{\mathrm{d}(\alpha,\beta)-1}(\beta) \subseteq G$ for all points α and β of G. Let $\alpha = (x_1, F_{i_1}^{(1)}, F_{j_1}^{(2)})$ and $\beta = (x_2, F_{i_2}^{(1)}, F_{j_2}^{(2)})$ denote two points of G and let γ denote an arbitrary element of $\Gamma_1(\alpha) \cap \Gamma_{\mathrm{d}(\alpha,\beta)-1}(\beta)$. Suppose that γ is of the form $(x_3, F_{i_3}^{(1)}, F_{j_1}^{(2)})$. Since $\mathcal{A}_1(F_{j_1}^{(2)})$ is classical, $\gamma \in \Gamma_1(\alpha) \cap \Gamma_{\mathrm{d}(\alpha,\beta')-1}(\beta')$ where β' denotes the unique point of $\mathcal{A}_1(F_{j_1}^{(2)})$ nearest to β. Since G_1 is convex, also $G_1' := \{(x, F_i^{(1)}, F_{j_1}^{(2)}) | x \in \mathcal{B}', i \in I\}$ is convex. Since $\alpha, \beta' \in G_1'$, also $\gamma \in G_1'$. Hence $x_3 \in \mathcal{B}'$, $i_3 \in I$ and $\gamma \in G$. In a similar way, one proves that $\gamma \in G$ if γ is of the form $(x_3, F_{i_1}^{(1)}, F_{j_3}^{(2)})$. Hence $\Gamma_1(\alpha) \cap \Gamma_{\mathrm{d}(\alpha,\beta)-1}(\beta) \subseteq G$ for all points α and β of G, proving that G is convex. □

4.7 Glued near polygons of type $\delta \in \{0,1\}$

As we already have mentioned, any known glued near polygon is either of type 0 or 1. In this section, we will give a more detailed study of glued near polygons of type 0 and 1.

4.7.1 Glued near polygons of type 0

Theorem 4.38. *If $\mathcal{A}_1$ and $\mathcal{A}_2$ are two near polygons of diameter at least 1, then $\mathcal{A}_1 \times \mathcal{A}_2$ is the (up to isomorphism) unique glued near polygon of type $\mathcal{A}_1 \otimes_0 \mathcal{A}_2$.*

Proof. Clearly, $\mathcal{A}_i$, $i \in \{1,2\}$, has a unique 0-spread S_i, namely the set $\binom{\mathcal{P}_i}{1}$ of all singletons of the point set $\mathcal{P}_i$. The near polygon $\mathcal{B}$ necessarily is the unique near 0-gon. By Theorem 4.16, we may choose arbitrary base elements in S_1 and S_2. For any choice of the base elements, θ_1 and θ_2 are uniquely determined. The glueing construction described in Section 4.3 now yields an incidence structure $\mathcal{A}_1 \otimes_0 \mathcal{A}_2$. Since $|\mathcal{B}| = 1$, there exists a natural bijection between the point set $\mathcal{B} \times S_1 \times S_2$ of $\mathcal{A}_1 \otimes_0 \mathcal{A}_2$ and the point set $\mathcal{P}_1 \times \mathcal{P}_2$ of $\mathcal{A}_1 \times \mathcal{A}_2$, namely $(*, \{p_1\}, \{p_2\}) \mapsto (p_1, p_2)$. This bijection clearly determines an isomorphism. □

4.7.2 Spreads of symmetry

Definitions.

(A) If K and L are two lines of a near polygon, then $\{K, L\}^\perp$ denotes the set of all lines meeting K and L and $\{K, L\}^{\perp\perp}$ denotes the set of all lines meeting every line of $\{K, L\}^\perp$.

(B) An admissible 1-spread S of a near polygon is called a *regular spread* if for all $K, L \in S$ with $\mathrm{d}(K, L) = 1$, (i) $\{K, L\}^{\perp\perp}$ cover the same set of points as $\{K, L\}^\perp$, and (ii) $\{K, L\}^{\perp\perp} \subseteq S$.

(C) A 1-spread S of a near polygon $\mathcal{A}$ is called a *spread of symmetry* if for every line $K \in S$ and for all points $k_1, k_2 \in K$, there exists an automorphism of $\mathcal{A}$ fixing each line of S and mapping k_1 to k_2. Every spread of symmetry is a regular spread (see e.g. the proof of Theorem 4.42 for the case of GQ's). Every trivial spread is a spread of symmetry.

Let S be an admissible 1-spread of a near $2d$-gon $\mathcal{A} = (\mathcal{P}, \mathcal{L}, \mathrm{I})$. The full group of automorphisms of $\mathcal{A}$ fixing each line of S is denoted by G_S. For every two lines K and L of S, let p_L^K denote the projection from K onto L. For a line $K \in S$, we call $\Pi_S(K) = \langle P_K^M \circ P_M^L \circ P_L^K \mid L, M \in S \rangle$ the *group of projectivities of* K *with respect to* S.

Theorem 4.39. (a) *The group $\Pi_S(K)$ is trivial if and only if the spread S is trivial.*

(b) *If $\Pi_S(K)$ is not the trivial group, then $\Pi_S(K)$ acts transitively on K.*

(c) *If S is not trivial, then G_S acts semiregularly on each line of S.*

Proof. (a) This follows from Theorem 4.12.

(b) Let $x \in K$ and $\theta \in \Pi_S(K)$ such that $x^\theta \neq x$. It is sufficient to prove that the orbit of x under $\Pi_S(K)$ is equal to K. So, let $\tilde{x}$ be an arbitrary point of K. There exists a path $x = x_0, x_1, \dots, x_k = x^\theta$ in $\mathcal{A}$ such that $\mathrm{d}(x_i, x_{i+1}) = 1$ and $x_i x_{i+1} \notin S$ for all $i \in \{0, \dots, k-1\}$. Take now the smallest i such that x is not the unique point of K nearest to x_i, and let y be the unique point of $x_{i-1}x_i$ nearest to $\tilde{x}$. Since $x_{i-1}x_i$ and K are parallel, $\tilde{x}$ is the unique point of K nearest to y. If L and M are the elements of S through x_{i-1} and y, respectively, then $P_K^M \circ P_M^L \circ P_L^K$ maps x to $\tilde{x}$, proving the result.

(c) If $\theta \in G_S$ fixes a point $x \in K$, then θ also fixes every point of $x^{\Pi_S(K)}$ and hence the whole point set of $\mathcal{A}$ if S is not trivial, see (b). Hence, if S is nontrivial, then only the trivial element of G_S has fixpoints. □

Theorem 4.40. *If $\theta \in G_S$, then θ induces a permutation $\bar{\theta}$ on the point set of $K \in S$ that commutes with each element of $\Pi_S(K)$. Conversely, if a permutation ϕ on the point set of K commutes with each element of $\Pi_S(K)$, then $\phi = \bar{\theta}$ for some $\theta \in G_S$.*

Proof. Let $\theta \in G_S$ and let $\bar{\theta}$ be the permutation on the point set of K induced by θ. If L and M are two arbitrary lines of S, then $\theta \circ p_M^L(x) = p_M^L \circ \theta(x)$ for every point x of L. Since $\theta = p_*^K \circ \bar{\theta} \circ p_K^*$, $p_K^M \circ p_M^L \circ p_L^K \circ \bar{\theta} = \bar{\theta} \circ p_K^M \circ p_M^L \circ p_L^K$ for all lines L and M of S. This proves that $\bar{\theta}$ commutes with every element of $\Pi_S(K)$. Conversely, suppose that a permutation ϕ on the point set of K commutes with every element of $\Pi_S(K)$. Then for every point x of a line L of S, we define $x^\theta := p_L^K \circ \phi \circ p_K^L(x)$. Suppose that x_1 and x_2 are two collinear points belonging to different lines K_1 and K_2 of S. Then $p_{K_2}^{K_1}(x_1^\theta) = p_{K_2}^{K_1} \circ p_{K_1}^K \circ \phi \circ p_K^{K_1}(x_1) = p_{K_2}^K \circ p_K^{K_2} \circ p_{K_2}^{K_1} \circ p_{K_1}^K \circ \phi \circ p_K^{K_1}(x_1) = p_{K_2}^K \circ \phi \circ p_K^{K_2} \circ p_{K_2}^{K_1} \circ p_{K_1}^K \circ p_K^{K_1}(x_1) = p_{K_2}^K \circ \phi \circ p_K^{K_2} \circ p_{K_2}^{K_1}(x_1) = p_{K_2}^K \circ \phi \circ p_K^{K_2}(x_2) = x_2^\theta$.

Hence, $x_1^\theta \sim x_2^\theta$. It is now easily seen that θ is an automorphism of $\mathcal{A}$ fixing each line of S. $\square$

Lemma 4.41 ([4], see also Theorem 1.9.1 of [82]). *Let θ be an automorphism of a finite generalized quadrangle of order (s,t). If f is the number of fixpoints of θ and if g is the number of points x for which $x^\theta \neq x \sim x^\theta$, then $(t+1)f + g \equiv 1 + st \pmod{s+t}$.*

Theorem 4.42. *If a finite generalized quadrangle Q of order (s,t) admits a spread of symmetry, then $s+1 \,|\, t(t-1)$ and $s+t \,|\, s(s+1)(t+1)$.*

Proof. Suppose S is a spread of symmetry of Q and let G_S denote the group of automorphisms of Q fixing each line of S. The theorem obviously holds if $t = 1$. So, suppose that $t \geq 2$. Let K be an arbitrary line of Q not contained in S and let $L_1, \ldots, L_{s+1}$ denote the lines of S intersecting K. Obviously, the lines $L_1, \ldots, L_{s+1}$ together with the lines K^θ, $\theta \in G_S$, determine a subgrid of Q. It follows that $|\{M_1, M_2\}^{\perp\perp}| = s+1$ and $\{M_1, M_2\}^{\perp\perp} \subseteq S$ for every two different lines M_1 and M_2 of S. The lines of S together with the spans $\{M_1, M_2\}^{\perp\perp}$, $M_1, M_2 \in S$ with $M_1 \neq M_2$, define a linear space $\mathcal{L}$. The total number of lines of this linear space is equal to $\frac{(st+1)(st)}{(s+1)s}$. The divisibility condition $s+1 \,|\, t(t-1)$ follows. Now, if we apply Lemma 4.41 to any nontrivial automorphism of G_S, then by Theorem 4.39, $f = 0$, $g = (s+1)(st+1)$ and $(s+t) \,|\, s(s+1)(t+1)$. $\square$

Theorem 4.43. *If K is a line of a nontrivial admissible spread S of a near polygon, then the following statements are equivalent:*

(1) *S is a spread of symmetry,*

(2) *$\Pi_S(K)$ acts regularly on the set of points of K,*

(3) *G_S acts regularly on the set of points of K.*

If one of the above statements holds, then $G_S \cong \Pi_S(K)$.

Proof. (3) $\Rightarrow$ (1): Trivial.
(1) $\Rightarrow$ (3): This follows from Theorem 4.39 (c).
(2) $\Leftrightarrow$ (3): To prove this, we will make use of the following elementary and well-known result in the theory of permutation groups (see e.g. [68]):

> If a group H acts regularly on a set X, then the group $\tilde{H}$ of permutations of X which commute with every element of H acts also regularly on X. Moreover, H and $\tilde{H}$ are isomorphic.

From Theorem 4.40, it follows that if $\Pi_S(K)$ acts regularly on K, then also G_S acts regularly on K. Conversely, suppose that G_S acts regularly on the line K, then by Theorem 4.40 and the above-mentioned result, $\Pi_S(K) \subseteq \tilde{H}$, where $\tilde{H}$ is some group acting regularly on K. Since $\Pi_S(K)$ acts transitively on K, see Theorem 4.39 (b), we must have that $\Pi_S(K) = \tilde{H}$. From the quoted property ($H \cong \tilde{H}$), it follows that $G_S \cong \Pi_S(K)$ if S is a spread of symmetry. $\square$

Theorem 4.44. *If $\Pi_S(K)$ is commutative, then S is a spread of symmetry.*

Proof. We may suppose that $\Pi_S(K)$ is not trivial by Theorem 4.39 (a). Since $\Pi_S(K)$ is commutative, every element of $\Pi_S(K)$ can be extended to an element of G_S by Theorem 4.40. Since $\Pi_S(K)$ acts transitively on the set of points of K (Theorem 4.39), S is a spread of symmetry. □

4.7.3 Glued near polygons of type 1

We will use the same notation as in Section 4.3. If $\delta_{\mathcal{B}} = 1$, then Theorem 4.11 can be strengthened.

Theorem 4.45. *If $\delta_{\mathcal{B}} = 1$, then $\mathcal{S}_\Gamma$ is a near polygon if and only if Π_1 and Π_2 commute.*

Proof. We will show that if $\delta_{\mathcal{B}} = 1$, then any element of $T_1 \cup T_2$ is convex. The theorem then immediately follows from Theorem 4.11. Let F denote an arbitrary element of $T_1 \cup T_2$. Without loss of generality, we may suppose that $F \in T_1$. So, $F = \mathcal{A}_1(F_j^{(2)})$ for a certain $j \in \{1, \ldots, n_2\}$. Let $\alpha = (x_1, F_{i_1}^{(1)}, F_j^{(2)})$ and $\beta = (x_2, F_{i_2}^{(1)}, F_j^{(2)})$ denote two arbitrary points of F. Since $F \cong \mathcal{A}_1$ and since the spread S_1 is admissible, $\mathrm{d}(\alpha, \beta) \leq \mathrm{d}_1(F_{i_1}^{(1)}, F_{i_2}^{(1)}) + 1$. If γ is a shortest path between α and β, then its length $l(\gamma)$ is equal to $N_A + N_B + N_C$, where N_A (N_B, respectively N_C) denotes the number of pairs of successive points in γ whose type is (A), (B), respectively (C). Obviously, $N_B \geq \mathrm{d}_1(F_{i_1}^{(1)}, F_{i_2}^{(1)})$. If γ is not completely contained in F, then $N_C \geq 2$ and $l(\gamma) \geq \mathrm{d}_1(F_{i_1}^{(1)}, F_{i_2}^{(1)}) + 2$, contradicting $\mathrm{d}(\alpha, \beta) \leq \mathrm{d}_1(F_{i_1}^{(1)}, F_{i_2}^{(1)}) + 1$. This proves that every element of $T_1 \cup T_2$ is convex. □

Theorem 4.46. *Let $\mathcal{S}$ be a glued near polygon of type 1 arising from a tuple $(\mathcal{A}_1, \mathcal{A}_2, \mathcal{B}, S_1, S_2, L_1^{(1)}, L_1^{(2)}, \theta_1, \theta_2)$. Let G_{S_i}, $i \in \{1, 2\}$, denote the group of automorphisms of $\mathcal{A}_i$ fixing each line of S_i. If none of the spreads S_1 and S_2 is trivial, then S_1 and S_2 are spreads of symmetry and G_{S_1} and G_{S_2} are isomorphic groups.*

Proof. We will use the same notation as in Section 4.3. Let $s+1$ denote the constant number of points on a line of $S_1 \cup S_2$. By Theorem 4.11, Π_1 and Π_2 commute. Since $\Pi_{S_i}(L_i^{(1)})$, $i \in \{1, 2\}$, acts transitively on $L_i^{(1)}$, it follows by Theorem 4.40 that G_{S_i} acts transitively on each line of S_i. So, S_i is a spread of symmetry. By (the proof of) Theorem 4.43 and the fact that Π_1 and Π_2 commute, the groups G_{S_1}, G_{S_2}, Π_1 and Π_2 are isomorphic. □

Theorem 4.47. *Let $\mathcal{A}_1$ and $\mathcal{A}_2$ be near polygons, let S_i, $i \in \{1, 2\}$, be a spread of symmetry of $\mathcal{A}_i$ and let G_i be the group of automorphisms of $\mathcal{A}_i$ fixing each line of S_i. Suppose that the spreads S_1 and S_2 are not trivial and that the groups G_1 and G_2 are isomorphic. Then there exists at least one glued near polygon of type 1 arising from $\mathcal{A}_1$, $\mathcal{A}_2$, S_1 and S_2.*

Proof. Let $\mathcal{B}$ denote an arbitrary line of size $|G_1| = |G_2|$. Let θ_1 and θ_2 denote arbitrary bijections between $\mathcal{B}$ and base lines $L_1^{(1)}$ and $L_1^{(2)}$ of S_1 and S_2, respectively. Let Π_1 and Π_2 denote the corresponding groups of permutations of L, see Section 4.3. Since S_i, $i \in \{1,2\}$ is a nontrivial spread of symmetry, Π_i acts regularly on $\mathcal{B}$ and is isomorphic to G_i, see Theorem 4.43. Hence, $\Pi_1 \cong \Pi_2$. Now, let $\tilde{\Pi}_1$ denote the group of permutations of $\mathcal{B}$ commuting with every element of Π_1. Then $\tilde{\Pi}_1$ is isomorphic to Π_1 and hence also with Π_2. So, there exists a permutation θ of $\mathcal{B}$ such that $\tilde{\Pi}_1 = \theta^{-1}\Pi_2\theta$. By Theorem 4.45, there arises now a glued near polygon of type 1 from the tuple $(\mathcal{A}_1, \mathcal{A}_2, \mathcal{B}, S_1, S_2, L_1^{(1)}, L_1^{(2)}, \theta_1, \theta_2\theta)$. □

Theorem 4.48. *Let $\mathcal{A}_1$ and $\mathcal{A}_2$ be slim near polygons, let S_i, $i \in \{1,2\}$, be a spread of symmetry of $\mathcal{A}_i$, let $L_1^{(i)}$ denote an arbitrary line of S_i and let $\mathcal{B}$ be a line of size 3. For every bijection $\theta_1 : \mathcal{B} \to L_1^{(1)}$ and every bijection $\theta_2 : \mathcal{B} \to L_1^{(2)}$, let $\mathcal{S}_{\theta_1,\theta_2}$ denote the incidence structure arising from the tuple $(\mathcal{A}_1, \mathcal{A}_2, \mathcal{B}, S_1, S_2, L_1^{(1)}, L_1^{(2)}, \theta_1, \theta_2)$, see Section 4.3. Then $\mathcal{S}_{\theta_1,\theta_2}$ is a glued near polygon of type 1. If the automorphism group of $\mathcal{A}_1$ fixing S_1 and the line $L_1^{(1)} \in S_1$ induces all six permutations of the line $L_1^{(1)}$, then all near polygons $\mathcal{S}_{\theta_1,\theta_2}$ are isomorphic.*

Proof. By the remark following Theorem 4.11, we may suppose that none of the spreads S_1 and S_2 is trivial. Then by Theorem 4.43 both Π_1 and Π_2 act regularly on $\mathcal{B}$. By Theorem 4.45 and the fact that $|\Pi_1| = |\Pi_2| = 3$, it then follows that $\mathcal{S}_{\theta_1,\theta_2}$ is a near polygon for any choice of θ_1 and any choice of θ_2. By reasons of symmetry, all these near polygons are isomorphic if the automorphism group of $\mathcal{A}_1$ fixing S_1 and the line $L_1^{(1)} \in S_1$ induces all six permutations of the line $L_1^{(1)}$. □

4.7.4 Admissible triples

Definition. An *admissible triple* is a triple $T = (\mathcal{L}, G, \Delta)$, where

- G is a nontrivial finite group. We put $s := |G| - 1$. Unless otherwise stated, we will always use the multiplicative notation for G.
- $\mathcal{L}$ is a linear space, different from a point, in which each line is incident with exactly $s+1$ points. We denote the point set of $\mathcal{L}$ by P.
- Δ is a map from $P \times P$ to G such that the following holds for any three points x, y and z of $\mathcal{L}$:

$$x, y \text{ and } z \text{ are collinear} \Leftrightarrow \Delta(x,y)\,\Delta(y,z) = \Delta(x,z).$$

Obviously, $\Delta(x,x) = 1$ and $\Delta(y,x) = [\Delta(x,y)]^{-1}$ for all points x and y of $\mathcal{L}$.

Coordinatization of generalized quadrangles with a spread of symmetry

In this section we will prove the following theorem.

Theorem 4.49. *Let $T = (\mathcal{L}, G, \Delta)$ be an admissible triple and let P denote the point set of $\mathcal{L}$. Let Γ be the graph with vertex set $G \times P$, with two vertices (g, p_1) and (g', p'_1) adjacent whenever either ($p_1 = p'_1$ and $g \neq g'$) or ($p_1 \neq p'_1$ and $g' = g\,\Delta(p_1, p'_1)$). Then, the vertices and maximal cliques of Γ define a generalized quadrangle Q. Moreover, $L_p := \{(g, p) \mid g \in G\}$ is a line of Q for every point p of $\mathcal{L}$ and the set of lines L_p, $p \in P$, defines a spread of symmetry S of Q. Conversely, if Q' is a finite generalized quadrangle with a spread of symmetry S', then the pair (Q', S') is derivable from an admissible triple in the above-described way.*

Proof. Put $|G| = s + 1$ and $|P| = 1 + st$.

(1) The graph Γ contains $(1 + s)(1 + st)$ vertices and every vertex is adjacent to $s(t + 1)$ others. We will prove that every two adjacent vertices are contained in a unique maximal clique and that this clique contains exactly $s + 1$ elements. The incidence structure Q then has order (s, t), and since the number of points at distance at most one from a given line is $(s + 1) + (s + 1)ts = (s + 1)(1 + st)$, Q must be a generalized quadrangle of order (s, t).

So, suppose that $p_1 = (g_1, x)$ and $p_2 = (g_2, y)$ are two adjacent vertices of Γ; we determine what the common neighbours (g_3, z) look like. If $x = y \neq z$, then $g_3 = g_1\,\Delta(x, z) = g_2\,\Delta(x, z)$, implying that $g_1 = g_2$, a contradiction. Hence if $x = y$, then p_1 and p_2 are in a unique maximal clique containing all the points (g, x) with $g \in G$. If $x \neq y$, then also $x \neq z \neq y$ and $g_3 = g_1\,\Delta(x, z) = g_2\,\Delta(y, z) = g_1\,\Delta(x, y)\,\Delta(y, z)$. This implies that $\Delta(x, z) = \Delta(x, y)\,\Delta(y, z)$ or that $z \in xy$. It now follows easily that p_1 and p_2 are contained in a unique maximal clique, namely $\{(g_1\,\Delta(x, z), z) \mid z \in xy\}$.

(2) By (1), L_p is a line of Q for every point p of $\mathcal{L}$. Obviously, the lines L_p determine a spread S of Q. For each $h \in G$, the map $\theta_h : (g, x) \mapsto (hg, x)$ defines an automorphism of Q that fixes each line of S. It is now easily seen that S is a spread of symmetry.

(3) Suppose that $S' = \{L_1, \ldots, L_{1+st}\}$ is a spread of symmetry of a generalized quadrangle Q' of order (s, t). If Q' is a grid, then the pair (Q', S') is derivable from an admissible triple: take for the linear space a line of size $s + 1$ and for G any group of order $s + 1$. Suppose that Q' is not a grid and let G denote the group of automorphisms of Q' fixing each line of S'. By Theorem 4.43, $|G| = s+1$. Let $\mathcal{L}$ be the linear space whose vertices are the elements of S' and whose lines are all the reguli $\{A, B\}^{\perp\perp}$, where A and B are two different lines of S' (natural incidence).

We will now construct the map Δ. Choose a point p in L_1. Take two points $x = L_i$ and $y = L_j$ of $\mathcal{L}$. Let x_1 be the projection of p on the line L_i (in the generalized quadrangle Q'), let x_2 be the projection of x_1 on the line L_j and finally let x_3 be the projection of x_2 on the line L_1. Now, there exists a unique element $\theta \in G$ such that $x_3 = p^\theta$ and we put $\Delta(x, y) := \theta^{-1}$. It remains to show that the points $x = L_i, y = L_j, z = L_k$ of $\mathcal{L}$ are collinear if and only if $\Delta(x, y)\,\Delta(y, z) = \Delta(x, z)$. Put $\Delta(x, y) = \alpha^{-1}$, $\Delta(y, z) = \beta^{-1}$ and $\Delta(x, z) = \gamma^{-1}$. Denote by p_l, $l \in \{1, i, j, k\}$, the projection on the line L_l in the generalized quadrangle Q'. Put $a = p_i(p)$, $b = p_j(a)$, $c = p_k(b)$, $d = p_k(a)$, then $p^\gamma = p_1(d)$. From $p \sim p_j(p) \sim p_k p_j(p) \sim p^\beta$ and $p^\alpha \sim b \sim c \sim p_1(c)$, it follows that $p^{\beta\alpha} = p_1(c)$. Now,

$$\begin{aligned} \Delta(x,y)\,\Delta(y,z) = \Delta(x,z) \quad &\Leftrightarrow \quad \gamma = \beta\alpha \\ &\Leftrightarrow \quad c = d \\ &\Leftrightarrow \quad a, b, c \text{ are on a line} \\ &\Leftrightarrow \quad x, y, z \text{ are collinear.} \end{aligned}$$

We will now show the isomorphism between Q' and the generalized quadrangle which is derived from $(\mathcal{L}, G, \Delta)$. Let z be an arbitrary point of Q'. Let L_z denote the unique line of S' incident with z and let z' denote the projection of z on the fixed line L_1. There exists a unique $g_z \in G$ such that $z' = p^{g_z}$. We prove now that the map $z \mapsto (g_z^{-1}, L_z)$ defines an isomorphism between the GQ's. It is clearly a bijection and since both geometries have the same order, it suffices to show that adjacency is preserved. So, let z_1 and z_2 be two adjacent points of Q' and put $x = L_{z_1}$ and $y = L_{z_2}$. We may suppose that $L_{z_1} \neq L_{z_2}$. From $p^{g_{z_1}} \sim z_1 \sim z_2 \sim p^{g_{z_2}}$, it follows that $g_{z_2} = [\Delta(x,y)]^{-1}\, g_{z_1}$ or $g_{z_2}^{-1} = g_{z_1}^{-1}\,\Delta(x,y)$. Hence $(g_{z_2}^{-1}, y) \sim (g_{z_1}^{-1}, x)$. □

Coordinatization of glued near hexagons

Using the connection between glued near hexagons, spreads of symmetry of generalized quadrangles and admissible triples, we are able to coordinatize glued near hexagons.

Theorem 4.50. *Suppose $\mathcal{S}$ is a finite glued near hexagon which is not a product near polygon, then there exist admissible triples $T_i = (\mathcal{L}_i, G_i, \Delta_i)$, $i \in \{1,2\}$, and an anti-isomorphism θ from G_1 to G_2 such that $\mathcal{S}$ is isomorphic to the geometry $\mathcal{S}_\theta$ which we will define now.*

Let P_i, $i \in \{1,2\}$, denote the point set of $\mathcal{L}_i$. Let Γ_θ be the graph with vertex set $G_1 \times P_1 \times P_2$ with two vertices (g, p_1, p_2) and (g', p_1', p_2') adjacent whenever one of the following conditions is satisfied:

(1) *$p_1 = p_1'$, $p_2 = p_2'$ and $g \neq g'$;*

(2) *$p_1 = p_1'$, $p_2 \neq p_2'$ and $g' = \theta^{-1}(\Delta_2(p_2, p_2'))\, g$;*

(3) *$p_1 \neq p_1'$, $p_2 = p_2'$ and $g' = g\, \Delta_1(p_1, p_1')$.*

The vertices and maximal cliques of Γ_θ define a geometry $\mathcal{S}_\theta$.

Proof. Let $\{R_1, R_2\} \in \Delta_1(\mathcal{S})$, let Q_i, $i \in \{1,2\}$, denote an arbitrary quad of R_i and let S_i be the spread of Q_i obtained by intersecting Q_i with all quads of R_{3-i}. Since $\mathcal{S}$ is not a product near polygon, none of the spreads S_1 and S_2 is trivial. Hence, S_i, $i \in \{1,2\}$, is a spread of symmetry of Q_i by Theorem 4.46. By Theorem 4.49, there exists an admissible triple $T_i = (\mathcal{L}_i, G_i, \Delta_i)$ coordinatizing the pair (Q_i, S_i).

Let x_i denote an arbitrary point of $\mathcal{L}_i$. Without loss of generality, see part (3) of the proof of Theorem 4.49, we may suppose that $\Delta_i(x_i, p)$ is the identity element of G_i for every point p of $\mathcal{L}_i$. We will consider the line L_{x_i} (see Theorem 4.49) as base line of S_i. The glued near hexagon $\mathcal{S}$ arises from the generalized quadrangles Q_1 and Q_2, the spreads S_1 and S_2, the base lines L_{x_1} and L_{x_2}, a line $\mathcal{B}$, and certain bijections $\theta_1 : \mathcal{B} \to L_{x_1}$, $\theta_2 : \mathcal{B} \to L_{x_2}$. Without loss of generality, we may suppose that $\mathcal{B} = G_1$ and $\theta_1(g_1) = (g_1, x_1)$ for every $g_1 \in \mathcal{B} = G_1$. Let θ be the bijection from G_1 to G_2 such that $\theta_2(g_1) = (\theta(g_1), x_2)$. The points of $\mathcal{S}$ are the elements of the set $G_1 \times S_1 \times S_2$. Two distinct points (g, L_{p_1}, L_{p_2}) and $(g', L_{p_1'}, L_{p_2'})$ of $\mathcal{S}$ are collinear if and only if one of the following conditions is satisfied:

(1) $p_1 = p_1'$, $p_2 = p_2'$ and $g \neq g'$;

(2) $p_1 = p_1'$, $p_2 \neq p_2'$ and $g' = \theta^{-1}[\theta(g)\,\Delta_2(p_2, p_2')]$;

(3) $p_1 \neq p_1'$, $p_2 = p_2'$ and $g' = g\,\Delta_1(p_1, p_1')$.

The conditions (1), (2) and (3) define a graph Γ_θ with vertex set $G_1 \times S_1 \times S_2$ and the vertices and maximal cliques of Γ_θ define a partial linear space $\mathcal{S}_\theta$ isomorphic to $\mathcal{S}$. Now, for every $a \in G_2$, let θ_a be the bijection from G_1 to G_2 which maps the element g_1 of G_1 to the element $a\,\theta(g_1)$ of G_2. Obviously, $\Gamma_\theta \cong \Gamma_{\theta_a}$ and $\mathcal{S}_\theta \cong \mathcal{S}_{\theta_a}$. So, without loss of generality, we may suppose that θ maps the identity element of G_1 to the identity element of G_2. Since $\mathcal{S}$ is a near hexagon, the condition in Theorem 4.45 must be fulfilled. This condition implies that

$$\theta^{-1}[\theta(\Delta_1(p_1, p_1'))\,\Delta_2(p_2, p_2')] = \theta^{-1}[\Delta_2(p_2, p_2')]\,\Delta_1(p_1, p_1') \qquad (*)$$

for all points p_1, p_1' of $\mathcal{L}_1$ and all points p_2, p_2' of $\mathcal{L}_2$. Since S_1 and S_2 are not trivial spreads, the groups $\Pi_{S_1}(L_{x_1})$ and $\Pi_{S_2}(L_{x_2})$ act regularly on the respective lines L_{x_1} and L_{x_2}, see Theorem 4.43. It follows that $\Delta_i(p_i, p_i')$, $i \in \{1, 2\}$, can take all values of G_i. Equation $(*)$ then implies that θ is an anti-isomorphism. The theorem now readily follows. □

4.7.5 The sets $\Upsilon_0(\mathcal{A})$ and $\Upsilon_1(\mathcal{A})$ for a dense near polygon $\mathcal{A}$

Let $\mathcal{A}$ be a dense near polygon and let $\Upsilon(\mathcal{A})$ denote the set of all partitions of $\mathcal{A}$ in convex subpolygons. We say that an element $T \in \Upsilon(\mathcal{A})$ belongs to $\Upsilon_i(\mathcal{A})$, $i \in \{0, 1\}$, if there exists a $T' \in \Upsilon(\mathcal{A})$ such that $\{T, T'\} \in \Delta_i(\mathcal{A})$.

Theorem 4.51. *If $T \in \Upsilon_0(\mathcal{A})$, then there exists a unique $T' \in \Upsilon(\mathcal{A})$ such that $\{T, T'\} \in \Delta_0(\mathcal{A})$.*

Proof. Let x denote an arbitrary point of $\mathcal{A}$, let F_x denote the unique element of T through x and let $L_1, \ldots, L_k$ denote all the lines through x not contained in T. If T' is an element of $\Upsilon(\mathcal{A})$ such that $\{T, T'\} \in \Delta_0(\mathcal{A})$, then the unique element of T' through x coincides with $\mathcal{C}(L_1, \ldots, L_k)$. The theorem now readily follows. □

Theorem 4.52. *If $T \in \Upsilon_1(\mathcal{A}) \setminus \Upsilon_0(\mathcal{A})$, then there exists a unique $T' \in \Upsilon(\mathcal{A})$ such that $\{T, T'\} \in \Delta_1(\mathcal{A})$.*

Proof. Let x denote an arbitrary point of $\mathcal{A}$, let F_x denote the unique element of T through x and let $L_1, \ldots, L_k$ denote the lines through x not contained in F_x. Let T' be an element of $\Upsilon(\mathcal{A})$ such that $\{T, T'\} \in \Delta_1(\mathcal{A})$ and let F'_x denote the unique element of T' through x. Clearly, $\text{diam}(F_x) + \text{diam}(F'_x) = \text{diam}(\mathcal{A}) + 1$, $\text{diam}(F_x \cap F'_x) = 1$ and $L_1, \ldots, L_k \subseteq F'_x$. There are two possibilities.

(a) $\mathcal{C}(L_1, \ldots, L_k) \cap F_x = \{x\}$.
In this case, we have $\text{diam}(\mathcal{C}(L_1, \ldots, L_k)) = \text{diam}(F'_x) - 1$ and hence $\text{diam}(\mathcal{A}) = \text{diam}(F_x) + \text{diam}(\mathcal{C}(L_1, \ldots, L_k))$. From Theorem 4.4, it readily follows that $T \in \Upsilon_0(\mathcal{A})$, a contradiction.

(b) $\mathcal{C}(L_1, \ldots, L_k) \cap F_x$ is a line.
In this case, we have $\mathcal{C}(L_1, \ldots, L_k) = F'_x$.

The proposition now readily follows. □

Remark. The previous proposition is not necessarily valid if $T \in \Upsilon_0(\mathcal{A})$. Suppose that $\{T, T'\} \in \Delta_0(\mathcal{A})$, let $F \in T$ and suppose that F has an admissible spread S. For every point x of F, define $\tilde{F}_x := \mathcal{C}(L_x, F'_x)$, where L_x denotes the unique line of S through x and F'_x denotes the unique element of T' through x. Put $\tilde{T} := \{\tilde{F}_x | x \in F\}$. Then $\{T, \tilde{T}\} \in \Delta_1(\mathcal{A})$. So, if the near polygon F has two different admissible spreads, there exist at least two $\tilde{T} \in \Upsilon(\mathcal{A})$ such that $\{T, \tilde{T}\} \in \Delta_1(\mathcal{A})$.

Definitions. Let $T \in \Upsilon_0(\mathcal{A}) \cup \Upsilon_1(\mathcal{A})$. If $T \in \Upsilon_0(\mathcal{A})$, then we denote by T^C the unique element of $\Upsilon_0(\mathcal{A})$ such that $\{T, T^C\} \in \Delta_0(\mathcal{A})$. If $T \in \Upsilon_1(\mathcal{A}) \setminus \Upsilon_0(\mathcal{A})$, then we denote by T^C the unique element of $\Upsilon(\mathcal{A})$ such that $\{T, T^C\} \in \Delta_1(\mathcal{A})$. We call T^C the *complementary partition* of T. If $T \in \Upsilon_0(\mathcal{A})$, then $(T^C)^C = T$. If $T \in \Upsilon_1(\mathcal{A}) \setminus \Upsilon_0(\mathcal{A})$, then $(T^C)^C$ is not necessarily equal to T (T^C might belong to $\Upsilon_0(\mathcal{A})$, see the previous remark). We denote by $\tilde{\Upsilon}_1(\mathcal{A})$ the set of all $T \in \Upsilon_1(\mathcal{A}) \setminus \Upsilon_0(\mathcal{A})$ for which $(T^C)^C = T$. We also define $\tilde{\Delta}_1(\mathcal{A}) := \Delta_1(\mathcal{A}) \cap \binom{\tilde{\Upsilon}_1(\mathcal{A})}{2}$.

4.7.6 Extensions of spreads and automorphisms

Let $\mathcal{A}$ be a dense near polygon, let T denote an element of $\Upsilon_0(\mathcal{A}) \cup \tilde{\Upsilon}_1(\mathcal{A})$, let F denote an arbitrary element of T and let F' denote an arbitrary element of T^C.

- For every 1-spread S of F, we define $\bar{S} := \{\pi_{F,E}(L) \,|\, E \in T \text{ and } L \in S\}$. Obviously, $\bar{S}$ is a spread of $\mathcal{A}$. We call $\bar{S}$ the *extension* of S.

- For every automorphism θ of F and for every point x of $\mathcal{A}$, we define $\bar{\theta}(x) := \pi_{F,F_x} \circ \theta \circ \pi_{F_x,F}(x)$. Here F_x denotes the unique element of T through x. Obviously, $\bar{\theta}$ is a permutation of the point set of $\mathcal{A}$. We call $\bar{\theta}$ the *extension* of θ.

Theorem 4.53. (a) *Let $T \in \Upsilon_0(\mathcal{A})$.*

(a1) *If θ is an automorphism of F, then $\overline{\theta}$ is an automorphism of $\mathcal{A}$.*

(a2) *If ϕ is an automorphism of $\mathcal{A}$ fixing each element of T, then $\phi = \bar{\theta}$ for some automorphism θ of F.*

(a3) *If ϕ_1 is an automorphism of $\mathcal{A}$ fixing each element of T and if ϕ_2 is an automorphism of $\mathcal{A}$ fixing each element of T^C, then ϕ_1 and ϕ_2 commute.*

(b) *Let $T \in \tilde{\Upsilon}_1(\mathcal{A})$, let S^* denote the spread of F obtained by intersecting F with every element of T^C and let G^* denote the group of automorphisms of F fixing each line of S^*. Let θ denote an automorphism of F. Then $\bar{\theta}$ is an automorphism of $\mathcal{A}$ if and only if θ commutes with every element of G^*.*

Proof. Properties (a1), (a2) and (a3) are straightforward. In order to prove property (b), it suffices to prove that $\bar{\theta}$ maps collinear points x and y to collinear points $\bar{\theta}(x)$ and $\bar{\theta}(y)$. There are two possibilities.

- $F_x = F_y$.
  The statement follows from the fact that the maps $\pi_{F_x,F}$, θ and π_{F,F_x} are isomorphisms.

- $F_x \neq F_y$.
  $\bar{\theta}(x)$ and $\bar{\theta}(y)$ are collinear if and only if $\pi_{F_x,F_y} \circ \bar{\theta}(x) = \bar{\theta}(y)$, i.e. if and only if $\pi_{F_x,F_y} \circ \pi_{F,F_x} \circ \theta \circ \pi_{F_x,F}(x) = \pi_{F,F_y} \circ \theta \circ \pi_{F_y,F} \circ \pi_{F_x,F_y}(x)$.

Hence, $\bar{\theta}$ is an automorphism if and only if $\pi_{F_x,F_y} \circ \pi_{F,F_x} \circ \theta = \pi_{F,F_y} \circ \theta \circ \pi_{F_y,F} \circ \pi_{F_x,F_y} \circ \pi_{F,F_x}$ for all $F_x, F_y \in T$, i.e. if and only if θ commutes with $\pi_{F_y,F} \circ \pi_{F_x,F_y} \circ \pi_{F,F_x}$ for all $F_x, F_y \in T$. Let S' denote the spread of F' obtained by intersecting F' with every element of T. Since $T \in \tilde{\Upsilon}_1(\mathcal{A})$, S' is not a trivial spread of F' and S^* is not a trivial spread of F. Put $H := \{\pi_{F_y,F} \circ \pi_{F_x,F_y} \circ \pi_{F,F_x} \mid F_x, F_y \in T\}$. Then $|H| = |\Pi_{S'}(F \cap F')|$ and $H \leq G^*$. By Theorem 4.39, H acts transitively and G^* acts semiregularly on every line of S^*. It follows that $H = G^*$. As a consequence, $\bar{\theta}$ is an isomorphism if and only if θ commutes with every element of G^*. □

By the following two theorems, all spreads of symmetry in product and glued near polygons are characterized.

Theorem 4.54. *Let S be a spread of symmetry of F and let G_S denote the group of automorphisms of F fixing each line of S.*

(a) *If $T \in \Upsilon_0(\mathcal{A})$, then $\bar{S}$ is a spread of symmetry of $\mathcal{A}$.*

(b) *If $T \in \tilde{\Upsilon}_1(\mathcal{A})$, let S^* denote the spread of F obtained by intersecting F with every element of T^C and let G^* denote the group of automorphisms of F fixing each line of S^*. Then $\bar{S}$ is a spread of symmetry of $\mathcal{A}$ if and only if $[G^*, G_S] = 0$.*

Proof. (a) The group $\bar{G}_S := \{\bar{\theta} \,|\, \theta \in G_S\}$ fixes each line of $\bar{S}$ and acts transitively on each line of $\bar{S}$. So, $\bar{S}$ is a spread of symmetry.

(b) If $[G^*, G_S] = 0$, then by Theorem 4.53, $\bar{G}_S := \{\bar{\theta} \,|\, \theta \in G_S\}$ is a group of automorphisms of $\mathcal{A}$. Since $\bar{G}_S$ fixes each line of $\bar{S}$ and acts transitively on each line of $\bar{S}$, $\bar{S}$ is a spread of symmetry. Conversely, suppose that $\bar{S}$ is a spread of symmetry, let $G_{\bar{S}}$ denote the group of automorphisms of $\mathcal{A}$ fixing each line of $\bar{S}$. Then $G_{\bar{S}} = \bar{G}$ for some subgroup G of G_S acting regularly on S. By Theorem 4.53, we have $[G^*, G] = 0$. If $G_S = G$, then we are done. If $G_S \neq G$, then G_S acts regularly, but not semiregularly on S. So, S is a trivial spread of F by Theorem 4.39. Since $T \in \tilde{\Upsilon}_1(\mathcal{A})$, $S \neq S^*$. By Theorem 4.5, it then follows that S and S^* have no line in common. (If L is a line of a trivial spread, then this trivial spread is the unique spread of symmetry containing L.) By Theorem 4.53 (a), it then follows that $[G^*, G_S] = 0$. □

Theorem 4.55. *Every admissible spread (spread of symmetry) S of $\mathcal{A}$ is the extension of an admissible spread (spread of symmetry) in F or F'.*

Proof. If $T \in \tilde{\Upsilon}_1(\mathcal{A})$, let $\tilde{S}$ denote the admissible spread of $\mathcal{A}$ obtained by intersecting each element of T with each element of T^C. Let W denote the set of all elements E of $T \cup T^C$ for which S_E is an admissible 1-spread (spread of symmetry) of E. Here, S_E denotes the set of lines of S contained in E, see Theorem 4.5.

Step 1. At least one of the following holds:

(i) every element of T belongs to W;

(ii) every element of T^C belongs to W.

Proof. Suppose that there exists an $E \in T$ such that $E \notin W$. Let E' denote an arbitrary element of T^C and let x denote a point of the intersection $E \cap E'$. By Theorem 4.5, the unique line of S through x does not belong to E. Hence, it belongs to E' and $E' \in W$ by Theorem 4.5. So, if there exists an $E \in T$ such that $E \notin W$, then every element of T^C belongs to W. Similarly, if there exists an $E \in T^C$ such that $E \notin W$, then every element of T belongs to W. The statement now readily follows. □

Without loss of generality we may suppose that every element of T belongs to W. Then we know that S_F is an admissible spread (spread of symmetry) of F. From Steps 2 and 3 below, it will follow that $S = \bar{S}_F$.

Step 2. If $F_1, F_2 \in T$ such that $\mathrm{d}(F_1, F_2) = 1$, then $S_{F_2} = \pi_{F_1,F_2}(S_{F_1})$.

Proof. Let L denote an arbitrary line of S_{F_1} and put $L' := \pi_{F_1,F_2}(L)$. We distinguish the following possibilities.

- $T \in \Upsilon_0(\mathcal{A})$ or ($T \in \tilde{\Upsilon}_1(\mathcal{A})$ and $L \notin \tilde{S}$).
  Then L and L' are contained in a quad Q which is isomorphic to a grid. By Theorem 4.5, S_Q is an admissible spread (spread of symmetry) of Q. Hence, $L' \in S$.

- $T \in \tilde{\Upsilon}_1(\mathcal{A})$ and $L \in \tilde{S}$.
  Let E denote the unique element of T^C through L. By Theorem 4.5, $E \in W$. Obviously, $L' = E \cap F_2 \in \tilde{S}$. Let x denote an arbitrary point of L'. The unique line of S through x lies in E and F_2 and hence coincides with L'. So, $L' \in S$. □

Step 3. If $F_0, F_1, \ldots, F_k \in T$ such that $\mathrm{d}(F_0, F_k) = k$ and $\mathrm{d}(F_{i-1}, F_i) = 1$ for every $i \in \{1, \ldots, k\}$, then $\pi_{F_0,F_k} = \pi_{F_{k-1},F_k} \circ \cdots \circ \pi_{F_1,F_2} \circ \pi_{F_0,F_1}$.

Proof. The map $\pi_{F_{k-1},F_k} \circ \cdots \circ \pi_{F_1,F_2} \circ \pi_{F_0,F_1}$ maps every point x of F_0 to a point of F_k at distance at most k from x_0. Since $\mathrm{d}(F_0, F_k) = k$, this point must coincide with $\pi_{F_0,F_k}(x)$. □

4.7.7 Compatible spreads of symmetry

Theorem 4.56. *Let $\mathcal{A} = (\mathcal{P}, \mathcal{L}, \mathrm{I})$ be a near polygon, let S_1 and S_2 denote two different spreads of symmetry in $\mathcal{A}$ and let G_i, $i \in \{1,2\}$, denote the group of automorphisms of $\mathcal{A}$ which fix each line of S_i. Then the following are equivalent:*

(i) $[G_1, G_2] = 0$;

(ii) *for every line $l \in S_1$ and every $g \in G_2$, $l^g \in S_1$;*

(iii) *for every line $l \in S_2$ and every $g \in G_1$, $l^g \in S_2$;*

(iv) *the partial linear space $\mathcal{B} = (\mathcal{P}, S_1 \cup S_2, \mathrm{I}_{|\mathcal{P} \times (S_1 \cup S_2)})$ is a disjoint union of lines and grids.*

Proof. (i) $\Rightarrow$ (ii) and (i) $\Rightarrow$ (iii): By symmetry, it suffices to prove the implication (i) $\Rightarrow$ (ii). Let l denote an arbitrary line of S_1, let x denote an arbitrary point of l and let g denote an arbitrary element of G_2. Then $l^g = (x^{G_1})^g = (x^g)^{G_1} \in S_1$.

(ii) $\Rightarrow$ (iv) and (iii) $\Rightarrow$ (iv): By symmetry, it suffices to prove the implication (ii) $\Rightarrow$ (iv). Suppose that the lines $K_1 \in S_1$ and $K_2 \in S_2$ intersect in a point x^*. For all $x_1 \in K_1$, $x_1^{G_2} \in S_2$ and for all $x_2 \in K_2$, $x_2^{G_1} \in S_1$. We will now prove that the lines $x_1^{G_2}$, $x_1 \in K_1$, and $x_2^{G_1}$, $x_2 \in K_2$, define a subgrid of $\mathcal{B}$. Obviously, $x_1^{G_2} \cap x_1'^{G_2} = \emptyset$ for all $x_1, x_1' \in K_1$ with $x_1 \neq x_1'$ and $x_2^{G_1} \cap x_2'^{G_1} = \emptyset$ for all $x_2, x_2' \in K_2$ with $x_2 \neq x_2'$. Now, consider arbitrary points $x_1 \in K_1$ and $x_2 \in K_2$ and let g_2 denote an arbitrary element of G_2 such that $x_2 = (x^*)^{g_2}$. The point x_2 lies on the line $K_1^{g_2}$ which, by our assumption, belongs to the spread S_1. So, $x_2^{G_1} = K_1^{g_2}$ and $x_2^{G_1} \cap x_1^{G_2} = K_1^{g_2} \cap x_1^{G_2} = \{x_1^{g_2}\}$. As a consequence, every two different intersecting lines of $S_1 \cup S_2$ are contained in a subgrid of $\mathcal{B}$. The implication now follows from the fact that every point of $\mathcal{B}$ is contained in at most two lines of $\mathcal{B}$.

(iv) $\Rightarrow$ (i): Let x be an arbitrary point of $\mathcal{A}$, let g_1 be an arbitrary element of G_1 and let g_2 be an arbitrary element of G_2. We will prove that $x^{g_1 g_2} = x^{g_2 g_1}$. We distinguish the following cases.

- Suppose that x is contained in a subgrid G of $\mathcal{B}$. Let l_i, $i \in \{1,2\}$, denote the unique line of S_i through x. Since $x^{g_i} \in l_i$, the unique line m_{3-i} of S_{3-i} through x^{g_i} is contained in G. Let y be the common point of the lines m_1 and m_2. Since $x \sim x^{g_1}$, $x^{g_2} \sim x^{g_1 g_2}$. So, $x^{g_1 g_2}$ is the unique point of m_2 collinear with x^{g_2}. Hence, $y = x^{g_1 g_2}$. In a similar way, one proves that $y = x^{g_2 g_1}$. As a consequence, $x^{g_1 g_2} = x^{g_2 g_1}$.

- Suppose that x is not contained in a subgrid of $\mathcal{B}$, i.e. x is contained in a line L of $S_1 \cap S_2$. Since $S_1 \neq S_2$, $\mathcal{B}$ has a subgrid G. Every line of G is parallel with L. Let $y \in G$ such that x is the unique point of L nearest to y. Then $x^{g_1 g_2}$ (respectively $x^{g_2 g_1}$) is the unique point of $L = L^{g_1 g_2} = L^{g_2 g_1}$ nearest to $y^{g_1 g_2}$ (respectively $y^{g_2 g_1}$). Since $y^{g_1 g_2} = y^{g_2 g_1}$, it follows that $x^{g_1 g_2} = x^{g_2 g_1}$. □

Definition. Let $\mathcal{A}$ be a near polygon, let S_1 and S_2 be two (possibly equal) spreads of symmetry of $\mathcal{A}$ and let G_i, $i \in \{1,2\}$, denote the group of automorphisms of $\mathcal{A}$ which fix each line of S_i. Then the spreads S_1 and S_2 are called *compatible* if $[G_1, G_2] = 0$. In the case that S_1 and S_2 are different, Theorem 4.56 provides some equivalent statements.

4.7.8 Compatible spreads of symmetry in product and glued near polygons

Theorem 4.57. *Let $\mathcal{A}$ be a product near polygon, let $\{T_1, T_2\} \in \Delta_0(\mathcal{A})$, let $F_1 \in T_1$ and let $F_2 \in T_2$. Let S_0 and S_1 denote two spreads of symmetry of F_1 and let S_2 denote a spread of symmetry of F_2. Then*

(i) *$\bar{S}_1$ and $\bar{S}_2$ are compatible spreads of symmetry of $\mathcal{A}$,*

(ii) *$\bar{S}_0$ and $\bar{S}_1$ are compatible spreads of symmetry of $\mathcal{A}$ if and only if S_0 and S_1 are compatible spreads of symmetry of F_1.*

Proof. For every $i \in \{0,1,2\}$, let G_i denote the group of automorphisms of F_i ($F_0 = F_1$) fixing each element of S_i. Then $\bar{G}_i = \{\bar{\theta} \,|\, \theta \in G_i\}$ is the group of automorphisms of $\mathcal{A}$ fixing each element of $\bar{S}_i$. By Theorem 4.53 (a3), every element of $\bar{G}_1$ commutes with every element of $\bar{G}_2$. This proves (i). If $\theta_0 \in G_0$ and $\theta_1 \in G_1$, then $\bar{\theta}_0 \bar{\theta}_1 = \overline{\theta_0 \theta_1}$ and $\bar{\theta}_1 \bar{\theta}_0 = \overline{\theta_1 \theta_0}$. So, θ_0 and θ_1 commute if and only if $\bar{\theta}_0$ and $\bar{\theta}_1$ commute. This proves (ii). □

Theorem 4.58. *Let $\mathcal{A}$ be a glued near polygon, let $\{T_1, T_2\} \in \tilde{\Delta}_1(\mathcal{A})$, let $F_1 \in T_1$ and let $F_2 \in T_2$. Let S_i^* denote the spread of symmetry of F_i obtained by intersecting F_i with the elements of T_{3-i}. Let S_0 and S_1 denote two spreads of symmetry of F_1 and let S_2 denote a spread of symmetry of F_2. Then*

(i) *$\bar{S}_0$ and $\bar{S}_1$ are compatible spreads of symmetry if and only if the spreads S_0, S_1 and S_1^* are mutually compatible.*

(ii) *$\bar{S}_1$ and $\bar{S}_2$ are compatible spreads of symmetry of $\mathcal{A}$ if and only if for every $i \in \{1,2\}$, S_i and S_i^* are compatible.*

Proof. (i) We may suppose that the pairs (S_0, S_1^*) and (S_1, S_1^*) are compatible (otherwise $\bar{S}_0$ and $\bar{S}_1$ would not be spreads of symmetry, see Theorem 4.54). For every $i \in \{0,1\}$, let G_i denote the group of automorphisms of F_1 fixing each element of S_i. Since (S_i, S_1^*) is compatible, $\bar{G}_i = \{\bar{\theta} \mid \theta \in G_i\}$ is the full group of automorphisms of $\mathcal{A}$ fixing each element of $\bar{S}_i$. If $\theta_0 \in G_0$ and $\theta_1 \in G_1$, then as before θ_0 and θ_1 commute if and only if $\bar{\theta}_0$ and $\bar{\theta}_1$ commute. So, $\bar{S}_0$ and $\bar{S}_1$ are compatible spreads of symmetry of $\mathcal{A}$ if and only if S_0 and S_1 are compatible spreads of symmetry of F. This proves (i).

(ii) For $\bar{S}_i$, $i \in \{1,2\}$, to be a spread of symmetry it is necessary that S_i and S_i^* are compatible. Conversely, suppose that for every $i \in \{1,2\}$, S_i and S_i^* are compatible. If $S_1 = S_1^*$ and $S_2 = S_2^*$, then by (i) it follows that $\bar{S}_1 = \bar{S}_2$ is compatible with itself. We will therefore suppose that $S_1 \neq S_1^*$ or $S_2 \neq S_2^*$. Then $\bar{S}_1 \neq \bar{S}_2$. Let $L_1 \in \bar{S}_1$ and $L_2 \in \bar{S}_2$ be two lines intersecting in a point x. Let U_i, $i \in \{1,2\}$, denote the unique element of T_i through x. Since $L_1 \subseteq U_1$, $L_2 \subseteq U_2$, $L_1 \neq L_2$, $\mathcal{C}(L_1, L_2)$ is a grid Q. By Theorem 4.5 the lines of Q disjoint from L_i, $i \in \{1,2\}$, belong to $\bar{S}_i$. So, the spreads $\bar{S}_1$ and $\bar{S}_2$ satisfy property (iv) of Theorem 4.56 and hence are compatible. This proves (ii). □

4.7.9 Near polygons of type $(F_1 * F_2) \circ F_3$

Theorem 4.59. *Let F_1, F_2 and F_3 be dense near polygons of diameter at least 2. Then every near polygon $\mathcal{A}$ of type $(F_1 \times F_2) \otimes_1 F_3$ is also of type $F_i \times (F_{3-i} \otimes_1 F_3)$ for a certain $i \in \{1,2\}$.*

Proof. Let x denote an arbitrary point of $\mathcal{A}$. Let $\{T, T_3\} \in \Delta_1(\mathcal{A})$ such that $F \cong F_1 \times F_2$ for every $F \in T$ and $G \cong F_3$ for every $G \in T_3$, and let $F(x)$, respectively $F_3(x)$, denote the unique element of T, respectively T_3, through x. Let $\{T_1, T_2\} \in \Delta_0(F(x))$ such that $F \cong F_1$ for every $F \in T_1$ and $G \cong F_2$ for every $G \in T_2$, and let $F_i(x)$, $i \in \{1,2\}$, denote the unique element of T_i through x. Without loss of generality we may assume that the line $F(x) \cap F_3(x)$ is contained in $F_2(x)$. By Theorems 4.36 and 4.37, $F_4(x) := \mathcal{C}(F_2(x), F_3(x))$ is a glued near polygon of type $F_2 \otimes_1 F_3$. Now, $\mathrm{diam}(\mathcal{A}) = (\mathrm{diam}(F_1(x)) + \mathrm{diam}(F_2(x))) + \mathrm{diam}(F_3(x)) - 1 = \mathrm{diam}(F_1(x)) + (\mathrm{diam}(F_2(x)) + \mathrm{diam}(F_3(x)) - 1) = \mathrm{diam}(F_1(x)) + \mathrm{diam}(F_4(x))$ and every line through x is contained in precisely one of the subpolygons $F_1(x)$ and $F_4(x)$. So, $\mathcal{A}$ satisfies the conditions of Theorem 4.4. Hence, $\mathcal{A} \cong F_1(x) \times F_4(x)$ and $\mathcal{A}$ is of type $F_1 \times (F_2 \otimes_1 F_3)$. □

Theorem 4.60. *Let F_1 and F_2 denote two dense near polygons of diameter at least 2, and let $\mathcal{A}$ be a near polygon of type $F_1 \otimes_1 F_2$. If $\mathcal{A}$ is a product near polygon, then at least one of the near polygons F_1 and F_2 is also a product near polygon.*

Proof. Let $\{T_1, T_2\} \in \Delta_1(\mathcal{A})$ such that every element of T_i, $i \in \{1,2\}$, is isomorphic to F_i, and let $\{T_3, T_4\}$ be an arbitrary element of $\Delta_0(\mathcal{A})$. Let x denote an arbitrary point of $\mathcal{A}$ and let G_i, $1 \leq i \leq 4$, be the unique element of T_i through

x. For every $i,j \in \{1,2,3,4\}$ we define $G_{i,j} = G_i \cap G_j$, $d_{i,j} := \mathrm{diam}(G_{i,j})$ and $d_i := d_{i,i}$. We then have $d := \mathrm{diam}(\mathcal{A}) = d_1 + d_2 - 1 = d_3 + d_4$, $d_{1,2} = 1$ and $d_{3,4} = 0$. Without loss of generality, we may suppose that the line $G_{1,2}$ is contained in G_3. By Theorems 4.36 and 4.37, it follows that $d_3 = d_{1,3} + d_{2,3} - 1$ and $d_4 = d_{1,4} + d_{2,4}$. Hence, $d = (d_{1,3} + d_{1,4}) + (d_{2,3} + d_{2,4}) - 1$. Now, $\mathcal{C}(G_{1,3}, G_{1,4}) = G_1$ and $\mathcal{C}(G_{2,3}, G_{2,4}) = G_2$; so $d_1 \le d_{1,3} + d_{1,4}$ and $d_2 \le d_{2,3} + d_{2,4}$. From $d_1 + d_2 - 1 = d = (d_{1,3} + d_{1,4}) + (d_{2,3} + d_{2,4}) - 1 \ge d_1 + d_2 - 1$, it then follows that $d_1 = d_{1,3} + d_{1,4}$ and $d_2 = d_{2,3} + d_{2,4}$. Since $d_4 = d_{1,4} + d_{2,4} \ge 1$, we have $d_{j,4} \ge 1$ for a $j \in \{1,2\}$. Now, since G_3 contains the line $G_{1,2}$, we also have $d_{j,3} \ge 1$. Now, the pair $\{G_{j,3}, G_{j,4}\}$ satisfies all conditions of Theorem 4.4 and hence $F_j \cong G_j \cong G_{j,3} \times G_{j,4}$. This proves the theorem. □

Definition. Let T and T' denote two partitions of a set. Then T' is called a *refinement* of T if each element of T is the union of some members of T'.

Theorem 4.61. *Let F_1, F_2 and F_3 denote three dense near polygons with diameter at least 2 and suppose that none of these near polygons is a product near polygon. If a near polygon $\mathcal{S}$ is of type $(F_1 \otimes_1 F_2) \otimes_1 F_3$ and if $\{T, T_3\} \in \Delta_1(\mathcal{S})$ such that every element of T is of type $F_1 \otimes_1 F_2$ and every element of T_3 is isomorphic to F_3, then there exists an element $\{\tilde{T}_1, \tilde{T}_2\} \in \Delta_1(\mathcal{S})$ and an $i \in \{1,2\}$ such that*

(i) *every element of $\tilde{T}_1$ is isomorphic to F_i,*

(ii) *every element of $\tilde{T}_2$ is of type $F_{3-i} \otimes_1 F_3$,*

(iii) *the partition $\tilde{T}_1$ is a refinement of the partition T,*

(iv) *the partition T_3 is a refinement of the partition $\tilde{T}_2$.*

Hence every dense near polygon $\mathcal{A}$ of type $(F_1 \otimes_1 F_2) \otimes_1 F_3$ is also of type $F_i \otimes_1 (F_{3-i} \otimes_1 F_3)$ for a certain $i \in \{1,2\}$.

Proof. By Theorem 4.60, $\mathcal{A}$ is not a product near polygon. Let F^* denote an arbitrary element of T, and let $\{T_1', T_2'\} \in \Delta_1(F^*)$ such that every element of T_i', $i \in \{1,2\}$, is isomorphic to F_i. The spread T is admissible. Put $T_i := \{\pi_F(G) | F \in T \text{ and } G \in T_i'\}$, $i \in \{1,2\}$, then T_i is a partition of $\mathcal{A}$ in convex subpolygons isomorphic to F_i. For every point x of $\mathcal{A}$ and every $i \in \{1,2,3\}$, let $F_i(x)$ denote the unique element of T_i through x and put $K(x) := F_1(x) \cap F_2(x)$ and $L(x) := \mathcal{C}(F_1(x), F_2(x)) \cap F_3(x)$. Since $F_1(x)$ and $F_2(x)$ are projections of elements of $T_1' \cup T_2'$ on the unique element $F(x)$ of T through x, we find that $K(x)$ is a line, that $\mathcal{C}(F_1(x), F_2(x)) = F(x)$ and that also $L(x)$ is a line. By Theorem 4.46 and the fact that $\mathcal{A}$ is not a product near polygon, it follows that $\{L(x) | x \in F^*\}$ is a spread of symmetry of F^* and so Theorem 4.55 applies. Without loss of generality, we may suppose that every line $L(x)$, $x \in F^*$, is contained in $F_2(x)$. Now, choose a point x outside F^* and let x' denote its projection on F^*. Since $L(x')$ is contained in $F_2(x')$, the projection $L(x)$ of $L(x')$ on $F(x)$ is contained in the projection $F_2(x)$ of $F_2(x')$ on $F(x)$. Hence, for every point x of $\mathcal{A}$, $F_3(x)$ intersects $F_2(x)$ in a line. By Theorems 4.36 and 4.37, it then follows that $\tilde{F}_2(x) := \mathcal{C}(F_3(x), F_2(x))$

is a glued near polygon of type $F_2 \otimes_1 F_3$. The set $\tilde{T}_2 := \{\tilde{F}_2(x) | x \in \mathcal{A}\}$ clearly determines a partition of $\mathcal{A}$ in convex subpolygons. Define $\tilde{T}_1 := T_1$. For every point $x \in \mathcal{A}$, $F_1(x) \cap \tilde{F}_2(x) = (F_1(x) \cap F(x)) \cap \tilde{F}_2(x) = F_1(x) \cap (F(x) \cap \tilde{F}_2(x)) = F_1(x) \cap F_2(x) = K(x)$. By Theorem 4.28, it now follows that $\{\tilde{T}_1, \tilde{T}_2\} \in \Delta_1(\mathcal{A})$. So, $\mathcal{A}$ is of type $F_1 \otimes_1 (F_2 \otimes_1 F_3)$. This proves the theorem. $\square$

Chapter 5

Valuations

In [60], we introduced the notion *valuation* for dense near polygons. Here we generalize this notion to a class of near polygons which we will call nice near polygons.

5.1 Nice near polygons

Definition. A pair $(\mathcal{S}, \mathcal{W})$ with $\mathcal{S}$ a near polygon and $\mathcal{W}$ a set of convex subpolygons of $\mathcal{S}$ is called *nice* if the following properties are satisfied:

(N1) the intersection of two elements of $\mathcal{W}$ is either empty or again an element of $\mathcal{W}$;

(N2) every two points x and y of $\mathcal{S}$ are contained in a unique convex sub-$2 \cdot \mathrm{d}(x, y)$-gon of $\mathcal{W}$; we will denote this convex subpolygon by $H(x, y)$.

Obviously, $\mathcal{C}(x, y) \subseteq H(x, y)$ for all points x and y of $\mathcal{S}$. A near polygon $\mathcal{S}$ is called *nice* if there exists a set $\mathcal{W}$ of convex subpolygons of $\mathcal{S}$ such that the pair $(\mathcal{S}, \mathcal{W})$ is nice.

Theorem 5.1. *Suppose that the pair $(\mathcal{S}, \mathcal{W})$ is nice. Let F denote an arbitrary element of $\mathcal{W}$ and let $\mathcal{W}'$ denote the set of all elements of $\mathcal{W}$ which are contained in F. Then $H(x, y) \subseteq F$ for all points x and y of F. As a consequence, also the pair $(F, \mathcal{W}')$ is nice.*

Proof. The two convex subpolygons $H(x, y)$ and $H(x, y) \cap F$ of $\mathcal{W}$ have both diameter $\mathrm{d}(x, y)$ and contain the points x and y. Hence $H(x, y) = H(x, y) \cap F$ or $H(x, y) \subseteq F$. It now immediately follows that the pair $(F, \mathcal{W}')$ satisfies properties (N1) and (N2). $\square$

By the results of Chapter 2, every dense near polygon is nice. In the case of dense near polygons the set $\mathcal{W}$ is uniquely determined: $\mathcal{W}$ is the set of all convex subpolygons. For all points x and y of $\mathcal{S}$, we then have $H(x, y) = \mathcal{C}(x, y)$. By [21] every dual polar space is nice. By the following theorem also every generalized polygon is nice.

Theorem 5.2. *Every generalized 2d-gon, $d \geq 2$, is nice.*

Proof. Let $\mathcal{S} = (\mathcal{P}, \mathcal{L}, \mathrm{I})$ be a generalized $2d$-gon, $d \geq 2$. For every two points x and y with $1 \leq \mathrm{d}(x,y) \leq d-1$, we define $H(x,y)$ as follows. Consider the unique geodesic path $x = x_0, x_1, \ldots, x_k = y$ between x and y ($k = \mathrm{d}(x,y)$). Then $H(x,y)$ is defined as the union of all the lines $x_{i-1}x_i$, $i \in \{1, \ldots, k\}$. If $x = y$, then we define $H(x,y) := \{x\}$ and if $\mathrm{d}(x,y) = d$, then we define $H(x,y) := \mathcal{S}$. With $\mathcal{W} := \{H(x,y) \,|\, x, y \in \mathcal{P}\}$, one easily verifies that properties (N1) and (N2) hold. □

5.2 Valuations of nice near polygons

Definition. Let $\mathcal{S} = (\mathcal{P}, \mathcal{L}, \mathrm{I})$ be a near $2n$-gon and let $\mathcal{W}$ be a set of convex subpolygons of $\mathcal{S}$ such that the pair $(\mathcal{S}, \mathcal{W})$ is nice. A function f from $\mathcal{P}$ to $\mathbb{N}$ is called a *valuation* of $(\mathcal{S}, \mathcal{W})$ (or shortly a valuation of $\mathcal{S}$ if no confusion is possible) if it satisfies the following properties (we call $f(x)$ the *value* of x):

(V_1) there exists at least one point with value 0;

(V_2) every line L of $\mathcal{S}$ contains a unique point x_L with smallest value and $f(x) = f(x_L) + 1$ for every point x of L different from x_L;

(V_3) every point x of $\mathcal{S}$ is contained in a convex subpolygon F_x of $\mathcal{W}$ such that the following properties are satisfied for every point y of F_x:

(i) $f(y) \leq f(x)$;

(ii) if $z \in \Gamma_1(y)$ such that $f(z) = f(y) - 1$, then $z \in F_x$;

(iii) $F_y \subseteq F_x$.

Theorem 5.3. *Let f be a valuation of a dense near polygon $\mathcal{S}$ (so $\mathcal{W}$ consists of all convex subpolygons of $\mathcal{S}$). Then for every point x of $\mathcal{S}$, there exists a unique convex subpolygon $F_x \in \mathcal{W}$ for which property (V_3) holds. In the case of valuations of dense near polygons, property (V_3)–(iii) follows from properties (V_2), (V_3)-(i) and (V_3)–(ii), i.e. property (V_3)–(iii) is superfluous.*

Proof. By properties (V_2), (V_3)–(i) and ($V3$)–(ii), the set $F_x \cap (\Gamma_0(x) \cup \Gamma_1(x))$ is uniquely determined by the valuation f. From Theorems 2.14 and 2.23, it then follows that F_x itself is completely determined by f. By properties (V_2), (V_3)–(i) and (V_3)–(ii), $(\Gamma_0(y) \cup \Gamma_1(y)) \cap F_y \subseteq F_x$ for every point y of F_x. Again by Theorems 2.14 and 2.23, $F_y \subseteq F_x$. So, property (V_3)–(iii) also follows from properties (V_2), (V_3)–(i) and (V_3)–(ii) in the case of dense near polygons. □

The next theorem immediately follows from properties (V_1) and (V_2).

Theorem 5.4. *Let f be a valuation of the nice pair $(\mathcal{S}, \mathcal{W})$. Then*

- $|f(x) - f(y)| \leq d(x, y)$ *for all points x and y of $\mathcal{S}$;*
- $f(x) \in \{0, \ldots, \mathrm{diam}(\mathcal{S})\}$ *for every point x of $\mathcal{S}$;*
- *if x is a point with value 0 and if $y \in \Gamma_1(x)$, then $f(y) = 1$.*

The following theorem says that the knowledge of the valuations of a certain near polygon can give information on how F can be embedded in a larger near polygon. This theorem will turn out to be very useful for obtaining several classification results about near polygons, see e.g. Chapter 9.

Theorem 5.5. *Let $\mathcal{S} = (\mathcal{P}, \mathcal{L}, \mathrm{I})$ be a near polygon and let $\mathcal{W}$ be a set of convex subpolygons of $\mathcal{S}$ such that $(\mathcal{S}, \mathcal{W})$ is nice. Let $F = (\mathcal{P}', \mathcal{L}', \mathrm{I}')$ be an element of $\mathcal{W}$ and let $\mathcal{W}'$ denote the set of all elements of $\mathcal{W}$ which are contained in F. For every point x of $\mathcal{S}$ and for every point y of F, we define $f_x(y) := \mathrm{d}(x, y) - \mathrm{d}(x, F)$. Then $f_x : \mathcal{P}' \to \mathbb{N}$ is a valuation of $(F, \mathcal{W}')$ for every point x of $\mathcal{S}$.*

Proof. Obviously, properties (V_1) and (V_2) are satisfied. For every point y of F, we define $F_y := H(x, y) \cap F$ where $H(x, y)$ is the unique convex sub-$2 \cdot \mathrm{d}(x, y)$-gon of $\mathcal{W}$ through the points x and y. Then $F_y \in \mathcal{W}'$. If $z \in F_y$, then $f_x(z) = \mathrm{d}(x, z) - \mathrm{d}(x, F) \leq \mathrm{d}(x, y) - \mathrm{d}(x, F) = f_x(y)$. By Theorem 5.1 $H(x, z) \subseteq H(x, y)$ and hence $F_z \subseteq F_y$. If z' is a neighbour of z in F such that $f(z') = f(z) - 1$, then $\mathrm{d}(x, z') = \mathrm{d}(x, z) - 1$, implying that $z' \in H(x, z) \cap F \subseteq H(x, y) \cap F = F_y$. So, also property (V_3) is satisfied. □

We can improve Theorem 5.5 in the following way.

Theorem 5.6. *Let $\mathcal{S} = (\mathcal{P}, \mathcal{L}, \mathrm{I})$ be a near polygon and let $\mathcal{W}$ be a set of convex subpolygons of $\mathcal{S}$ such that $(\mathcal{S}, \mathcal{W})$ is nice. Let $F = (\mathcal{P}', \mathcal{L}', \mathrm{I}')$ be a subpolygon of $\mathcal{S}$ satisfying the following properties:*

- *F is a subspace of $\mathcal{S}$;*
- $d_F(x, y) = d_{\mathcal{S}}(x, y)$ *for all points x and y of F;*
- *$(F, \mathcal{W}')$ is nice where $\mathcal{W}' := \{F \cap F' \mid F' \in \mathcal{W}$ and $F \cap F' \neq \emptyset\}$.*

For every point x of $\mathcal{S}$ and for every point y of F, we define $f_x(y) := d_{\mathcal{S}}(x, y) - d_{\mathcal{S}}(x, F)$. Then $f_x : \mathcal{P}' \to \mathbb{N}$ is a valuation of $(F, \mathcal{W}')$ for every point x of $\mathcal{S}$.

Proof. Since $H(x, y) \cap F$ belongs to $\mathcal{W}'$ for every point x and every point y of F, we can follow a similar reasoning as in the proof of Theorem 5.5. □

Valuations can also be induced in subpolygons.

Theorem 5.7. *Let $\mathcal{S} = (\mathcal{P}, \mathcal{L}, \mathrm{I})$ be a near polygon and let $\mathcal{W}$ be a set of convex subpolygons of $\mathcal{S}$ such that $(\mathcal{S}, \mathcal{W})$ is nice. Let $F = (\mathcal{P}', \mathcal{L}', \mathrm{I}')$ be a subpolygon of $\mathcal{S}$ satisfying the following properties:*

- *F is a subspace of $\mathcal{S}$;*

- *$d_F(x,y) = d_{\mathcal{S}}(x,y)$ for all points x and y of F;*
- *$(F, \mathcal{W}')$ is nice where $\mathcal{W}' := \{F \cap F' \mid F' \in \mathcal{W}$ and $F \cap F' \neq \emptyset\}$.*

If f is a valuation of $(\mathcal{S}, \mathcal{W})$, then the map $f_F : \mathcal{P}' \to \mathbb{N}; x \mapsto f(x) - \min\{f(x)|x \in \mathcal{P}'\}$ is a valuation of F.

Proof. Obviously, f_F satisfies properties (V_1) and (V_2). For every point x of $\mathcal{S}$, let F_x denote the unique convex subpolygon of $\mathcal{W}$ for which property (V_3) holds (with respect to the valuation f). The map f_F also satisfies (V_3) if for every point x of F one takes $F'_x := F_x \cap F$ as convex subpolygon of F through x. □

Definition. The valuation f_F defined in Theorem 5.7 is called an *induced valuation.*

We will now define two important classes of valuations.

Theorem 5.8. *Let $\mathcal{S} = (\mathcal{P}, \mathcal{L}, \mathrm{I})$ be a near polygon and let $\mathcal{W}$ be a set of convex subpolygons such that the pair $(\mathcal{S}, \mathcal{W})$ is nice.*

(a) *If y is a point of $\mathcal{S}$, then $f_y : \mathcal{P} \to \mathbb{N};\ x \mapsto d(x,y)$ is a valuation of $(\mathcal{S}, \mathcal{W})$.*

(b) *If O is an ovoid of $\mathcal{S}$, then $f_O : \mathcal{P} \to \mathbb{N};\ x \mapsto d(x,O)$ is a valuation of $(\mathcal{S}, \mathcal{W})$.*

Proof. In both cases, (V_1) and (V_2) are satisfied. In case (a), we put F_x equal to $H(x,y)$, i.e. the unique convex sub-$[2 \cdot \mathrm{d}(x,y)]$-gon of $\mathcal{W}$ through x and y. In case (b), we put $F_x := \{x\}$ if $x \in O$ and $F_x := \mathcal{S}$ otherwise. For these choices of F_x, also (V_3) holds. □

Definition. A valuation of a nice pair $(\mathcal{S}, \mathcal{W})$ is called *classical*, respectively *ovoidal*, if it is obtained as in (a), respectively (b), of Theorem 5.8.

5.3 Characterizations of classical and ovoidal valuations

In this section, we will characterize classical and ovoidal valuations in dense near polygons. From the first characterization, we immediately obtain that every valuation of a dense generalized quadrangle is either classical or ovoidal. Notice that valuations of dense near 0-gons and dense near 2-gons are trivial objects. There is a unique point with value 0 and all other points (in the case of dense near 2-gons) have value 1.

Theorem 5.9. *Let f be a valuation of a dense near $2n$-gon $\mathcal{S} = (\mathcal{P}, \mathcal{L}, \mathrm{I})$ with $n \geq 1$. Then*

(a) $\max\{f(u)|u \in \mathcal{P}\} \leq n$ *with equality if and only if f is classical;*

(b) $\max\{f(u)|u \in \mathcal{P}\} \geq 1$ *with equality if and only if f is ovoidal.*

Proof. Obviously, the above inequalities hold and become equalities if f is classical, respectively ovoidal.

(a) Suppose that $\max\{f(u)|u \in \mathcal{P}\} = n$. Let x be a point of $\mathcal{S}$ with value 0 and let y be a point with value n. By Theorem 5.4, $\mathrm{d}(x,y) = n$. Let y' be an arbitrary point of $\Gamma_n(x) \cap \Gamma_1(y)$ and let y'' denote the unique point of the line yy' at distance $n-1$ from x. By Theorem 5.4, $f(y'') = f(y'') - f(x) \leq n-1$ and $f(y'') = f(y) + f(y'') - f(y) \geq n-1$. Hence, $f(y'') = n-1$ and by property (V_2), $f(y') = n$. So, every point of $\Gamma_n(x) \cap \Gamma_1(y)$ has value n. By the connectedness of $\Gamma_n(x)$, see Theorem 2.7, it then follows that every point of $\Gamma_n(x)$ has value n. Now, let z be an arbitrary point of $\mathcal{S}$. Then, by Theorem 2.14, there exists a path of length $n - \mathrm{d}(x,z)$ between z and a point z' of $\Gamma_n(x)$. From $\mathrm{d}(x,z) \geq |f(z) - f(x)| = f(z)$ and $n - f(z) = |f(z') - f(z)| \leq \mathrm{d}(z,z') = n - \mathrm{d}(x,z)$, it follows that $f(z) = \mathrm{d}(x,z)$. This proves that f is classical.

(b) Suppose now that $\max\{f(x)|x \in \mathcal{P}\} = 1$. By property (V_2), every line of $\mathcal{S}$ contains a unique point with value 0. So, the points with value 0 determine an ovoid of $\mathcal{S}$ and f is ovoidal. □

Corollary 5.10. *Every valuation of a dense generalized quadrangle is either classical or ovoidal.*

Theorem 5.11. *Let f be a valuation of a dense near polygon $\mathcal{S}$.*

(a) *If every induced quad valuation is classical, then the valuation f itself is classical.*

(b) *If every induced quad valuation is ovoidal, then the valuation f itself is ovoidal.*

Proof. (a) Suppose that f is a nonclassical valuation of $\mathcal{S}$. Let x denote an arbitrary point with value 0 and let i be the smallest nonnegative integer for which there exists a point y satisfying $i = \mathrm{d}(x,y) \neq f(y)$. Obviously, $i \geq 2$. Choose points $y' \in \Gamma_1(y) \cap \Gamma_{i-1}(x)$ and $y'' \in \Gamma_1(y') \cap \Gamma_{i-2}(x)$. Then $f(y'') = i-2$, $f(y') = i-1$ and $f(y) \in \{i-1, i-2\}$. Every point of $Q := \mathcal{C}(y'', y)$ collinear with y'' has distance $i-1$ from x and hence has value $i-1$. Since the valuation induced in Q is classical, y'' is the unique point of $\mathcal{C}(y, y'')$ with smallest value and $f(y) = f(y'') + \mathrm{d}(y'', y) = i - 2 + 2 = i$, a contradiction.

(b) Suppose that f is a nonovoidal valuation of $\mathcal{S}$. Let x denote an arbitrary point with value 0 and let i be the smallest nonnegative integer for which there exists a point y satisfying $i = \mathrm{d}(x,y)$ and $f(y) \geq 2$. Obviously, $i \geq 2$. Choose points $y' \in \Gamma_1(y) \cap \Gamma_{i-1}(x)$ and $y'' \in \Gamma_1(y') \cap \Gamma_{i-2}(x)$. Clearly every point of the line through y' and y'' has value 0 or 1. But then the valuation induced in the quad $\mathcal{C}(y, y'')$ cannot be ovoidal, a contradiction. □

5.4 The partial linear space G_f

Definition. For every valuation f of a near polygon $\mathcal{S}$, we define O_f as the set of all points of $\mathcal{S}$ with value 0.

Theorem 5.12. *Let f be a valuation of a dense near polygon $\mathcal{S}$ and let x be a point of $\mathcal{S}$. If $\mathrm{d}(x, O_f) \leq 2$, then $f(x) = \mathrm{d}(x, O_f)$.*

Proof. Obviously, this holds if $\mathrm{d}(x, O_f) \leq 1$. So, suppose that $\mathrm{d}(x, O_f) = 2$ and let x' denote a point of O_f at distance 2 from x. If the valuation induced in the quad $\mathcal{C}(x, x')$ is ovoidal, then x would be collinear with a point of $O_f \cap \mathcal{C}(x, x')$, a contradiction. So, the valuation induced in $\mathcal{C}(x, x')$ is classical and $f(x) = f(x') + \mathrm{d}(x, x') = 2$. □

If $x, y \in O_f$, then by Theorem 5.4, $d(x, y) \geq 2$. A quad Q of $\mathcal{S}$ is called *special* if it contains at least two points of O_f. Let G_f be the partial linear space whose points are the points of O_f and whose lines are the special quads of $\mathcal{S}$ (natural incidence). If f is a valuation of a dense near polygon and if x and y are two different collinear points of G_f, then the line of G_f through x and y corresponds with an ovoid in the special quad through x and y. So, in this case every line of G_f contains at least three points.

5.5 A property of valuations

Theorem 5.13. *Let f be a valuation of a finite dense near $2n$-gon $\mathcal{S}$ and let let m_i, $i \in \mathbb{N}$, denote the number of points with value i. If $\mathcal{S}$ contains lines of size $s + 1$, then $\sum_{i=0}^{\infty} \frac{m_i}{(-s)^i} = 0$.*

Proof. Notice that $m_i = 0$ if $i \geq n + 1$.

(a) Suppose first that $\mathcal{S}$ has order (s, t). For every line L of $\mathcal{S}$,

$$\sum_{x \in L} \frac{1}{(-s)^{f(x)}} = \frac{1}{(-s)^{f(x_L)}} + s \frac{1}{(-s)^{f(x_L)+1}} = 0,$$

with x_L the unique point of L with smallest value. Hence,

$$0 = \sum_{L \in \mathcal{L}} \sum_{x \in L} \frac{1}{(-s)^{f(x)}} = \sum_{x \in \mathcal{P}} \sum_{L I x} \frac{1}{(-s)^{f(x)}} = (t+1) \sum_{x \in \mathcal{P}} \frac{1}{(-s)^{f(x)}} = (t+1) \sum_{i=0}^{\infty} \frac{m_i}{(-s)^i}.$$

This proves that the theorem holds if $\mathcal{S}$ has an order.

(b) Suppose next that not every line of $\mathcal{S}$ is incident with the same number of points, then by Theorem 1.12, $\mathcal{S}$ has a partition in isomorphic convex subpolygons of order (s, t') for some $t' \geq 0$. By (a), the theorem holds for each valuation induced in one of the subpolygons of the partition. If we add all obtained equations (after multiplying with a suitable power of $-s$), then we obtain the required equation. □

Corollary 5.14. *Let f be a valuation of a finite dense near polygon $\mathcal{S} = (\mathcal{P}, \mathcal{L}, \mathrm{I})$. If k different line sizes $s_1 + 1, \ldots, s_k + 1$ occur in $\mathcal{S}$, then* $\max\{f(x) \,|\, x \in \mathcal{P}\} \geq k$.

Proof. Put $M := \max\{f(x) \,|\, x \in \mathcal{P}\}$. By Theorem 5.13, the polynomial $p(s) := \sum_{i=0}^{M} m_i(-s)^{M-i} = 0$ has at least k different roots. Hence, $k \leq \deg(f(s)) = M$. □

5.6 Some classes of valuations

The two most important classes of valuations are the classical and ovoidal valuations. In this section, we will describe several other classes of valuations.

5.6.1 Hybrid valuations

Let $\mathcal{S} = (\mathcal{P}, \mathcal{L}, \mathrm{I})$ be a near $2n$-gon, $n \geq 2$, and let $\mathcal{W}$ be a set of convex subpolygons of $\mathcal{S}$ such that the pair $(\mathcal{S}, \mathcal{W})$ is nice. Choose a point x in $\mathcal{S}$ and a $\delta \in \{2, \ldots, n\}$. Let $\mathcal{A}_{x,\delta}$ be the incidence structure with points the points of $\mathcal{S}$ at distance at least δ from x and with lines the lines of $\mathcal{S}$ at distance at least $\delta - 1$ from x (natural incidence). For every ovoid O of $\mathcal{A}_{x,\delta}$, we can define the following function $f_{x,O}$ from $\mathcal{P}$ to $\mathbb{N}$: if y is a point of $\mathcal{S}$ at distance at most $\delta - 1$ from x, then $f_{x,O}(y) := \mathrm{d}(x, y)$; if y is a point of $\mathcal{S}$ at distance at least δ from x, then $f_{x,O}(y) = \delta - 2$ if $y \in O$ and $f_{x,O}(y) = \delta - 1$ otherwise.

Theorem 5.15. *The map $f_{x,O}$ is a valuation of $\mathcal{S}$.*

Proof. Since $f(x) = 0$, property (V_1) holds. Now, let L be an arbitrary line of $\mathcal{S}$. If $\mathrm{d}(x, L) \leq \delta - 2$, then the unique point on L nearest to x is also the unique point on L with smallest value. If $\mathrm{d}(x, L) \geq \delta - 1$, then the unique point of O on L is the unique point of L with smallest value. This proves property (V_2). Now, property (V_3) also holds if we make the following choices for F_y, $y \in \mathcal{P}$: we put $F_y := H(x, y)$ if $\mathrm{d}(x, y) \leq \delta - 2$, $F_y := \{y\}$ if $y \in O$ and $F_y := \mathcal{S}$ otherwise. Here, $H(x, y)$ denotes the unique convex sub-$2 \cdot \mathrm{d}(x, y)$-gon of $\mathcal{W}$ through the points x and y. □

Definition. Any valuation which can be obtained as in Theorem 5.15 is called a *hybrid valuation of type δ*. A hybrid valuation of type 2 is just an ovoidal valuation. A hybrid valuation of type n is also called a *semi-classical valuation.* We could regard classical valuations as hybrid valuations of type $n + 1$.

Theorem 5.16. *If f is a valuation of a dense near $2n$-gon and if x is a point of $\mathcal{S}$ such that $f(y) = d(x, y)$ for every point y at distance at most $n - 1$ from y, then f is either classical or semi-classical.*

Proof. Suppose that f is not classical and consider a point $z \in \Gamma_n(x)$. Every point of $\Gamma_1(z) \cap \Gamma_{n-1}(x)$ has value $n - 1$. Hence by property (V_2) and Theorem 5.9, $f(z) \in \{n - 2, n - 1\}$. By property (V_2), it now follows that the points of $\Gamma_n(x)$ with value $n - 2$ form an ovoid in $\mathcal{A}_{x,n}$. This proves that f is semi-classical. □

Theorem 5.17. *Let $\mathcal{S}$ be a dense near $2n$-gon, $n \geq 2$, of order $(2,t)$ and let x be a point of $\mathcal{S}$ for which $\Gamma_n(x) \neq \emptyset$. Then there exists a semi-classical valuation f with $f(x) = 0$ if and only if $\Gamma_n(x)$ is bipartite. In that case, there are precisely 2 semi-classical valuations with $f(x) = 0$.*

Proof. Every line of $\mathcal{A}_{x,n}$ contains two points. So, $\mathcal{A}_{x,n}$ has ovoids if and only if the graph induced by $\Gamma_n(x)$ is bipartite. It follows that there exist semi-classical valuations with $f(x) = 0$ if and only if $\Gamma_n(x)$ is bipartite. If $\Gamma_n(x)$ is bipartite, then $\Gamma_n(x)$ contains two ovoids by Theorem 2.7. □

5.6.2 Product valuations

Let $\mathcal{S}_i = (\mathcal{P}_i, \mathcal{L}_i, \mathrm{I}_i)$, $i \in \{1,2\}$, be a near polygon and let $\mathcal{W}_i$ denote a set of convex subpolygons of $\mathcal{S}_i$ such that the pair $(\mathcal{S}_i, \mathcal{W}_i)$ is nice. For every $F_1 \in \mathcal{W}_1$ and every $F_2 \in \mathcal{W}_2$, $F_1 \times F_2 := \{(a,b) \,|\, a \in F_1, b \in F_2\}$ is a convex subpolygon of $\mathcal{S}_1 \times \mathcal{S}_2$. We define $\mathcal{W}_1 \times \mathcal{W}_2 := \{F_1 \times F_2 \,|\, F_1 \in \mathcal{W}_1, F_2 \in \mathcal{W}_2\}$.

Theorem 5.18. *The pair $(\mathcal{S}_1 \times \mathcal{S}_2, \mathcal{W}_1 \times \mathcal{W}_2)$ is nice.*

Proof.

- Let $F_1, F_1' \in \mathcal{W}_1$ and $F_2, F_2' \in \mathcal{W}_2$, then $F_1 \cap F_1' \in \mathcal{W}_1$, $F_2 \cap F_2' \in \mathcal{W}_2$ and $(F_1 \times F_2) \cap (F_1' \times F_2') = (F_1 \cap F_1') \times (F_2 \cap F_2') \in \mathcal{W}_1 \times \mathcal{W}_2$. This proves property (N1).
- Let $H_i(a,b)$, $i \in \{1,2\}$, denote the unique convex sub-$2 \cdot \mathrm{d}_i(a,b)$-gon of $\mathcal{W}_i$ through the points a and b of $\mathcal{S}_i$. If (x_1,y_1) and (x_2,y_2) are two points of $\mathcal{S}_1 \times \mathcal{S}_2$ and if $F_1 \times F_2$ is a convex subpolygon of $\mathcal{W}_1 \times \mathcal{W}_2$ through (x_1,y_1) and (x_2,y_2), then F_1 contains the points x_1, x_2 and F_2 contains the points y_1, y_2. Hence, $H_1(x_1,x_2) \subseteq F_1$, $H_2(x_1,x_2) \subseteq F_2$, $\mathrm{diam}(F_1) \geq \mathrm{d}_1(x_1,x_2)$, $\mathrm{diam}(F_2) \geq \mathrm{d}_2(y_1,y_2)$ and $\mathrm{diam}(F_1 \times F_2) = \mathrm{diam}(F_1) + \mathrm{diam}(F_2) \geq \mathrm{d}_1(x_1,x_2) + \mathrm{d}_2(y_1,y_2) = \mathrm{d}[(x_1,y_1),(x_2,y_2)]$. So, if $\mathrm{diam}(F_1 \times F_2) = \mathrm{d}[(x_1,y_1),(x_2,y_2)]$, then $F_1 = H_1(x_1,x_2)$ and $F_2 = H_2(y_1,y_2)$. As a consequence, $H_1(x_1,x_2) \times H_2(y_1,y_2)$ is the unique convex sub-$2 \cdot \mathrm{d}[(x_1,y_1),(x_2,y_2)]$-gon of $\mathcal{W}_1 \times \mathcal{W}_2$ through the points (x_1,y_1) and (x_2,y_2). This proves property (N2). □

Theorem 5.19. *If f_i, $i \in \{1,2\}$, is a valuation of $(\mathcal{S}_i, \mathcal{W}_i)$, then the map $f : \mathcal{P}_1 \times \mathcal{P}_2 \rightarrow \mathbb{N}, (x_1,x_2) \mapsto f_1(x_1) + f_2(x_2)$ is a valuation of $(\mathcal{S}_1 \times \mathcal{S}_2, \mathcal{W}_1 \times \mathcal{W}_2)$.*

Proof. If x_i, $i \in \{1,2\}$, is a point of $\mathcal{S}_i$ for which $f_i(x_i) = 0$, then $f[(x_1,x_2)] = 0$. This proves property (V_1). If L is a line of $\mathcal{S}_1 \times \mathcal{S}_2$, then without loss of generality, we may suppose that L is of the form $K \times \{y\}$, with K a line of $\mathcal{S}_1$ and y a point of $\mathcal{S}_2$. Now, $f[(k,y)] = f_1(k) + f_2(y)$ for every point k of K. Property (V_2) now immediately follows: the unique point of L with smallest f-value is the point (x_K, y), where x_K denotes the unique point of K with smallest f_1-value. It remains to check property (V_3). For every point x_i, $i \in \{1,2\}$, of $\mathcal{S}_i$, let F_{x_i}, $i \in \{1,2\}$, denote the subpolygon of $\mathcal{S}_i$ satisfying (V_3). For every point (x_1,x_2) of $\mathcal{S}_1 \times \mathcal{S}_2$, we define $F_{(x_1,x_2)} := F_{x_1} \times F_{x_2}$. Let (a_1,a_2) be a point of $F_{(x_1,x_2)}$. Then $f[(a_1,a_2)] = f_1(a_1) + f_2(a_2) \leq f_1(x_1) + f_2(x_2) = f[(x_1,x_2)]$ and

$F_{(a_1,a_2)} = F_{a_1} \times F_{a_2} \subseteq F_{x_1} \times F_{x_2}$. If (b_1, b_2) is a point of $\mathcal{S}_1 \times \mathcal{S}_2$ collinear with (a_1, a_2) and satisfying $f[(b_1, b_2)] = f(a_1, a_2) - 1$, then without loss of generality, we may suppose that $a_2 = b_2$ and $a_1 \sim b_1$ (in $\mathcal{S}_1$). Then $f_1(b_1) = f[(b_1, b_2)] - f_2(b_2) = f[(a_1, a_2)] - 1 - f_2(a_2) = f_1(a_1) - 1$. Since $a_1 \in F_{x_1}$, also the point b_1 belongs to F_{x_1}. Hence, the point (b_1, b_2) belongs to $F_{(x_1,x_2)}$. This proves property (V_3). □

Definition. Any valuation of a product near polygon which can be obtained as in Theorem 5.19 is called a *product valuation.*

5.6.3 Diagonal valuations

Theorem 5.20. *Let $\mathcal{S} = (\mathcal{P}, \mathcal{L}, \mathrm{I})$ be a near polygon and let $\mathcal{W}$ be a set of convex subpolygons of $\mathcal{S}$ such that the set $(\mathcal{S}, \mathcal{W})$ is nice. Define $X := \{(x, x) \,|\, x \in \mathcal{P}\}$. Then the function $f : \mathcal{P} \times \mathcal{P} \to \mathbb{N}; p \mapsto d(p, X)$ is a valuation of $(\mathcal{S} \times \mathcal{S}, \mathcal{W} \times \mathcal{W})$.*

Proof. For every point (u, v) of $\mathcal{S}\times\mathcal{S}$, we have $f[(u, v)] = \mathrm{d}(u, v)$. So, every point of $\mathcal{S}\times\mathcal{S}$ has value at most $\mathrm{diam}(\mathcal{S})$. Obviously, there exists a point with value 0. Let L denote a line of $\mathcal{S}\times\mathcal{S}$. Without loss of generality, we may suppose that $L = \{u\}\times M$ for some point u and some line M of $\mathcal{S}$. If u' denotes the unique point of M nearest to u, then (u, u') is the unique point of L with smallest value. Now, for every point (u, v) of $\mathcal{S} \times \mathcal{S}$, we define $F_{(u,v)} := H(u, v) \times H(u, v)$, where $H(u, v)$ denotes the unique convex sub-$2 \cdot \mathrm{d}(u, v)$-gon of $\mathcal{W}$ through u and v. Let (u_1, v_1) denote an arbitrary point of $F_{(u,v)}$. Then $f[(u_1, v_1)] = \mathrm{d}(u_1, v_1) \leq \mathrm{d}(u, v) = f[(u, v)]$ and $F_{(u_1,v_1)} = H(u_1, v_1) \times H(u_1, v_1) \subseteq H(u, v) \times H(u, v) = F_{(u,v)}$. Let (u_2, v_2) be a point of $\mathcal{S} \times \mathcal{S}$ collinear with (u_1, v_1) such that $f[(u_2, v_2)] = f[(u_1, v_1)] - 1$. Without loss of generality, we may suppose that $u_1 = u_2$. Then $v_2 \sim v_1$ and $\mathrm{d}(u_1, v_2) = \mathrm{d}(u_1, v_1) - 1$. So, $v_2 \in H(u_1, v_1) \subseteq H(u, v)$. As a consequence, $(u_2, v_2) \in H(u, v)\times H(u, v) = F_{(u,v)}$. This proves that f is a valuation of $(\mathcal{S}\times\mathcal{S}, \mathcal{W}\times\mathcal{W})$. □

Definition. Any valuation which can be obtained as in Theorem 5.20 is called a *diagonal valuation.*

Remark. With every set Y of points in $F \times F$, we can associate a matrix M_Y whose rows and columns are indexed by the points of F. If $(u, v) \in Y$, then the (u, v)-th entry of M_Y is equal to 1; otherwise it is equal to 0. The matrix M_X corresponding with the above-mentioned set X gives rise to a matrix with all 1's on the diagonal. This explains the name we have given to these valuations.

5.6.4 Semi-diagonal valuations

Theorem 5.21. *Let $\mathcal{S} = (\mathcal{P}, \mathcal{L}, \mathrm{I})$ be a dense glued near hexagon, let $\{T_1, T_2\} \in \Delta_1(\mathcal{S})$ and let X be a set of points of $\mathcal{S}$ satisfying the following properties:*

(1) *every element of $T_1 \cup T_2$ has a unique point in common with X;*

(2) *$d(x_1, x_2) = 2$ for all $x_1, x_2 \in X$ with $x_1 \neq x_2$.*

Then the map $f : \mathcal{P} \to \mathbb{N}; x \mapsto d(x, X)$ is a valuation of $\mathcal{S}$.

Proof. (a) *Let Q denote an arbitrary quad of $T_1 \cup T_2$, let x_Q denote the unique point of X in Q and let y denote an arbitrary point of Q. Then $f(y) = \mathrm{d}(y, x_Q)$.*

Obviously, this holds if $\mathrm{d}(y, x_Q) \leq 1$. So, suppose that $\mathrm{d}(y, x_Q) = 2$. Since $y \notin X$, $\mathrm{d}(y, X) \in \{1, 2\}$. If $\mathrm{d}(y, X) = 1$, then y is collinear with a point $y' \in X$ which is not contained in Q. Then $\mathrm{d}(y', x_Q) = \mathrm{d}(y', y) + \mathrm{d}(y, x_Q) = 3$, a contradiction. So, $\mathrm{d}(y, X) = 2 = \mathrm{d}(y, x_Q)$.

(b) *Every line L contains a unique point with smallest value.*

Let Q denote a quad of $T_1 \cup T_2$ through L and let x_Q denote the unique point of X in Q. Then $f(y) = \mathrm{d}(y, x_Q)$ for every point y of L. So, L contains a unique point with smallest value, namely the point of L nearest to x_Q.

If x is a point of $\mathcal{S}$ with value 0, then we define $F_x := \{x\}$. If x is a point with value 2, then we define $F_x := \mathcal{S}$. Suppose now that x is a point with value 1, let Q_i, $i \in \{1, 2\}$, denote the unique element of T_i through x and let y_i denote the unique point of $X \cap Q_i$. Then $\mathrm{d}(x, y_1) = \mathrm{d}(x, y_2) = 1$. Put $F_x := \mathcal{C}(x, y_1, y_2)$. If $y_1 = y_2$, then F_x is a line. Suppose now that $y_1 \neq y_2$; then F_x is a grid. Let L be a line of F_x not containing y_1 and y_2, let Q_L denote the unique element of $T_1 \cup T_2$ through L and let x_L denote the unique point of X in Q_L. Since $\mathrm{d}(y_1, x_L) = \mathrm{d}(y_2, x_L) = 2$, the points $\pi_{Q_L}(y_1)$ and $\pi_{Q_L}(y_2)$ of the line L are collinear with x_L. Hence, the point x_L itself is also contained in L. This proves that $F_x \cap X$ is an ovoid.

By the above discussion, it is now clear that f is a valuation of $\mathcal{S}$. □

Definition. Any valuation which can be obtained as described in Theorem 5.21 is called a *semi-diagonal valuation*. Semi-diagonal valuations were introduced in [48]. They have similar properties as diagonal valuations, see e.g. the following lemma.

Lemma 5.22. *Let $\mathcal{S} = (\mathcal{P}, \mathcal{L}, \mathrm{I})$ be a dense product near polygon, let $\{T_1, T_2\} \in \Delta_0(\mathcal{S})$ and let F_i, $i \in \{1, 2\}$, denote an arbitrary element of T_i . For every point x of $\mathcal{S}$ and every $i \in \{1, 2\}$, let $F_i(x)$ denote the unique element of T_i through x. Let X be a set of points of $\mathcal{S}$ satisfying the following properties.*

- *Every element of $T_1 \cup T_2$ has a unique point in common with X.*
- *For every point x of $\mathcal{S}$, $d(x, x_1) = d(x, x_2)$, with x_i, $i \in \{1, 2\}$, the unique point of X in $F_i(x)$.*

Then F_1 and F_2 are isomorphic. Moreover, there exists an isomorphism ϕ from $F_1 \times F_1$ to $\mathcal{S}$ such that $X = \{\phi(x, x) \,|\, x \in F_1\}$. As a consequence, the function $f : \mathcal{P} \to \mathbb{N}; x \to d(x, X)$ is a diagonal valuation of $\mathcal{S}$.

Proof. For every point x of F_1, let $\theta(x)$ denote the unique point of F_2 such that the unique point $\tilde{x}$ in $F_2(x) \cap F_1(\theta(x))$ belongs to X. Obviously, θ is a bijection. Now, suppose that x and y are collinear points of F_1. The unique point in $F_2(y) \cap F_1(\theta(x))$ has distance 1 from the point $\tilde{x} \in X$ and hence has distance 1 from another point in X. This point necessarily coincides with the unique point $\tilde{y}$ in $F_2(y) \cap F_1(\theta(y))$, proving that $\theta(x)$ and $\theta(y)$ are collinear. So, θ preserves collinearity. By symmetry, also θ^{-1} preserves collinearity. So, θ is an isomorphism from F_1 to F_2. Now, for every point $x \in F_1$ and every point $y \in F_2$, let $\phi'(x, y)$ denote the unique point in $F_1(y) \cap F_2(x)$. The map $\phi : F_1 \times F_1; (x, y) \mapsto \phi'(x, \theta(y))$ is an isomorphism from $F_1 \times F_1$ to $\mathcal{S}$. Moreover, $X = \{\phi(x, x) \,|\, x \in F_1\}$. This proves the lemma. □

Property (V)

Let $T_1 = (\mathcal{L}_1, G_1, \Delta_1)$ and $T_2 = (\mathcal{L}_2, G_2, \Delta_2)$ be two admissible triples where G_1 and G_2 are isomorphic additive groups. Let P_i, $i \in \{1, 2\}$, be the point set of $\mathcal{L}_i$. If $x_1, x_2, \ldots, x_k$ are points of P_i, $i \in \{1, 2\}$, then we define $\Delta_i(x_1, x_2, \ldots, x_k) := \Delta_i(x_1, x_2) + \Delta_i(x_2, x_3) + \cdots + \Delta_i(x_{k-1}, x_k)$.

Definition. Let x be a point of $\mathcal{L}_2$, let α be an isomorphism between $\mathcal{L}_2$ and $\mathcal{L}_1$ and let g_x be an element of G_1. An anti-isomorphism θ between G_1 and G_2 is said to satisfy *property* (V_x) *(with respect to* (α, g_x)*)* if

$$g_x = \theta^{-1}(\Delta_2(x, y, z, x)) + g_x + \Delta_1(x^\alpha, y^\alpha, z^\alpha, x^\alpha)$$

holds for all points y and z of $\mathcal{L}_2$.

Lemma 5.23. *Suppose that $\mathcal{L}_2$ is not a line. There for every point x of $\mathcal{L}_2$, for every isomorphism α from $\mathcal{L}_2$ to $\mathcal{L}_1$ and for every $g_x \in G_1$, there exists at most one anti-isomorphism θ satisfying property (V_x) with respect to (α, g_x).*

Proof. It suffices to prove that $\Delta_2(x, y, z, x)$ can take all values of G_2. Let L denote a line of $\mathcal{L}_2$ not containing x and let y denote an arbitrary point of L. If

$$\Delta_2(x, y) + \Delta_2(y, z_1) + \Delta_2(z_1, x) = \Delta_2(x, y) + \Delta_2(y, z_2) + \Delta_2(z_2, x)$$

for two different points z_1 and z_2 of L, then

$$\Delta_2(z_2, z_1) + \Delta_2(z_1, x) = \Delta_2(z_2, y) + \Delta_2(y, z_1) + \Delta_2(z_1, x) = \Delta_2(z_2, x),$$

contradicting the fact that x, z_1 and z_2 are not collinear. So, the elements $\Delta_2(x, y, z, x)$, $z \in L$, of G_2 are mutually different. Since there are as many points on L as elements in G_2, $\Delta_2(x, y, z, x)$ will take all values of G_2. This proves the lemma. □

Lemma 5.24. *Let x and y be points of $\mathcal{L}_2$, let α be an isomorphism from $\mathcal{L}_2$ to $\mathcal{L}_1$ and let $g_x \in G_1$. If an anti-isomorphism θ between G_1 and G_2 satisfies property (V_x) with respect to (α, g_x), then θ satisfies property (V_y) with respect to $(\alpha, \theta^{-1}(\Delta_2(x, y)) + g_x + \Delta_1(x^\alpha, y^\alpha))$.*

Proof. Let z_1 and z_2 be arbitrary points of $\mathcal{L}_2$. Then we have

$$\begin{aligned} g_x &= \theta^{-1}(\Delta_2(x, y, z_1, x)) + g_x + \Delta_1(x^\alpha, y^\alpha, z_1^\alpha, x^\alpha), && (5.1) \\ g_x &= \theta^{-1}(\Delta_2(x, z_1, z_2, x)) + g_x + \Delta_1(x^\alpha, z_1^\alpha, z_2^\alpha, x^\alpha), && (5.2) \\ g_x &= \theta^{-1}(\Delta_2(x, z_2, y, x)) + g_x + \Delta_1(x^\alpha, z_2^\alpha, y^\alpha, x^\alpha). && (5.3) \end{aligned}$$

From equations (5.1) and (5.2), we have

$$g_x = \theta^{-1}(\Delta_2(x, y, z_1, z_2, x)) + g_x + \Delta_1(x^\alpha, y^\alpha, z_1^\alpha, z_2^\alpha, x^\alpha). \tag{5.4}$$

From equations (5.3) and (5.4), we have

$$g_x = \theta^{-1}(\Delta_2(x, y, z_1, z_2, y, x)) + g_x + \Delta_1(x^\alpha, y^\alpha, z_1^\alpha, z_2^\alpha, y^\alpha, x^\alpha),$$

or

$$g_y = \theta^{-1}(\Delta_2(y, z_1, z_2, y)) + g_y + \Delta_1(y^\alpha, z_1^\alpha, z_2^\alpha, y^\alpha),$$

with $g_y = \theta^{-1}(\Delta_2(x, y)) + g_x + \Delta_1(x^\alpha, y^\alpha)$. This proves the lemma. □

Definition. Let α be an isomorphism from $\mathcal{L}_2$ to $\mathcal{L}_1$. We say that an anti-isomorphism θ between G_1 and G_2 satisfies *property* (V) *with respect to* α if there exists a point x in $\mathcal{L}_2$ and an element $g_x \in G_1$ such that θ satisfies property (V_x) with respect to (α, g_x). We say that an anti-isomorphism between G_1 and G_2 satisfies *property* (V) if it satisfies property (V) with respect to some isomorphism between $\mathcal{L}_2$ and $\mathcal{L}_1$.

Necessary and sufficient conditions for the existence of semi-diagonal valuations

We will now determine necessary and sufficient conditions for the existence of semi-diagonal valuations. We will use the coordinatization of glued near hexagons as presented in Theorem 4.50. Let $T_1 = (\mathcal{L}_1, G_1, \Delta_1)$ and $T_2 = (\mathcal{L}_2, G_2, \Delta_2)$ denote two admissible triples where G_1 and G_2 are two isomorphic additive groups. For any anti-isomorphism θ between G_1 and G_2, let $\mathcal{S}_\theta$ denote the corresponding glued near hexagon. Let $\{R_1, R_2\}$ be the element of $\Delta_1(\mathcal{S}_\theta)$ which arises in the natural way from the admissible triples T_1 and T_2. If (g, p_1, p_2), (g', p_1', p_2') and (g'', p_1'', p_2'') are three points of $\mathcal{S}_\theta$ at mutual distance 2, then

$$\begin{aligned} g' &= \theta^{-1}(\Delta_2(p_2, p_2')) + g + \Delta_1(p_1, p_1'), \\ g'' &= \theta^{-1}(\Delta_2(p_2', p_2'')) + g' + \Delta_1(p_1', p_1''), \\ g &= \theta^{-1}(\Delta_2(p_2'', p_2)) + g'' + \Delta_1(p_1'', p_1). \end{aligned}$$

As a consequence,

$$g = \theta^{-1}(\Delta_2(p_2, p_2', p_2'', p_2)) + g + \Delta_1(p_1, p_1', p_1'', p_1).$$

We have: p_1, p_1' and p_1'' are collinear points of $\mathcal{L}_1 \Leftrightarrow \Delta_1(p_1, p_1', p_1'', p_1) = 0 \Leftrightarrow \Delta_2(p_2, p_2', p_2'', p_2) = 0 \Leftrightarrow p_2$, p_2' and p_2'' are collinear points of $\mathcal{L}_2$.

Proposition 5.25. *The following are equivalent:*

(1) *θ satisfies property (V);*

(2) *there exists a set X of points in $\mathcal{S}_\theta$ satisfying the following properties:*

(i) *$d(x,y)=2$ for any two different points x and y of X;*

(ii) *every quad of $R_1 \cup R_2$ has a unique point in common with X.*

Proof. By the remark preceding this proposition, we know that if X is a set of points satisfying (2i) and (2ii), then there exists an isomorphism α from $\mathcal{L}_2$ to $\mathcal{L}_1$ and a function λ from P_2 to G_1 such that $X = X_{(\alpha,\lambda)} := \{(\lambda(p), p^\alpha, p) \,|\, p \in P_2\}$. We will now determine the conditions that need to be satisfied by α and λ so that $X_{(\alpha,\lambda)}$ satisfies properties (2i) and (2ii). Choose a point $x \in P_2$ and put $g_x := \lambda(x)$. Then for every point y of P_2, the unique point $(*, y^\alpha, y)$ at distance 2 from (g_x, x^α, x) is the point $(\theta^{-1}(\Delta_2(x,y)) + g_x + \Delta_1(x^\alpha, y^\alpha), y^\alpha, y)$. Now, the points $(\theta^{-1}(\Delta_2(x,y)) + g_x + \Delta_1(x^\alpha, y^\alpha), y^\alpha, y)$ and $(\theta^{-1}(\Delta_2(x,z)) + g_x + \Delta_1(x^\alpha, z^\alpha), z^\alpha, z)$ are at distance 2 if and only if $g_x = \theta^{-1}\Big(\Delta_2(x,y) + \Delta_2(y,z) + \Delta_2(z,x)\Big) + g_x + \Delta_1(x^\alpha, y^\alpha) + \Delta_1(y^\alpha, z^\alpha) + \Delta_1(z^\alpha, x^\alpha)$. So, $X_{(\alpha,\lambda)}$ satisfies properties (2i) and (2ii) if and only if θ satisfies property (V_x) with respect to (α, g_x). The proposition now readily follows. □

Corollary 5.26. *Let θ be an anti-isomorphism between G_1 and G_2. Then $\mathcal{S}_\theta$ has semi-diagonal valuations if and only if θ satisfies property (V).*

Examples. Suppose that T_1 and T_2 are equal to some admissible triple $T = (\mathcal{L}, G, \Delta)$ where G is a commutative group of size at least three. Let $1_{\mathcal{L}}$ denote the trivial automorphism of $\mathcal{L}$ and let -1_G be the (anti-)automorphism of G mapping g to $-g$. Put $\mathcal{S} := \mathcal{S}_{-1_G}$. The map -1_G satisfies property (V_x) with respect to $(1_{\mathcal{L}}, g)$ for every point x of $\mathcal{L}$ and every $g \in G$. Hence, $\mathcal{S}$ has a semi-diagonal valuation f_g with $O_{f_g} := \{(g, p, p) \,|\, p \in \mathcal{L}\}$. All these semi-diagonal valuations are equivalent since the map $(g, p_1, p_2) \mapsto (g+h, p_1, p_2)$ defines an automorphism of $\mathcal{S}$ for every $h \in G$. For each known admissible triple $(\mathcal{L}, G, \Delta)$ for which the linear space $\mathcal{L}$ is not a line, the group G is commutative, see [29, Section 8].

5.6.5 Distance-j-ovoidal valuations

Definition. Let $\mathcal{S}$ be a near $2n$-gon, $n \geq 2$. A distance-j-ovoid $(2 \leq j \leq n)$ of $\mathcal{S}$ is a set X of points satisfying:

(1) $\mathrm{d}(x,y) \geq j$ for every two different points x and y of X;

(2) for every point a of $\mathcal{S}$, there exists a point $x \in X$ such that $\mathrm{d}(a,x) \leq \frac{j}{2}$;

(3) for every line L of $\mathcal{S}$, there exists a point $x \in X$ such that $\mathrm{d}(L,x) \leq \frac{j-1}{2}$.

A distance-2-ovoid is just an ovoid. From (1), (2) and (3), we immediately have:

- If j is odd, then for every point a of $\mathcal{S}$, there exists a unique point $x \in X$ such that $\mathrm{d}(a,x) \leq \frac{j-1}{2}$.
- If j is even, then for every line L of $\mathcal{S}$, there exists a unique point $x \in X$ such that $\mathrm{d}(L,x) \leq \frac{j-2}{2}$.

Theorem 5.27. *Let $\mathcal{S} = (\mathcal{P}, \mathcal{L}, \mathrm{I})$ be a near $2n$-gon, $n \geq 2$, and let $\mathcal{W}$ be a set of convex subpolygons of $\mathcal{S}$ such that the pair $(\mathcal{S}, \mathcal{W})$ is nice. Let X be a distance-j-ovoid of $\mathcal{S}$ ($2 \leq j \leq n$ and j even). Then the map $f : \mathcal{P} \to \mathbb{N}, x \mapsto d(x, X)$ is a valuation of $(\mathcal{S}, \mathcal{W})$.*

Proof. Since $f(x) = 0$ for every point $x \in X$, property (V_1) holds.

Let L be a line of $\mathcal{S}$. Then there exists a unique point $x^* \in X$ such that $\mathrm{d}(x^*, L) \leq \frac{j-2}{2} = \frac{j}{2} - 1$. So, $\mathrm{d}(a, x^*) \leq \frac{j}{2}$ for every point a of L. By property (1), we then have that $\mathrm{d}(a, X) = \mathrm{d}(a, x^*)$ for every point a of L. It is now easily seen that property (V_2) holds: the point x_L of L with smallest value is the unique point of L nearest to x^*.

Let x denote an arbitrary point of $\mathcal{S}$. If $\mathrm{d}(x, X) = \frac{j}{2}$, then we define $F_x := \mathcal{S}$. If $\mathrm{d}(x, X) < \frac{j}{2}$, then by property (1), there exists a unique point $x' \in X$ at distance $\mathrm{d}(x, X)$ from x and we define $F_x := H(x, x')$, where $H(x, x')$ denotes the unique convex sub-$2\cdot\mathrm{d}(x, x')$-gon of $\mathcal{W}$ through the points x and x'. Clearly, property (V_3) holds for any point x for which $\mathrm{d}(x, X) = \frac{j}{2}$. Suppose therefore that $\mathrm{d}(x, X) < \frac{j}{2}$ and let x' denote the unique point of X at distance $\mathrm{d}(x, X)$ from x. Then for every point y of F_x, $\mathrm{d}(y, x') \leq \mathrm{d}(x, x') < \frac{j}{2}$. So, $f(y) = \mathrm{d}(y, X) = \mathrm{d}(y, x') \leq f(x)$ and $F_y = H(y, x') \subseteq H(x, x') = F_x$. If z is a point of $\mathcal{S}$ collinear with y such that $f(z) = f(y) - 1$, then there exists a point $x'' \in X$ such that $\mathrm{d}(z, x'') = \mathrm{d}(y, x') - 1$. Since y has distance at most $\mathrm{d}(y, x')$ to x'', x' coincides with x''. So, $\mathrm{d}(z, x') = \mathrm{d}(y, x') - 1$ and $z \in H(x', y) \subseteq F_x$. This proves that also (V_3) holds. □

Definition. Any valuation f which can be obtained as in Theorem 5.27 is called a *distance-j-ovoidal valuation.* A distance-2-ovoidal valuation is the same as an ovoidal valuation.

5.6.6 Extended valuations

Theorem 5.28. *Let $\mathcal{S} = (\mathcal{P}, \mathcal{L}, \mathrm{I})$ be a near $2n$-gon and let $\mathcal{W}$ be a set of convex subpolygons of $\mathcal{S}$ such that $(\mathcal{S}, \mathcal{W})$ is nice. Let F be a convex subpolygon of $\mathcal{S}$ which is classical in $\mathcal{S}$. Put $\mathcal{W}' := \{F \cap F' \,|\, F' \in \mathcal{W}$ and $F \cap F' \neq \emptyset\}$. Suppose that the following holds:*

(a) *$(F, \mathcal{W}')$ is nice;*

(b) *for every $G \in \mathcal{W}'$ and every $x \in G$, there exists an $x' \in G$ such that $d(x, x') = \mathrm{diam}(G)$.*

If f' is a valuation of $(F, \mathcal{W}')$, then the map $f : \mathcal{P} \to \mathbb{N}, x \mapsto f(x) := d(x, \pi_F(x)) + f'(\pi_F(x))$, is a valuation of $(\mathcal{S}, \mathcal{W})$. If f' is a classical valuation, then also f is classical.

Proof. Obviously, property (V_1) is satisfied. From Theorem 1.9, it easily follows that also property (V_2) is satisfied.

For every point x of F, let G_x denote a convex subpolygon of $\mathcal{W}'$ for which property (V_3) holds (with respect to the valuation f'). Put $G_x := G_{\pi_F(x)}$ for every point x of $\mathcal{S}$. For all points x and y of $\mathcal{S}$, let $H(x,y)$ denote the unique convex sub-$2 \cdot \mathrm{d}(x,y)$-gon of $\mathcal{W}$ through the points x and y. For every point x of $\mathcal{S}$, we define $F_x := H(x, x_1)$, where x_1 is a point of G_x at distance $\mathrm{diam}(G_x)$ from $\pi_F(x)$. This is a good definition. Let x_1' denote another point of G_x at distance $\mathrm{diam}(G_x)$ from $\pi_F(x)$. By our assumption (b) and the fact that F is classical in $\mathcal{S}$, $H(x, x_1)$ contains a point at distance $\mathrm{d}(x, \pi_F(x)) + \mathrm{diam}(H(x, x_1) \cap F)$ from x. Hence, $\mathrm{diam}(H(x, x_1) \cap F) \leq \mathrm{d}(x, x_1) - \mathrm{d}(x, \pi_F(x)) = \mathrm{d}(\pi_F(x), x_1)$. Since $H(x, x_1) \cap F$ contains the points $\pi_F(x)$ and x_1, $H(x, x_1) \cap F = G_x$. Hence, $H(x, x_1)$ contains the point x_1' and $H(x, x_1') \subseteq H(x, x_1)$. In a similar way one shows that $H(x, x_1) \subseteq H(x, x_1')$. So, $H(x, x_1)$ is equal to $H(x, x_1')$ and F_x is well-defined. By the above reasoning we also know that $F_x \cap F = G_x$ for every point x of $\mathcal{S}$. The following properties hold:

- *For every $y \in F_x$, $\pi_F(y) \in G_x$.* Clearly every shortest path between y and a point $z \in G_x$ is contained in F_x. Since the point $\pi_F(y)$ is contained in a shortest path between y and z, the point $\pi_F(y)$ belongs to $F_x \cap F = G_x$.

- *For every point y of F_x, $d(y, \pi_F(y)) \leq d(x, \pi_F(x))$.* Let y' denote a point of G_x at maximal distance $\mathrm{diam}(G_x)$ from the point $\pi_F(y)$. Then $\mathrm{d}(x, \pi_F(x)) + \mathrm{diam}(G_x) = \mathrm{diam}(F_x) \geq \mathrm{d}(y, y') = \mathrm{d}(y, \pi_F(y)) + \mathrm{d}(\pi_F(y), y') = \mathrm{d}(y, \pi_F(y)) + \mathrm{diam}(G_x)$. Hence, $\mathrm{d}(y, \pi_F(y)) \leq \mathrm{d}(x, \pi_F(x))$.

Let u be a point of F_x. Since $\pi_F(u) \in G_x$, $f'(\pi_F(u)) \leq f'(\pi_F(x))$. Hence, $f(u) = \mathrm{d}(u, \pi_F(u)) + f'(\pi_F(u)) \leq \mathrm{d}(x, \pi_F(x)) + f'(\pi_F(x)) = f(x)$. Since $\pi_F(u) \in G_{\pi_F(x)}$, $G_u \subseteq G_x$ and $F_u \subseteq F_x$. Let v be a neighbour of u with value $f(u) - 1$. In order to prove property (V_3), we distinguish two possibilities.

- $\mathrm{d}(v, \pi_F(v)) \neq \mathrm{d}(u, \pi_F(u))$. Then $\pi_F(u) = \pi_F(v)$ by Lemma 1.8. In this case we have $\mathrm{d}(v, \pi_F(v)) = \mathrm{d}(u, \pi_F(u)) - 1$. So, v is on a shortest path between u and $\pi_F(u) = \pi_F(v)$. Since $u, \pi_F(u) \in F_x$, also v belongs to F_x.

- $\mathrm{d}(v, \pi_F(v)) = \mathrm{d}(u, \pi_F(u))$. In this case we have $f'(\pi_F(v)) = f'(\pi_F(u)) - 1$. By Theorem 1.9, $\mathrm{d}(\pi_F(u), \pi_F(v)) = 1$. From $\pi_F(u) \in G_x$, it then follows that also $\pi_F(v) \in G_x$. Now, v lies on a shortest path between $\pi_F(v)$ and u. Since $\pi_F(v) \in F_x$ and $u \in F_x$, also v belongs to F_x.

If f' is classical valuation of F, then we have $f(x) = \mathrm{d}(x, \pi_F(x)) + f'(\pi_F(x)) = \mathrm{d}(x, \pi_F(x)) + \mathrm{d}(\pi_F(x), x^*) = \mathrm{d}(x, x^*)$, where x^* denotes the unique point of F for which $f'(x^*) = 0$. Hence f is classical if f' is classical. □

Definition. The valuation f is called an *extension* of f'. If $F = \mathcal{S}$, then f is called a *trivial extension* of f'.

5.6.7 SDPS-valuations

Let $\mathcal{A} = (\mathcal{P}, \mathcal{L}, \mathrm{I})$ be a possibly infinite thick dual polar space of rank $2n$, $n \in \mathbb{N}$. A set X of points of $\mathcal{A}$ is called an *SDPS-set* (SDPS = sub dual polar space) of $\mathcal{A}$ if it satisfies the following properties:

(1) No two points of X are collinear in $\mathcal{A}$.

(2) If $x, y \in X$ such that $\mathrm{d}(x, y) = 2$, then $X \cap \mathcal{C}(x, y)$ is an ovoid of the quad $\mathcal{C}(x, y)$.

(3) The partial linear space $\widetilde{\mathcal{A}}$ whose points are the elements of X and whose lines are the quads of $\mathcal{A}$ containing at least two points of X (natural incidence) is a dual polar space of rank n.

(4) For all $x, y \in X$, $\mathrm{d}(x, y) = 2 \cdot \delta(x, y)$. Here, $\mathrm{d}(x, y)$ and $\delta(x, y)$ denote the distances between x and y in the respective dual polar spaces $\mathcal{A}$ and $\widetilde{\mathcal{A}}$.

(5) If $x \in X$ and if L is a line of $\mathcal{A}$ through x, then L is contained in a quad of $\mathcal{A}$ which contains at least two points of X.

An SDPS-set of the near 0-gon consists of the unique point of the near 0-gon. An SDPS-set of a thick generalized quadrangle is just an ovoid of that generalized quadrangle. We will now describe two classes of SDPS-sets in thick dual polar spaces of rank $2n \geq 4$ (see also Section 9 of [90]).

Consider the finite field $\mathbb{F}_{q^2}$ with q^2 elements and let $\mathbb{F}_q$ denote the unique subfield of order q of $\mathbb{F}_{q^2}$. Let η denote an arbitrary element of $\mathbb{F}_{q^2} \setminus \mathbb{F}_q$. Then $\mathbb{F}_{q^2} = \{x_1 + x_2\eta \,|\, x_1, x_2 \in \mathbb{F}_q\}$; define $\tau : \mathbb{F}_{q^2} \to \mathbb{F}_q, x_1 + x_2\eta \mapsto x_1$.

- Consider the following bijection ϕ between the vector spaces $\mathbb{F}_q^{4n}$ and $\mathbb{F}_{q^2}^{2n}$:

$$\phi(x_1, x_2, \ldots, x_{4n}) = (x_1 + \eta x_2, \ldots, x_{4n-1} + \eta x_{4n}).$$

  If $\langle \cdot, \cdot \rangle$ is a nondegenerate symplectic form of $\mathbb{F}_{q^2}^{2n}$, then $\tau(\langle \phi(\cdot), \phi(\cdot) \rangle)$ is a nondegenerate symplectic form in $\mathbb{F}_q^{4n}$. If α is a totally isotropic n-dimensional subspace of $\mathbb{F}_{q^2}^{2n}$, then $\phi^{-1}(\alpha)$ is a $2n$-dimensional totally isotropic subspace of $\mathbb{F}_q^{4n}$. In this way we obtain an "embedding" of $DW(2n-1, q^2)$ in $DW(4n-1, q)$, giving rise to an SDPS-set.

- Consider the following bijection ϕ between the vector spaces $\mathbb{F}_q^{4n+2}$ and $\mathbb{F}_{q^2}^{2n+1}$:

$$\phi(x_1, x_2, \ldots, x_{4n+2}) = (x_1 + \eta x_2, \ldots, x_{4n+1} + \eta x_{4n+2}).$$

  Let $\langle \cdot, \cdot \rangle$ be a nondegenerate hermitian form of $\mathbb{F}_{q^2}^{2n+1}$. For every $x \in \mathbb{F}_{q^2}^{2n+1}$, we define $h(x) := \langle x, x \rangle$ and for every $x \in \mathbb{F}_q^{4n+2}$, we define $q(x) := \langle \phi(x), \phi(x) \rangle$. The equation $h(x) = 0$, respectively $q(x) = 0$, defines a nonsingular hermitian variety $H(2n, q^2)$ in $\mathrm{PG}(2n, q^2)$, respectively a nonsingular elliptic quadric $Q^-(4n+1,q)$ in $\mathrm{PG}(4n+1,q)$. With every generator of $H(2n,q^2)$, there corresponds a generator of $Q^-(4n+1,q)$. In this way we obtain an "embedding" of $DH(2n,q^2)$ in $DQ^-(4n+1,q)$, giving rise to an SDPS-set.

Theorem 5.29 (Section 5.8). *Let X be an SDPS-set of a thick dual polar space $\mathcal{A}$ of rank $2n \geq 0$. For every point x of $\mathcal{A}$, we define $f(x) := d(x, X)$. Then f is a valuation of $\mathcal{A}$.*

Definition. A valuation which can be obtained from an SDPS-set as described in Theorem 5.29 is called an *SDPS-valuation.* SDPS-valuations were introduced in [61] for finite thick dual polar spaces.

Theorem 5.30 (Section 5.9). *Let $\mathcal{A}$ be a thick dual polar space and let f be a valuation of $\mathcal{A}$ which is the (possibly trivial) extension of an SDPS-valuation in a convex subpolygon $\mathcal{B}$ of $\mathcal{A}$. Then the following holds:*

(i) *$f(x) = d(x, O_f)$ for every point x of $\mathcal{A}$;*

(ii) *if H is a hex of $\mathcal{A}$, then the valuation induced in H is either classical or the extension of an ovoidal valuation in a quad of H.*

Theorem 5.31 (Section 5.10). *If X is an SDPS-set of a finite thick dual polar space $\mathcal{A}$ of rank $2n \geq 4$ and if $\widetilde{\mathcal{A}}$ denotes the associated dual polar space of rank n, then one of the following cases occurs:*

(1) *$\mathcal{A} \cong DW(4n-1,q)$ and $\widetilde{\mathcal{A}} \cong DW(2n-1,q^2)$ for some prime power q. If Q is a quad containing two points of X, then $Q \cap X$ is a classical ovoid of Q, i.e., an elliptic quadric $Q^-(3,q)$ on $Q \cong Q(4,q)$.*

(2) *$\mathcal{A} \cong DQ^-(4n+1,q)$ and $\widetilde{\mathcal{A}} \cong DH(2n,q^2)$ for some prime power q. If Q is a quad containing two points of X, then $Q \cap X$ is a classical ovoid of Q, i.e., a unital $H(2,q^2)$ on $Q \cong H(3,q^2)$.*

Also the converse of Theorem 5.30 is true.

Theorem 5.32 (Section 5.11). *Let $\mathcal{A}$ be a thick dual polar space and let f be a valuation of $\mathcal{A}$ with the property that every induced hex valuation is either classical or the extension of an ovoidal valuation in a quad. Then f is the (possibly trivial) extension of an SDPS-valuation in a convex subpolygon of $\mathcal{A}$.*

5.7 Valuations of dense near hexagons

Let f be a valuation of a dense near hexagon $\mathcal{S} = (\mathcal{P}, \mathcal{L}, \mathrm{I})$. We know that f is classical if and only if $\max\{f(x) | x \in \mathcal{P}\} = 3$ and that f is ovoidal if and only if $\max\{f(x) | x \in \mathcal{P}\} = 1$.

Theorem 5.33. *If $|O_f| = 1$, then f is a classical or a semi-classical valuation.*

Proof. This follows directly from Theorems 5.12 and 5.16. $\square$

Theorem 5.34. *Suppose that $|O_f| \geq 2$ and that f is not ovoidal. Then every two points of O_f lie at distance 2 from each other. As a consequence, G_f is a linear space.*

Proof. Let x and y denote two distinct points of O_f. Then $\mathrm{d}(x,y) \in \{2,3\}$. Suppose that $\mathrm{d}(x,y)=3$ and consider a shortest path x, x_1, x_2, y from x to y. By property (V_2), the points x_1 and x_2 have value 1, and there exists a point p on x_1x_2 with value 0. Let F_{x_1} denote the subpolygon through x_1 satisfying property (V_3). Since x and p are points with value 0 collinear with x_1, we have $x, p \in F_{x_1}$. Since x_1 and p belong to F_{x_1}, also the point x_2 belongs to F_{x_1}. Since y is a point with value 0 collinear with x_2, we also have $y \in F_{x_1}$. Hence, $x, y \in F_{x_1}$ and $\mathcal{C}(x,y) \subseteq F_{x_1}$. Since $\mathrm{d}(x,y)=3$, $\mathcal{S} = \mathcal{C}(x,y) = F_{x_1}$, a contradiction, since every point of F_{x_1} has value at most 1 and $\mathcal{S}$ contains points with value 2. □

Theorem 5.35. *If not every line of a dense near hexagon $\mathcal{S}$ is incident with the same number of points, then f is either a classical valuation or the extension of an ovoidal valuation in a quad of $\mathcal{S}$.*

Proof. By Corollary 1.13, the near hexagon $\mathcal{S}$ is the direct product of a generalized quadrangle Q and a line L such that no line of Q is incident with the same number of points as L.

Let x denote an arbitrary point of O_f, let Q_x denote the unique quad through x isomorphic to Q and let L_x denote the line through x not contained in Q_x. Every quad intersecting Q_x in a line is a nonsymmetrical grid. As a consequence, the valuation induced in each such quad is classical. We distinguish two cases.

(i) The valuation induced in Q_x is classical. Since the valuation induced in every quad through x is classical, $f(y) = \mathrm{d}(x,y)$ for every point y at distance at most 2 from x. By Theorem 5.16, f is either classical or semi-classical. Let Q denote a grid-quad not containing x intersecting Q_x in a line. If f were semi-classical, then the valuation induced in Q would be ovoidal, which is impossible. Hence, f is a classical valuation.

(ii) The valuation induced in Q_x is ovoidal. Then $\mathrm{d}(y, O_f) \leq 2$ for every point y of $\mathcal{S}$. By Theorem 5.12, $f(y) = \mathrm{d}(y, O_f)$. Since the valuation induced in every quad intersecting Q_x in a line is classical, O_f is an ovoid of Q_x. It follows that f is the extension of the ovoidal valuation of Q_x associated with the ovoid O_x. □

Theorem 5.36. *If $\mathcal{S}$ is classical, then f is either classical, ovoidal, semi-classical or the extension of an ovoidal valuation in a quad of $\mathcal{S}$.*

Proof. By Theorem 5.12, $f(x) = \mathrm{d}(x, O_f)$ for every point x at distance at most 2 from O_f. If $|O_f| = 1$, then f must be classical or semi-classical by Theorem 5.16. Suppose therefore that $|O_f| \geq 2$. Suppose also that f is not ovoidal, then every two different points of O_f lie at distance 2 from each other by Theorem 5.34. Let o_1 and o_2 be two different points of O_f and let Q denote the unique quad through o_1 and o_2. Then $Q \cap O_f$ is an ovoid of Q. For every point x outside Q, there exists a point in $O_f \cap Q$ at distance 3 from x. Hence $O_f = O_f \cap Q$. Now, every point of $\mathcal{S}$ lies at distance at most 2 from O_f. So, $f(x) = \mathrm{d}(x, O_f) = \mathrm{d}(x, \pi_Q(x)) + \mathrm{d}(\pi_Q(x), O_f)$ and f is the extension of an ovoidal valuation of Q. □

Theorem 5.37. *Let f be a valuation of a dense glued near hexagon $\mathcal{S}$ of type 1 and let $\{T_1, T_2\} \in \Delta_1(\mathcal{S})$. Then the following are equivalent:*

(i) *the valuation induced in every quad of $T_1 \cup T_2$ is classical;*

(ii) *f is either classical or semi-diagonal.*

Proof. Obviously, (ii) implies (i). Suppose that the valuation induced in every quad of $T_1 \cup T_2$ is classical. If there exists a point with value 3, then f is classical by Theorem 5.9. Suppose now that $f(x) \leq 2$ for every point x of $\mathcal{S}$. The valuation induced in an element $Q \in T_1 \cup T_2$ is classical and hence Q contains a unique point of O_f. By Theorem 5.34, every two points of O_f lie at distance 2 from each other. Now, every point x of $\mathcal{S}$ lies at distance at most 2 from O_f. By Theorem 5.12, $f(x) = \mathrm{d}(x, O_f)$. Hence, f is a semi-diagonal valuation. □

5.8 Proof of Theorem 5.29

We will prove Theorem 5.29 in several lemmas. We will use the same notation as in Section 5.6.7. We call a quad of $\mathcal{A}$ *special* if it contains at least two points of X. If Q is a special quad, then $Q \cap X$ is an ovoid of Q.

Lemma 5.38. *No two special quads intersect in a line.*

Proof. Suppose the contrary. Let Q_1 and Q_2 denote two special quads which intersect in a line L. Let x denote the unique point of X on L and let x_1, x_2 denote two other points of L. Let y_i, $i \in \{1, 2\}$, denote a point of $(Q_i \cap X) \setminus \{x\}$ collinear with x_i. Then $\mathrm{d}(y_1, y_2) = \mathrm{d}(y_1, x_1) + \mathrm{d}(x_1, y_2) = 1 + 2 = 3$, contradicting property (4) in the definition of SDPS-set. □

Corollary 5.39. *If x is a point of X, then every line through x is contained in a unique special quad.*

Lemma 5.40. *Let x_1 and x_2 be two distinct points of X and let L be a line through x_1 contained in $\mathcal{C}(x_1, x_2)$. Then the unique special quad Q through L is contained in $\mathcal{C}(x_1, x_2)$.*

Proof. Suppose the contrary. Let y denote the unique point on L nearest to x_2 and let x_3 denote a point of $Q \cap O_f$ at distance 2 from y. Then $\mathrm{d}(x_2, x_3) = \mathrm{d}(x_2, y) + \mathrm{d}(y, x_3) = \mathrm{d}(x_1, x_2) - 1 + 2 = \mathrm{d}(x_1, x_2) + 1$. Now, by property (4) in the definition of SDPS-set, $\mathrm{d}(x_2, x_3)$ and $\mathrm{d}(x_1, x_2)$ must be even. A contradiction follows. □

Lemma 5.41. *Let F be a convex subpolygon of $\mathcal{A}$. Then $F \cap X$ is either empty or a convex subspace of $\widetilde{\mathcal{A}}$.*

Proof. We suppose that $F \cap X$ is nonempty. Let x_1 and x_2 denote two points of $F \cap X$ such that $\delta(x_1, x_2) = 1$. Then $\mathrm{d}(x_1, x_2) = 2$. Since F is convex, $\mathcal{C}(x_1, x_2) \subseteq F$ and hence $\mathcal{C}(x_1, x_2) \cap X \subseteq F \cap X$. This proves that $F \cap X$ is a subspace of $\widetilde{\mathcal{A}}$. Now, let a, b, c denote points of X such that $a, b \in F \cap X$, $\delta(a, c) = \delta(a, b) - 1$ and $\delta(c, b) = 1$. Then $\mathrm{d}(a, c) = \mathrm{d}(a, b) - 2$ and $\mathrm{d}(c, b) = 2$. Since F is convex and $\mathrm{d}(a, c) + \mathrm{d}(c, b) = \mathrm{d}(a, b)$, $c \in F \cap X$. Hence $F \cap X$ is also convex (in $\widetilde{\mathcal{A}}$). □

Lemma 5.42. *Let $\tilde{F}$ be a convex subpolygon of $\widetilde{\mathcal{A}}$. Then there exists a unique convex subpolygon F of $\mathcal{A}$ such that* $\mathrm{diam}_{\mathcal{A}}(F) = 2 \cdot \mathrm{diam}_{\widetilde{\mathcal{A}}}(\tilde{F})$ *and* $\tilde{F} = F \cap X$. *Moreover, $\tilde{F}$ is an SDPS-set of F.*

Proof. Let x and y be two points of $\tilde{F}$ such that $\delta(x, y) = \mathrm{diam}_{\widetilde{\mathcal{A}}}(\tilde{F})$. If F is a convex subpolygon of $\mathcal{A}$ such that $\mathrm{diam}_{\mathcal{A}}(F) = 2 \cdot \mathrm{diam}_{\widetilde{\mathcal{A}}}(\tilde{F})$ and $\tilde{F} = F \cap X$, then F necessarily equals $\mathcal{C}(x, y)$ since $x, y \in F$ and $\mathrm{d}(x, y) = 2 \cdot \delta(x, y) = 2 \cdot \mathrm{diam}_{\widetilde{\mathcal{A}}}(\tilde{F}) = \mathrm{diam}_{\mathcal{A}}(F)$. We will now show that $F := \mathcal{C}(x, y)$ satisfies all required properties. By Lemma 5.41, $F \cap X$ is a convex subpolygon of $\widetilde{\mathcal{A}}$ containing x and y and hence also $\tilde{F}$. On the other hand, we have $\mathrm{diam}_{\widetilde{\mathcal{A}}}(F \cap X) = \frac{1}{2} \mathrm{diam}_{\mathcal{A}}(F \cap X) \leq \frac{1}{2} \mathrm{diam}(F) = \frac{1}{2}\, \mathrm{d}(x, y) = \delta(x, y) = \mathrm{diam}_{\widetilde{\mathcal{A}}}(\tilde{F})$. This proves that $F \cap X = \tilde{F}$. From Lemma 5.40, it readily follows that $\tilde{F}$ is an SDPS-set of F. □

Definition. A convex subpolygon F of $\mathcal{A}$ is called *special* if $F \cap X$ is an SDPS-set of F. The special subpolygons of $\mathcal{A}$ are those subpolygons of the form $\mathcal{C}(x_1, x_2)$ where $x_1, x_2 \in X$.

From Lemmas 5.41 and 5.42, we have the following corollary.

Corollary 5.43. *If F is a convex subpolygon of $\mathcal{A}$, then $F \cap X$ is either empty or an SDPS-set in a convex subpolygon of F.*

Lemma 5.44. *Let $n \geq 1$. If F is a convex sub-$(4n-2)$-gon of $\mathcal{A}$, then $F \cap X$ is a convex sub-$(2n-2)$-gon of $\widetilde{\mathcal{A}}$.*

Proof. We first show that $F \cap X$ is nonempty. Let x be an arbitrary point of X. If $x \in F$, then we are done. If $x \notin F$, let Q denote the unique special quad through the line $x\pi_F(x)$. Then $F \cap Q$ is a line. Since this line contains a point of X, also F contains a point of X.

Now, let x^* denote a point of $F \cap X$ and let y denote a point of X at maximal distance from x^*. Then $\mathrm{d}(x^*, y) = 2n$ and $y \notin F$. Let Q denote the unique special quad through the line $y\pi_F(y)$. The quad Q intersects F in a line which contains a point y'' of X. Now, $\mathrm{d}(x^*, y'') \geq 2n - 2$ and hence $\delta(x^*, y'') \geq n - 1$. This proves that $\mathrm{diam}_{\widetilde{\mathcal{A}}}(F \cap X) \geq n - 1$. On the other hand, we have $\mathrm{diam}_{\widetilde{\mathcal{A}}}(F \cap X) = \frac{1}{2} \mathrm{diam}_{\mathcal{A}}(F \cap X) \leq \frac{1}{2} \mathrm{diam}_{\mathcal{A}}(F) = \frac{2n-1}{2}$. Hence, $\mathrm{diam}_{\widetilde{\mathcal{A}}}(F \cap X) = n - 1$. In Lemma 5.41, we have already shown that $F \cap X$ is convex (in $\widetilde{\mathcal{A}}$). □

Lemma 5.45. *Let $n \geq 1$. Let x be a point of $\mathcal{A}$ and let F denote a convex sub-$(4n-2)$-gon through x. Then $d(x, X) = d(x, F \cap X)$.*

Proof. Let y be a point of $X \setminus F$, let Q denote the unique special quad through the line $y\pi_F(y)$ and let y' denote the unique point of X on the line $Q \cap F$. Then $\mathrm{d}(x,y) = \mathrm{d}(x,\pi_F(y)) + 1 \geq \mathrm{d}(x,y')$. As a consequence, $\mathrm{d}(x, X \setminus F) \geq \mathrm{d}(x, F \cap X)$ and $\mathrm{d}(x,X) = \mathrm{d}(x, F \cap X)$. □

Lemma 5.46. *Let x be a point of $\mathcal{A}$ and let F denote a special subpolygon through x. Then $d(x,X) = d(x, F \cap X)$.*

Proof. The lemma holds if $F = \mathcal{A}$. So, suppose that $F \neq \mathcal{A}$ and that the lemma holds for any special subpolygon of diameter $\mathrm{diam}(F) + 2$. Let F'' denote a special subpolygon of diameter $\mathrm{diam}(F) + 2$ through F and let F' denote a convex subpolygon of diameter $\mathrm{diam}(F)+1$ such that $F \subset F' \subset F''$. By our assumption, $\mathrm{d}(x,X) = \mathrm{d}(x, F'' \cap X)$. By Lemma 5.45, $\mathrm{d}(x, F'' \cap X) = \mathrm{d}(x, F'' \cap X \cap F')$ and by Lemma 5.44, $F'' \cap X \cap F' = F \cap X$. The lemma now follows. □

Lemma 5.47. *For every point x of $\mathcal{A}$, $d(x,X) \leq n$. Moreover, there exists a point x^* of $\mathcal{A}$ such that $d(x^*,X) = n$.*

Proof. We will prove this by induction on n. Obviously, the lemma holds if n is equal to 0 or 1. Suppose therefore that $n \geq 2$. Let F denote a convex sub-$(4n-2)$-gon through x and let F' denote the unique convex sub-$(4n-4)$-gon of $\mathcal{A}$ containing all points of $F \cap X$. Then $\mathrm{d}(x, \pi_{F'}(x)) \leq 1$. Since $F \cap X$ is an SDPS-set in F', we have $\mathrm{d}(\pi_{F'}(x), F \cap X) \leq n-1$ and hence $\mathrm{d}(x,X) \leq n$. By the induction hypothesis, we know that there exists a point $y \in F'$ such that $\mathrm{d}(y, X \cap F) = n-1$. If x^* denotes a point of $F \setminus F'$ collinear with y, then $\mathrm{d}(x^*,X) = \mathrm{d}(x^*, X \cap F) = 1 + \mathrm{d}(y, X \cap F) = n$. This proves the lemma. □

Lemma 5.48. *Let x be a point of $\mathcal{A}$. Then there exist two points $x_1, x_2 \in X$ such that $d(x,x_1) = d(x,x_2) = d(x,X)$ and $d(x_1,x_2) = 2 \cdot d(x,X)$. As a consequence, every point is contained in a special convex sub-$[4 \cdot d(x,X)]$-gon.*

Proof. We will prove this by induction on the distance $\mathrm{d}(x,X)$. Obviously, the property holds if $\mathrm{d}(x,X) = 0$. Suppose therefore that $\mathrm{d}(x,X) \geq 1$ and that the property holds for any point at distance at most $\mathrm{d}(x,X)-1$ from X. Take a point x' collinear with x at distance $\mathrm{d}(x,X)-1$ from X and let x_1' and x_2' denote two points of X such that $\mathrm{d}(x',x_1') = \mathrm{d}(x',x_2') = \mathrm{d}(x,X) - 1$ and $\mathrm{d}(x_1',x_2') = 2 \cdot \mathrm{d}(x,X) - 2$. Put $F' := \mathcal{C}(x_1',x_2')$. Then $x' \in F'$ since x' is on a shortest path between x_1' and x_2'. If $x \in F'$, then by Lemma 5.47, x would have distance at most $\mathrm{d}(x,X)-1$ from $X \cap F'$, a contradiction. So, x is not contained in F' and $\mathrm{d}(x,x_1') = \mathrm{d}(x,x_2') = 1 + \mathrm{d}(x',x_1') = 1 + \mathrm{d}(x',x_2') = \mathrm{d}(x,X)$. Now, the special convex sub-$[4 \cdot \mathrm{d}(x,X)]$-gons through F' partition the set of lines through x_2' not contained in F' and hence also the set of lines through x' not contained in F'. As a consequence, there exists a unique special convex sub-$[4 \cdot \mathrm{d}(x,X)]$-gon F through F' containing the line xx'. Let L_2 denote a line of F through x_2' not contained in F' and containing a point y_2 at distance $\mathrm{d}(x,X)-1$ from x. Let Q_2 denote the unique special quad through L_2. Then $Q_2 \subseteq F$ (recall Lemma 5.40). Now, put $x_1 := x_1'$ and let x_2 denote a point of $(X \cap Q_2) \setminus \{x_2'\}$ collinear with y_2. Then $\mathrm{d}(x,x_2) \leq \mathrm{d}(x,y_2) + \mathrm{d}(y_2,x_2) = \mathrm{d}(x,X)$

and hence $\mathrm{d}(x, x_2) = \mathrm{d}(x, X)$. Since the quad Q_2 intersects F' only in the point x'_2, we have $\mathrm{d}(x_1, x_2) = \mathrm{d}(x_1, x'_2) + \mathrm{d}(x'_2, x_2) = 2 \cdot \mathrm{d}(x, X)$. We have already noticed that $\mathrm{d}(x, x_1) = \mathrm{d}(x, x'_1) = \mathrm{d}(x, X)$. This proves the lemma. □

Lemma 5.49. *Every line L of $\mathcal{A}$ contains a unique point at smallest distance from X.*

Proof. We prove the lemma by induction on n. Obviously, the lemma holds if n is equal to 0 or 1. Suppose therefore that $n \geq 2$. Let F denote a convex sub-$(4n-2)$-gon through L and let F' denote the unique special sub-$(4n-4)$-gon contained in F. For every point x of L, $\mathrm{d}(x, X) = \mathrm{d}(x, F' \cap X)$ by Lemma 5.45. If L is contained in F', then the lemma holds for the line L by the induction hypothesis. If L is disjoint from F', then the lemma holds for the line $\pi_{F'}(L)$ and hence also for the line L since $\mathrm{d}(x, F' \cap X) = 1 + \mathrm{d}(\pi_{F'}(x), F' \cap X)$ for every point x of L. If L intersects F' in a unique point, then this point is the unique point of L at smallest distance from X. □

Lemma 5.50. *Let x be a point of $\mathcal{A}$ and let F denote a special sub-$[4 \cdot d(x, X)]$-gon through x. Then a line L through x contains a point at distance $d(x, X) - 1$ from X if and only if L is contained in F. As a consequence, x is contained in a unique special sub-$[4 \cdot d(x, X)]$-gon F_x.*

Proof. Suppose that L is contained in F. By Lemma 5.49, L contains a unique point nearest to X and by Lemmas 5.46 and 5.47 every point of F has distance at most $\mathrm{d}(x, X)$ to X. Hence L contains a unique point at distance $\mathrm{d}(x, X) - 1$ from x. Conversely, suppose that L is not contained in F and contains a point y at distance $\mathrm{d}(x, X) - 1$ from X. Let F' denote the unique special $(4 \cdot \mathrm{d}(x, X) + 4)$-gon through $\mathcal{C}(y, F)$. Then $\mathrm{d}(y, X) = \mathrm{d}(y, F' \cap X)$ by Lemma 5.46. Let y' denote a point of $X \cap F'$ at distance $\mathrm{d}(x, X) - 1$ from y. Then $y' \notin F$ and hence also $y' \notin \mathcal{C}(y, F)$. Let y'' denote the unique point of $\mathcal{C}(y, F)$ collinear with y' and let Q denote the unique special quad through the line $y'y''$. Then the line $Q \cap \mathcal{C}(y, F)$ contains a point z of X which necessarily belongs to F. We have $\mathrm{d}(y, z) \geq \mathrm{d}(x, X) + 1$ and $\mathrm{d}(y, z) \leq \mathrm{d}(y, y') + \mathrm{d}(y', z) = \mathrm{d}(x, X) - 1 + 2 = \mathrm{d}(x, X) + 1$. Hence y' is contained in a shortest path between y and z. But this is impossible since $y' \notin \mathcal{C}(y, F)$. □

Lemma 5.51. *For every point x of $\mathcal{A}$ and every point $y \in F_x$, $F_y \subseteq F_x$.*

Proof. In the near polygon F_x, there exists a unique special sub-$[4 \cdot \mathrm{d}(y, F_x \cap X)]$-gon F'_y through y. By Lemma 5.46, $\mathrm{d}(y, F_x \cap X) = \mathrm{d}(y, X)$. Hence F'_y must coincide with F_y. This proves the lemma. □

The following lemma completes the proof of Theorem 5.29.

Lemma 5.52. *The map $f : \mathcal{P} \to \mathbb{N}; x \mapsto d(x, X)$ is a valuation of $\mathcal{A}$.*

Proof. Obviously, there exists a point with value 0. By Lemma 5.49, every line contains a unique point with smallest value. We will now show that f also satisfies property (V_3). For every point y of F_x, we have $\mathrm{d}(y, X) \leq \mathrm{d}(x, X)$ by Lemma 5.47. If y is a point of F_x and if z is a point of $\mathcal{A}$ collinear with y for which $\mathrm{d}(z, X) = \mathrm{d}(y, X) - 1$, then $z \in F_y$ by Lemma 5.50. By Lemma 5.51, it now follows that $z \in F_x$. Hence f is a valuation of $\mathcal{A}$. □

5.9 Proof of Theorem 5.30

For every point x of $\mathcal{A}$, $f(x) = \mathrm{d}(x, \pi_{\mathcal{B}}(x)) + \mathrm{d}(\pi_{\mathcal{B}}(x), O_f) = \mathrm{d}(x, O_f)$. This proves (i). We will prove property (ii) by induction on the diameter n of $\mathcal{A}$. Obviously, the property holds if $n \leq 3$. So, suppose $n \geq 4$ and that the property holds for any thick dual polar space of diameter less than n (= induction hypothesis). We distinguish the following possibilities:

- $\mathcal{B} \neq \mathcal{A}$.
  Let $\mathcal{A}'$ denote a convex sub-$(2n-2)$-gon of $\mathcal{A}$ through $\mathcal{B}$. For every point x of $\mathcal{A}$, we have

  $$f(x) = \mathrm{d}(x, O_f) = \mathrm{d}(x, \pi_{\mathcal{A}'}(x)) + \mathrm{d}(\pi_{\mathcal{A}'}(x), O_f). \qquad (*)$$

  If H is contained in $\mathcal{A}'$, then the valuation induced in H is either classical or the extension of an ovoidal valuation in a quad of H since property (ii) holds for $\mathcal{A}'$. If H meets $\mathcal{A}'$ in a quad Q, then by $(*)$, the valuation induced in H is the extension of the valuation induced in Q. So, the property also holds in this case. If H is disjoint from $\mathcal{A}'$, then by the induction hypothesis applied to the dual polar space $\mathcal{A}'$, the property holds for the hex $\pi_{\mathcal{A}'}(H)$. Hence, the property also holds for H by $(*)$.

- $\mathcal{B} = \mathcal{A}$.
  Let $\mathcal{A}'$ denote a convex sub-$2(n-1)$-gon through H and let $\mathcal{A}''$ denote the unique special convex sub-$(2n-4)$-gon of $\mathcal{A}'$. The valuation f' induced by f in $\mathcal{A}'$ is the extension of an SDPS-valuation in $\mathcal{A}''$. By the induction hypothesis applied to $\mathcal{A}'$ it follows that the valuation f'' induced by f' in H is classical or the extension of an SDPS-valuation in a quad of H. The theorem then holds since f'' is also the valuation induced by f in H.

5.10 Proof of Theorem 5.31

Let (s, t), respectively (s', t'), denote the order of $\mathcal{A}$, respectively $\widetilde{\mathcal{A}}$. Let (s, t_2), respectively (s', t_2'), denote the order of the quads of $\mathcal{A}$, respectively $\widetilde{\mathcal{A}}$. We call a quad Q of $\mathcal{A}$ special if it contains two points of X, or equivalently, if $Q \cap X$ is an ovoid of Q. By Corollary 5.39, $(1+t) = (1+t_2)(1+t')$ with $t = t_2 + t_2^2 + \cdots + t_2^{2n-1}$

and $t' = t_2' + {t_2'}^2 + \cdots + {t_2'}^{n-1}$. Hence, $t_2' + {t_2'}^2 + \cdots + {t_2'}^{n-1} = t_2^2 + t_2^4 + \cdots + t_2^{2n-2}$. It follows that $t_2' = t_2^2$. Obviously, $s' = st_2$.

Suppose $n = 2$. By the classification of finite thick dual polar spaces, see Section 1.9.5, one of the following cases occurs:

(1) $\mathcal{A} \cong DW(7,q)$ and $\widetilde{\mathcal{A}}$ is a generalized quadrangle of order q^2;

(2) $\mathcal{A} \cong DQ(8,q)$ and $\widetilde{\mathcal{A}}$ is a generalized quadrangle of order q^2;

(3) $\mathcal{A} \cong DQ^-(9,q)$ and $\widetilde{\mathcal{A}}$ is a generalized quadrangle of order (q^3, q^2);

(4) $\mathcal{A} \cong DH(8,q^2)$ and $\widetilde{\mathcal{A}}$ is a generalized quadrangle of order (q^5, q^4);

(5) $\mathcal{A} \cong DH(7,q^2)$ and $\widetilde{\mathcal{A}}$ is a generalized quadrangle of order (q^3, q^4).

Case (5) cannot occur, since the quads of $DH(7,q^2)$ are isomorphic to $Q(5,q)$ and do not have ovoids, see e.g. [82, 1.8.3]. Pralle and Shpectorov [84] have studied the automorphism group of a special quad Q stabilizing $Q \cap X$. From their treatment, it follows that either ($\mathcal{A} \cong DW(7,q)$ and $\widetilde{\mathcal{A}} \cong DW(3,q^2)$) or ($\mathcal{A} \cong DQ^-(9,q)$ and $\widetilde{\mathcal{A}} \cong DH(4,q^2)$). Moreover, they have shown the claims regarding the intersection of a special quad Q with the SDPS-set.

Suppose now that $n \geq 3$. By Lemma 5.42, every special quad Q is contained in a convex suboctagon F such that $F \cap X$ is an SDPS-set of F. From the discussion of the case $n = 2$, the theorem then readily follows.

5.11 Proof of Theorem 5.32

Let n be the diameter of $\mathcal{A}$. We will prove the theorem by induction on n. Obviously, the theorem holds if $n \leq 3$. So, suppose that $n \geq 4$ and that the theorem holds for any thick dual polar space of diameter less than n (= induction hypothesis). Recall that a quad of $\mathcal{A}$ is special if it contains two points of O_f, or equivalently, if it intersects O_f in an ovoid.

Lemma 5.53. *No two special quads intersect in a line.*

Proof. Suppose that Q_1 and Q_2 are two special quads intersecting in a line. Then $\mathcal{C}(Q_1, Q_2)$ is a hex H. Now, the valuation induced in H is either classical or the extension of an ovoidal valuation in a quad. It is easily seen that none of these two possibilities can occur. □

Lemma 5.54. *For all $x_1, x_2 \in O_f$, $\mathrm{d}(x_1, x_2)$ is even.*

Proof. We distinguish three possibilities.

(1) $\mathrm{d}(x_1, x_2) < n$. If f' denotes the valuation induced in $\mathcal{C}(x_1, x_2)$, then by the induction hypothesis, f' is an SDPS-valuation in $\mathcal{C}(x_1, x_2)$. Since $x_1, x_2 \in O_{f'}$, $\mathrm{d}(x_1, x_2)$ is even.

(2) $\mathrm{d}(x_1, x_2) = n$ and n is even.

(3) $\mathrm{d}(x_1, x_2) = n$ and n is odd. Let F denote a convex sub-$(2n-2)$-gon through x_1 and let x_2' denote the unique point of F collinear with x_2. Then $f(x_2') = 1$. By the induction hypothesis, the valuation f' induced in F is the (possibly trivial) extension of an SDPS-valuation in a convex subpolygon of F. Since $f(x_2') = 1$, x_2' is collinear with a point x_2'' of $O_f \cap F$, see Theorem 5.30. Since $\mathrm{d}(x_1, x_2'')$ is even and $\mathrm{d}(x_1, x_2'') \geq n-2$, $\mathrm{d}(x_1, x_2'') = n-1$. So, f' is an SDPS-valuation of $F = \mathcal{C}(x_1, x_2'')$. Now, the line $x_2'x_2''$ is contained in at least two special quads, namely $\mathcal{C}(x_2, x_2'')$ and the unique special quad through $x_2'x_2''$ contained in F, see Corollary 5.39. This contradicts Lemma 5.53. □

Lemma 5.55. *If $x_1, x_2 \in O_f$ with $d(x_1, x_2)$ as big as possible, then $O_f \subseteq \mathcal{C}(x_1, x_2)$.*

Proof. Obviously, this holds if $\mathrm{d}(x_1, x_2) = n$. Suppose therefore that $\mathrm{d}(x_1, x_2) < n$ and let F denote a convex sub-$(2n-2)$-gon through $\mathcal{C}(x_1, x_2)$. By the induction hypothesis applied to F, it follows that every point of $O_f \cap F$ is contained in $\mathcal{C}(x_1, x_2)$. Suppose now that there exists a point $y \in O_f$ not contained in F and let y' denote the unique point of F collinear with y. Since $f(y') = 1$, there exists a point $x_3 \in O_f \cap F$ collinear with y'. Then $x_3 \in \mathcal{C}(x_1, x_2)$. Since the valuation induced in $\mathcal{C}(x_1, x_2)$ is an SDPS-valuation, there exists a point $x_4 \in O_f \cap \mathcal{C}(x_1, x_2)$ at distance $\mathrm{d}(x_1, x_2)$ from x_3. We now distinguish two possibilities.

(1) The quad $\mathcal{C}(y, x_3)$ intersects $\mathcal{C}(x_1, x_2)$ in a line L. Then L would be contained in two special quads, namely $\mathcal{C}(y, x_3)$ and the unique special quad through L contained in $\mathcal{C}(x_1, x_2)$. This contradicts Lemma 5.53.

(2) The quad $\mathcal{C}(y, x_3)$ intersects $\mathcal{C}(x_1, x_2)$ in the point x_3. Then $\mathrm{d}(y, x_4) = 2 + \mathrm{d}(x_1, x_2)$, contradicting the maximality of $\mathrm{d}(x_1, x_2)$. □

Lemma 5.56. *The valuation f is not semi-classical.*

Proof. Suppose the contrary. Let x denote the unique point of $\mathcal{A}$ with value 0 and let H denote a hex containing a point at maximal distance n from x. Then it is easily seen that the valuation induced in H is semi-classical, contradicting our assumptions. □

Lemma 5.57. *If $|O_f| = 1$, then f is a classical valuation.*

Proof. Let x denote the unique point of O_f. If y is a point at distance at most $n-1$ from x, then by the induction hypothesis, the valuation induced in $\mathcal{C}(x, y)$ is the (possibly trivial) extension of an SDPS-valuation in a convex subpolygon of $\mathcal{C}(x, y)$. Since $|O_f| = 1$, this induced valuation must be classical. Hence, $f(y) = \mathrm{d}(x, y)$ for every point y at distance at most $n-1$ from x. The lemma now follows from Theorem 5.16 and Lemma 5.56. □

Lemma 5.58. *If $|O_f| \geq 2$, then every point x of O_f is contained in a special quad.*

Proof. Suppose the contrary. Then, by the induction hypothesis, the valuation induced in every convex sub-$(2n-2)$-gon through x must be classical. Hence, $f(y) = \mathrm{d}(x, y)$ for every point y at distance at most $n-1$ from x. By Theorem 5.16 and Lemma 5.56, it now follows that f is classical, contradicting $|O_f| \geq 2$. □

Lemma 5.59. *For every point x of $\mathcal{A}$, $f(x) = d(x, O_f)$.*

Proof. By Lemma 5.57, this holds if $|O_f| = 1$. Suppose therefore that $|O_f| \geq 2$. Let x^* denote a point of O_f nearest to x and let Q denote a special quad through x^*. Then either $\pi_Q(x) = x^*$ or $\pi_Q(x) \sim x^*$. Since $\mathrm{d}(x, \pi_Q(x)) \leq n-2$, we have $\mathrm{d}(x, x^*) \leq n-1$. By the induction hypothesis, the valuation f' induced in $\mathcal{C}(x, x^*)$ is the (possibly trivial) extension of an SDPS-valuation in a convex subpolygon of $\mathcal{C}(x, x^*)$. So, $f(x) = f'(x) = \mathrm{d}(x, O_f \cap \mathcal{C}(x, x^*)) = \mathrm{d}(x, x^*) = \mathrm{d}(x, O_f)$. □

Lemma 5.60. *If the maximal distance between two points of O_f is smaller than n, then f is the extension of an SDPS-valuation in a convex subpolygon of $\mathcal{A}$.*

Proof. Let $x_1, x_2 \in O_f$ with $\mathrm{d}(x_1, x_2)$ as big as possible. Since $\mathrm{d}(x_1, x_2) < n$, the valuation induced in $F := \mathcal{C}(x_1, x_2)$ is an SDPS-valuation f' with $O_{f'} = O_f$, see Lemma 5.55. For every point x of $\mathcal{A}$, we have $f(x) = \mathrm{d}(x, O_f) = \mathrm{d}(x, \pi_F(x)) + \mathrm{d}(\pi_F(x), O_{f'})$, proving the lemma. □

In the sequel, we will suppose that the maximal distance between two points of O_f is equal to n. This implies that n is even, see Lemma 5.54. Let $\widetilde{\mathcal{A}}$ denote the following partial linear space:

- the points of $\widetilde{\mathcal{A}}$ are the elements of O_f,
- the lines of $\widetilde{\mathcal{A}}$ are the special quads,
- incidence is containment.

If x_1 and x_2 are two points of O_f, then we denote by $\mathrm{d}(x_1, x_2)$ the distance between x_1 and x_2 in the geometry $\mathcal{A}$ and by $\delta(x_1, x_2)$ the distance between x_1 and x_2 in the geometry $\widetilde{\mathcal{A}}$.

Lemma 5.61. *For all points $x_1, x_2 \in O_f$, $d(x_1, x_2) = 2 \cdot \delta(x_1, x_2)$. As a consequence, the diameter of $\widetilde{\mathcal{A}}$ is half the diameter of $\mathcal{A}$.*

Proof. Every path of $\widetilde{\mathcal{A}}$ between x_1 and x_2 can be turned into a path of $\mathcal{A}$ with double length. This proves that $\mathrm{d}(x_1, x_2) \leq 2 \cdot \delta(x_1, x_2)$ for all points $x_1, x_2 \in O_f$.

We will prove the lemma by induction on the distance $\mathrm{d}(x_1, x_2)$ which is always even by Lemma 5.54. Obviously, the lemma holds if $\mathrm{d}(x_1, x_2)$ is 0 or 2. Suppose therefore that $\mathrm{d}(x_1, x_2) = 2k \geq 4$. Then we already know that $\delta(x_1, x_2) \geq k$. We will now show that there exists a special quad Q through x_2 containing a point x_3 at distance $2k-2$ from x_1. If $2k < n$, this follows from the fact that the valuation induced in $\mathcal{C}(x_1, x_2)$ is an SDPS-valuation. If $2k = n$, this follows from Lemma 5.58. If $x_3 \notin O_f$, then there exists a point in $Q \cap O_f$ at distance $2k-1$ from x_1, contradicting Lemma 5.54. Hence $x_3 \in O_f$. Since $\mathrm{d}(x_1, x_3) = 2k-2$, $\delta(x_1, x_3) = k-1$ and $\delta(x_1, x_2) \leq k$. Together with $\delta(x_1, x_2) \geq k$, this implies that $\mathrm{d}(x_1, x_2) = 2 \cdot \delta(x_1, x_2)$. □

Lemma 5.62. *$\widetilde{\mathcal{A}}$ is a near polygon (and hence a near n-gon).*

Proof. Let x denote a point of O_f and let Q denote a special quad. If $\pi_Q(x) \notin O_f$, then there exists a point in $O_f \cap Q$ at distance $\mathrm{d}(x, \pi_Q(x)) + 1$ from x and a point in $O_f \cap Q$ at distance $\mathrm{d}(x, \pi_Q(x)) + 2$ from x, contradicting Lemma 5.54. So, $\pi_Q(x) \in O_f$ and $\pi_Q(x)$ is the unique point of $O_f \cap Q$ nearest to x. The claim now follows from Lemma 5.61. □

Lemma 5.63. *$\widetilde{\mathcal{A}}$ is a dense near polygon.*

Proof. Obviously, every line of $\widetilde{\mathcal{A}}$ is incident with at least three points. Now, let x_1 and x_2 be two points of $\widetilde{\mathcal{A}}$ at distance 2 from each other, i.e., at distance 4 in $\mathcal{A}$. We will show that every line L of $\mathcal{C}(x_1, x_2)$ through x_1 is contained in a (necessarily unique) special quad $Q_L \subseteq \mathcal{C}(x_1, x_2)$. Let y denote the point of L nearest to x_2. By the induction hypothesis, the valuation induced in $\mathcal{C}(y, x_2)$ is the possibly trivial extension of an SDPS-valuation. Since $f(y) = 1$, there exists a point $x_3 \in \mathcal{C}(y, x_2) \cap O_f$ collinear with y. Obviously, L is contained in a special quad $\mathcal{C}(x_1, x_3)$ and x_3 is a common neighbour of x_1 and x_2 in the near polygon $\widetilde{\mathcal{A}}$. Repeating this argument for every line L of $\mathcal{C}(x_1, x_2)$ through x_1, we see that x_1 and x_2 have at least two common neighbours (in $\widetilde{\mathcal{A}}$). □

If x_1 and x_2 are two points of O_f at distance 4 from each other, then $O_f \cap \mathcal{C}(x_1, x_2)$ is a convex subspace and hence a quad of $\widetilde{\mathcal{A}}$. Conversely, every quad is obtained in this way.

Lemma 5.64. *The near polygon $\widetilde{\mathcal{A}}$ is classical.*

Proof. The lemma holds trivially if $\widetilde{\mathcal{A}}$ is a generalized quadrangle. So, suppose $n > 4$. Let x denote a point of $\widetilde{\mathcal{A}}$ and let $\widetilde{Q}$ denote a quad of $\widetilde{\mathcal{A}}$. Let F denote the convex suboctagon of $\mathcal{A}$ containing all points of $\widetilde{Q}$. Then, by the induction hypothesis, the valuation induced in F is an SDPS-valuation. Let H denote an arbitrary hex of F through $\pi_F(x)$. Then $H \cap O_f$ is an ovoid in a quad Q. Let x' denote the unique point of Q nearest to $\pi_F(x)$. Then $\mathrm{d}(\pi_F(x), O_f \cap F) = \mathrm{d}(\pi_F(x), H \cap O_f) = \mathrm{d}(\pi_F(x), x') + \mathrm{d}(x', H \cap O_f) \leq 2$. So, we have the following possibilities.

(a) $\pi_F(x) \in O_f \cap F$. Then $\pi_F(x)$ is indeed the unique point of $\widetilde{Q}$ nearest to x.

(b) $\mathrm{d}(\pi_F(x), O_f \cap F) = 1$. Then $\pi_F(x)$ is contained in a special quad Q of F, see Corollary 5.39. Then there exists a point in $O_f \cap Q$ at distance $\mathrm{d}(x, \pi_F(x)) + 1$ from x and a point in $O_f \cap Q$ at distance $\mathrm{d}(x, \pi_F(x)) + 2$ from x, contradicting Lemma 5.54.

(c) $\mathrm{d}(\pi_F(x), O_f \cap F) = 2$. Let x' denote a neighbour of $\pi_F(x)$ collinear with a point of $O_f \cap F$, then x' is contained in a special quad Q of F. Then there exists a point in $O_f \cap Q$ at distance $\mathrm{d}(x, \pi_F(x)) + 2$ from x and a point in $O_f \cap Q$ at distance $\mathrm{d}(x, \pi_F(x)) + 3$ from x, contradicting Lemma 5.54. □

The following lemma, in combination with Lemma 5.59, completes the proof of Theorem 5.32.

Lemma 5.65. *The set O_f is an SDPS-set of $\mathcal{A}$.*

Proof. We still must check property (5) in the definition of SDPS-set. Let x denote an arbitrary point of O_f and let L denote an arbitrary line of $\mathcal{A}$ through x. Since $\widetilde{\mathcal{A}}$ is a dense near n-gon, there exists a point x' in O_f at distance n from x. Let y denote the unique point of L nearest to x'. By the induction hypothesis, the valuation induced in $\mathcal{C}(y, x')$ is the possibly trivial extension of an SDPS-valuation. So, there exists a point $x'' \in O_f \cap \mathcal{C}(y, x')$ collinear with y. Hence, the line L is contained in the special quad $\mathcal{C}(x, x'')$. $\square$

Chapter 6

The known slim dense near polygons

In this chapter, we will discuss five infinite classes of slim dense near polygons and three "exceptional" slim dense near hexagons. We will determine the convex subpolygons and the spreads of symmetry of these near polygons. For the near hexagons, we will determine all valuations. The above-mentioned near polygons cover all known examples of slim dense near polygons which are not glued. At the end of this chapter, we will also discuss the glued near polygons which can be derived from the five infinite classes and the three exceptional examples.

6.1 The classical near polygons $DQ(2n, 2)$ and $DH(2n-1, 4)$

Let $n \in \mathbb{N} \setminus \{0, 1\}$. Let $\mathcal{F}$ be a nonsingular parabolic quadric in $\mathrm{PG}(2n, 2)$ or a nonsingular hermitian variety in $\mathrm{PG}(2n-1, 4)$ and let $\mathcal{F}^D$ denote its corresponding dual polar space, i.e.

- the points of $\mathcal{F}^D$ are the maximal subspaces (i.e. generators) of $\mathcal{F}$,
- the lines of $\mathcal{F}^D$ are the next-to-maximal subspaces of $\mathcal{F}$,
- incidence is reverse containment.

The generalized quadrangle $Q(4, 2)^D = DQ(4, 2)$ is isomorphic to $W(2)$ and the generalized quadrangle $H(3, 4)^D = DH(3, 4)$ is isomorphic to $Q(5, 2)$.

Theorem 6.1. (a) *If π_1 and π_2 are two generators of $\mathcal{F}$ intersecting each other in an $(n-1-i)$-dimensional subspace, then the distance $d(\pi_1, \pi_2)$ between π_1 and π_2 in the near polygon $\mathcal{F}^D$ is equal to i.*

(b) *$\mathcal{F}^D$ is a near $2n$-gon.*

(c) *The near polygon $\mathcal{F}^D$ is dense.*

(d) *Suppose π is an $(n-1-i)$-dimensional subspace lying on $\mathcal{F}$. Then the generators through π form a convex sub-$2i$-gon of $\mathcal{F}^D$.*

(e) *If π_1 and π_2 are two generators of $\mathcal{F}$, then $\mathcal{C}(\pi_1, \pi_2)$ consists of all generators through $\pi_1 \cap \pi_2$.*

(f) *Every convex sub-$2(n-1)$-gon of $\mathcal{F}^D$ is big in $\mathcal{F}^D$.*

(g) *Every convex subpolygon of $\mathcal{F}^D$ is classical in $\mathcal{F}^D$. As a consequence, $\mathcal{F}^D$ is classical.*

Proof. (a) Suppose $\mathrm{d}(\pi_1, \pi_2) = k$. Then there exist $k+1$ generators $\alpha_0, \alpha_1, \ldots, \alpha_k$ such that $\alpha_0 = \pi_1$, $\alpha_k = \pi_2$ and $\dim(\alpha_j \cap \alpha_{j+1}) = n-2$ for every $j \in \{0, \ldots, k-1\}$. It easily follows (by induction) that $\dim(\alpha_0 \cap \alpha_j) \geq n-1-j$ for every $j \in \{0, \ldots, k\}$. In particular, $n+1-i \geq n-1-k$ or $k \geq i$. We will now show by induction on i that $k \leq i$. Obviously, this holds if $i \leq 1$. So, suppose that $i \geq 2$. Let x be a point of $\pi_1 \setminus \pi_2$ and let π_3 be the unique generator through x such that $\dim(\pi_3 \cap \pi_2) = n-2$. Then $\pi_1 \cap \pi_3 = \langle x, \pi_1 \cap \pi_2 \rangle$ and $\dim(\pi_1 \cap \pi_3) = n-1-(i-1)$. By the induction hypothesis, $\mathrm{d}(\pi_1, \pi_3) \leq i-1$. Since $\mathrm{d}(\pi_2, \pi_3) = 1$, $\mathrm{d}(\pi_1, \pi_2) \leq i$ or $k \leq i$.

(b) Let π be a point and let α be a line of $\mathcal{F}^D$. First, suppose that $\alpha \cap \pi = \emptyset$. Then there exists a unique generator through α intersecting π in a point. This proves that the line α of $\mathcal{F}^D$ contains a unique point nearest to π. Next, suppose that $\alpha \cap \pi \neq \emptyset$. Then the generators through $\alpha \cap \pi$ define a polar space $\mathcal{F}'$ of type $Q(2k, 2)$ or $H(2k-1, 4)$. (We take the following convention: a dual polar space of type $Q(2, 2)$ or $H(1, 4)$ is a line of size 3.) The point π of $\mathcal{F}$ corresponds with a point π' of $\mathcal{F}'^D$ and the line α of $\mathcal{F}$ corresponds with a line α' of $\mathcal{F}'^D$. In $\mathcal{F}'^D$, the line α' contains a unique point nearest to π'. It follows that the line α of $\mathcal{F}^D$ contains a unique point nearest to π. Hence, $\mathcal{F}^D$ is a near polygon. By (a) the maximal distance between two points of $\mathcal{F}^D$ is equal to n. So, $\mathcal{F}^D$ is a near $2n$-gon.

(c) Obviously, every line of $\mathcal{F}^D$ is incident with precisely three points. Now, let π_1 and π_2 be two points of $\mathcal{F}^D$ at mutual distance 2. Let x be a point of $\pi_1 \setminus \pi_2$, let π_3 be the generator through x intersecting π_2 in an $(n-2)$-dimensional subspace, let y be a point of $\pi_2 \setminus \pi_3$ and let π_4 be the generator of $\mathcal{F}$ through y intersecting π_1 in an $(n-2)$-dimensional subspace. Then π_3 and π_4 are two common neighbours of π_1 and π_2. This proves that $\mathcal{F}^D$ is dense.

(d) Let X denote the set of generators through π. Let α be a line of $\mathcal{F}^D$ containing two different points π_1 and π_2 of X. Then $\alpha = \pi_1 \cap \pi_2$ contains π and hence all generators through α contain π. This proves that X is a subspace. Now, let π_1 and π_2 be two generators of X and let π_3 be a third generator such that $\mathrm{d}(\pi_1, \pi_3) = \mathrm{d}(\pi_1, \pi_2) - 1$ and $\mathrm{d}(\pi_3, \pi_2) = 1$. Then $\dim(\pi_3 \cap \pi_1) = \dim(\pi_1 \cap \pi_2) + 1$. So, there exists a point $x \in \pi_3 \cap \pi_1$ not contained in π_2.

Since $\mathrm{d}(\pi_3,\pi_2)=1$, π_3 is the unique generator through x intersecting π_2 in an $(n-2)$-dimensional subspace. It follows that $\pi_1\cap\pi_2\subseteq\pi_3$ and hence also that $\pi\subseteq\pi_3$. Hence, X is convex. Through π, there are now two generators α_1 and α_2 such that $\alpha_1\cap\alpha_2=\pi$. By (a), X induces a convex sub-$2i$-gon.

(e) Let G denote the convex subpolygon which consists of all generators through $\pi_1\cap\pi_2$. By (d), $\mathrm{diam}(G)=n-1-\dim(\pi_1\cap\pi_2)$. Since G and $\mathcal{C}(\pi_1,\pi_2)$ are two convex subpolygons of diameter $n-1-\dim(\pi_1\cap\pi_2)=\mathrm{d}(\pi_1,\pi_2)$ through π_1 and π_2, we must have that $G=\mathcal{C}(\pi_1,\pi_2)$.

(f) Let G denote an arbitrary convex sub-$2(n-1)$-gon of $\mathcal{F}^D$. Then G consists of all generators through a certain point x of $\mathcal{F}$. Let α be an arbitrary generator of $\mathcal{F}$. If $x\in\alpha$, then the point α of $\mathcal{F}^D$ lies in G. If $x\not\in\alpha$, then there exists a unique generator α' through x intersecting α in an $(n-2)$-dimensional subspace. Obviously, α' is the unique point of G collinear with α.

(g) Let G be a proper convex subpolygon of $\mathcal{F}^D$. If $\mathrm{diam}(G)=n-1$, then G is classical in $\mathcal{F}^D$ by (f). If $\mathrm{diam}(G)\leq n-2$, then G can be obtained as the intersection of convex subpolygons of diameter $n-1$. The statement then follows from Theorem 1.6. □

Corollary 6.2. (a) *Every convex sub-2i-gon, $i\geq 2$, of $DQ(2n,2)$ is isomorphic to $DQ(2i,2)$. In particular, every quad is isomorphic to $W(2)$. Every local space of $DQ(2n,2)$ is isomorphic to* $\mathrm{PG}(n-1,2)$.

(b) *Every convex sub-2i-gon, $i\geq 2$, of $DH(2n-1,4)$ is isomorphic to $DH(2i-1,4)$. In particular, every quad is isomorphic to $Q(5,2)$. Every local space of $DH(2n-1,4)$ is isomorphic to* $\mathrm{PG}(n-1,4)$.

Theorem 6.3. *The near polygon $DQ(2n,2)$, $n\geq 2$, has no spreads of symmetry.*

Proof. Suppose that S is a spread of symmetry of $DQ(2n,2)$. Let L be a line of S and let Q be a $W(2)$-quad through L. By Theorem 4.5, S_Q is a spread of symmetry of Q, a contradiction, since $W(2)$ has no spreads of symmetry by Theorem 4.42. □

Theorem 6.4. *If x is a point of the near hexagon $DQ(6,2)$, then there exists a path of length 7 in $\Gamma_3(x)$.*

Proof. Let x_1 and x_2 be points collinear with x such that $\mathrm{d}(x_1,x_2)=2$. Let Q_i, $i\in\{1,2\}$, denote a quad through x_i such that (i) $x\not\in Q_i$ and (ii) $Q_1\cap Q_2=\emptyset$. We will use Sylvester's model for Q_1 and Q_2. Suppose that the points of Q_1 are the subsets $\{i,j\}$ of size 2 of $\{1,2,3,4,5,6\}$. We denote the unique point of Q_2 collinear with $\{i,j\}$ by $\{i^*,j^*\}$. Without loss of generality, we may suppose that $x_1=\{1,2\}$ and $x_2=\{3^*,4^*\}$. (Note that $x_2\sim\pi_{Q_2}(x_1)$ since $\mathrm{d}(x_1,x_2)=2$.) Then $\{1,3\},\{2,4\},\{1,5\},\{2,3\},\{2^*,3^*\},\{4^*,5^*\},\{1^*,3^*\},\{1,3\}$ is a path of length 7 in $\Gamma_3(x)$. □

Corollary 6.5. *The near hexagon $DQ(6,2)$ has no semi-classical valuations.*

Lemma 6.6. *If a slim dense near octagon $\mathcal{S}$ contains two disjoint big $W(2)$-quads, then $\mathcal{S}$ has no ovoids.*

Proof. Suppose the contrary. Let Q_1 and Q_2 be two disjoint big $W(2)$-quads and let O be an ovoid of $\mathcal{S}$. Then $O_i := O \cap Q_i$, $i \in \{1,2\}$, is an ovoid of Q. The ovoids $\pi_{Q_1}(O_2)$ and O_1 of Q_1 have at least one point in common. So, O must contain two collinear points, a contradiction. □

Corollary 6.7. *Every valuation of $DQ(6,2)$ is either classical or the extension of an ovoidal valuation in a quad of $DQ(6,2)$.*

Proof. This follows from Theorem 5.36, Corollary 6.5 and Lemma 6.6. □

Theorem 6.8. *All valuations of $DH(2n-1,4)$ are classical.*

Proof. Let f be a valuation of $DH(2n-1,4)$. Since $Q(5,2)$ has no ovoids, every induced quad-valuation is classical. Hence, f is classical by Theorem 5.11. □

We will now determine all spreads of symmetry of $DH(2n-1,4)$. Let $(\cdot,\cdot)$ denote the hermitian form of $V(2n,4)$ associated with $H(2n-1,4)$ and let ζ denote the corresponding hermitian polarity. We will suppose that $(\cdot,\cdot)$ is linear in the first argument and semi-linear in the second.

Lemma 6.9. *Let $\mathcal{L}$ be the linear space whose points are the points of $H(2n-1,4)$ and whose lines are the lines of* $\mathrm{PG}(2n-1,4)$ *which are not tangent to $H(2n-1,4)$. Then the subspaces of $\mathcal{L}$ are precisely the intersections of $H(2n-1,4)$ with subspaces of* $\mathrm{PG}(2n-1,4)$.

Proof. For every set X of points on $H(2n-1,4)$, let $\langle X\rangle$ denote the subspace of $\mathrm{PG}(2n-1,4)$ generated by all points of X and let $\overline{X}$ denote the smallest subspace of $\mathcal{L}$ through X. Since $\langle X\rangle \cap H(2n-1,4)$ is a subspace of $\mathcal{L}$, we have $\overline{X} \subseteq \langle X\rangle \cap H(2n-1,4)$. We will now prove that $\overline{X} = \langle X\rangle \cap H(2n-1,4)$ $(*)$. This property obviously holds if $m := \dim(\langle X\rangle) \leq 1$, and since every intersection of a hermitian variety with a plane is either the plane itself, a unital or a cone pB with p a point and B a Baer subline, it also holds if $m=2$. So, suppose that $m \geq 3$ and consider a subset Y of X such that $\dim(\langle Y\rangle) = m-1$. We may suppose that $\langle Y\rangle \cap H(2n-1,4) = \overline{Y} \subset \overline{X}$ (otherwise use induction). Now consider a fixed point x in $X \setminus \langle Y\rangle$. For every point x' of $\langle X\rangle \cap H(2n-1,4)$ different from x, the line xx' intersects $\langle Y\rangle$ in a point x'' and one of the following possibilities occurs.

- There exists a Baer subline $L \subseteq H(2n-1,4)$ in $\langle Y\rangle$ through the point x''. Since property $(*)$ holds if $m=2$, we have $x' \in \langle x, L\rangle \cap H(2n-1,4) = \overline{L \cup \{x\}} \subseteq \overline{X}$.

- Every line of $\langle Y\rangle$ through the point x'' is a tangent line. Then x'' is a singular point of the hermitian variety $\langle Y\rangle \cap H(2n-1,4)$. Hence $x'' \in H(2n-1,4)$ and $x' \in xx'' \cap H(2n-1,4) \subseteq \overline{X}$.

In any case, we have $x' \in \overline{X}$. Since x' was an arbitrary point of $\langle X \rangle \cap H(2n-1,4)$ different from x and since $x \in \overline{X}$, we have $\langle X \rangle \cap H(2n-1,4) \subseteq \overline{X}$ and hence $\overline{X} = \langle X \rangle \cap H(2n-1,4)$. As a consequence the subspaces of $\mathcal{L}$ are precisely the intersections of $H(2n-1,4)$ with subspaces of $\mathrm{PG}(2n-1,4)$. □

Theorem 6.10. *For every subspace α of $H(2n-1,4)$, let α^ϕ denote the corresponding convex subpolygon of $DH(2n-1,4)$. If V is the set of all $(n-2)$-dimensional subspaces of $H(2n-1,4)$ which lie in a nontangent hyperplane Π, then $V^\phi := \{\alpha^\phi | \alpha \in V\}$ is a spread of symmetry of $DH(2n-1,4)$. Conversely, every spread of symmetry is obtained in this way.*

Proof. (a) Every generator of $H(2n-1,4)$ contains a unique element of V, or equivalently, every point of $DH(2n-1,4)$ is incident with a unique line of V^ϕ. So, V^ϕ is a spread. We can choose our reference system in such a way that $H(2n-1,4)$ has equation $X_0^3 + \cdots + X_{2n-1}^3 = 0$ and that Π has equation $X_{2n-1} = 0$. The group $G := \{\theta_\lambda : (x_0, \ldots, x_{2n-2}, x_{2n-1}) \to (x_0, \ldots, x_{2n-2}, \lambda x_{2n-1}) \,|\, \lambda^3 = 1\}$ of automorphisms of $\mathrm{PG}(2n-1,4)$ fixes $H(2n-1,4)$ setwise and Π pointwise. So, G determines a group G^ϕ of automorphisms of $DH(2n-1,4)$ which fixes every element of V^ϕ. This group G^ϕ acts regularly on each line of V^ϕ, proving that V^ϕ is a spread of symmetry.

(b) Now consider a spread of symmetry S of $DH(2n-1,4)$ and let X denote the set of the points x of $H(2n-1,4)$ for which the convex sub-$2(n-1)$-gon x^ϕ contains a line of S. Take two different points x_1 and x_2 in X, then one of the following possibilities occurs:

- $|x_1x_2 \cap H(2n-1,4)| = 5$. Let y denote an arbitrary point of $x_1^\phi \cap x_2^\phi = (x_1x_2)^\phi$. By Theorem 4.5, the unique line of S through y is contained in x_1^ϕ and x_2^ϕ and hence in $x_1^\phi \cap x_2^\phi$. As a consequence, each of the five convex sub-$2(n-1)$-gons through $(x_1x_2)^\phi$ belongs to X^ϕ, or equivalently, each of the five points of x_1x_2 belongs to X.

- $|x_1x_2 \cap H(2n-1,4)| = 3$. In this case x_1^ϕ and x_2^ϕ are two disjoint convex sub-$2(n-1)$-gons of $DH(2n-1,4)$. Let x_3 denote the third point of $H(2n-1,4)$ on the line x_1x_2. Let L denote an arbitrary line of S contained in x_1^ϕ and let Q denote the unique quad through L which intersects x_2^ϕ and x_3^ϕ in a line. Now, let y denote an arbitrary point of $Q \cap x_2^\phi$. The unique line of S through y is contained in Q and in x_2^ϕ and hence coincides with the line $Q \cap x_2^\phi$. Since $Q \cap (x_1^\phi \cup x_2^\phi \cup x_3^\phi)$ is a subgrid of Q and since S is a regular spread, we now see that also $Q \cap x_3^\phi$ belongs to S. Hence $x_1, x_2, x_3 \in X$.

As a consequence, the set X is a subspace of $\mathcal{L}$ and hence the intersection of $H(2n-1,4)$ with a subspace π of $\mathrm{PG}(2n-1,4)$. The elements of S are precisely the elements α^ϕ, where α is an $(n-2)$-dimensional subspace contained in $\pi \cap H(2n-1,4)$. For, if $L \in S$, then every point of $L^{\phi^{-1}}$ belongs to X and hence $L^{\phi^{-1}}$ is contained in $\pi \cap H(2n-1,4)$. Conversely, let α be an $(n-2)$-dimensional subspace contained in $\pi \cap H(2n-1,4)$, let $x_1, \dots, x_{n-1}$ denote $n-1$ points of X generating α and let u denote an arbitrary point of α^ϕ. By Theorem 4.5 the unique line K of S through u is contained in each convex subpolygon x_i^ϕ and hence coincides with the line $\alpha^\phi = x_1^\phi \cap \cdots \cap x_{n-1}^\phi$. If $\pi \cap H(2n-1,4)$ contains a subspace β of dimension $n-1$, then every line through the point β^ϕ would belong to S, which is impossible. As a consequence $n-2$ is the maximal dimension of the subspaces contained in $\pi \cap H(2n-1,4)$. If x is a singular point of $\pi \cap H(2n-1,4)$, then x is contained in all $(n-2)$-dimensional subspaces of $\pi \cap H(2n-1,4)$ and so all lines of S would be contained in the convex sub-$2(n-1)$-gon x^ϕ, a contradiction. So, $\pi \cap H(2n-1,4)$ is a nonsingular hermitian variety of type $H(2n-2,4)$ or $H(2n-3,4)$, but since S must have the right amount of lines, i.e. $\frac{|DH(2n-1,4)|}{3}$, we know that $\pi \cap H(2n-1,4)$ is of type $H(2n-2,4)$ and that π is a nontangent hyperplane. □

If p is a point of $\mathrm{PG}(2n-1,4)$ not contained in $H(2n-1,4)$, then by Theorem 6.10, p^ζ is a nontangent plane which determines a spread of symmetry S_p of $DH(2n-1,4)$. Let $p = \langle \bar{e} \rangle$ for a certain vector $\bar{e}$ of $V(2n,4)$. For every $\lambda \in \mathbb{F}_4^*$, let $\theta_{\bar{e},\lambda}$ denote the following linear map of $V(2n,4)$: $\bar{x} \mapsto \bar{x} + (\lambda - 1)\frac{(\bar{x},\bar{e})}{(\bar{e},\bar{e})}\bar{e}$. The map $\theta_{\bar{e},\lambda}$ induces an automorphism of $\mathrm{PG}(2n-1,4)$ fixing the hermitian variety $H(2n-1,4)$ and every point of the hyperplane p^ζ. Hence, $\theta_{\bar{e},\lambda}$ also induces an automorphism of $DH(2n-1,4)$ fixing each line of S_p. Since $\theta_{\bar{e},\lambda_1} \circ \theta_{\bar{e},\lambda_2} = \theta_{\bar{e},\lambda_1\lambda_2}$ for all $\lambda_1, \lambda_2 \in \mathbb{F}_4^*$, $H := \{\theta_{\bar{e},\lambda} \mid \lambda \in \mathbb{F}_4^*\}$ is a subgroup of $\mathrm{GL}(2n,4)$. The group of automorphisms of $DH(2n-1,4)$ induced by the elements of H is the whole group of automorphisms of $DH(2n-1,4)$ fixing each line of S_p.

Theorem 6.11. *Let p_1 and p_2 denote two points of $\mathrm{PG}(2n-1,4)$ not contained in $H(2n-1,4)$. Then S_{p_1} is compatible with S_{p_2} if and only if either $p_1 = p_2$ or $p_1 \in p_2^\zeta$.*

Proof. Let $\bar{e}_1$ and $\bar{e}_2$ be vectors of $V(2n,4)$ such that $p_1 = \langle \bar{e}_1 \rangle$ and $p_2 = \langle \bar{e}_2 \rangle$ and let λ_1 and λ_2 denote arbitrary elements of $\mathbb{F}_4^* \setminus \{1\}$. Then one calculates that $\theta_{\bar{e}_2,\lambda_2} \circ \theta_{\bar{e}_1,\lambda_1}$ is the following map:

$$\bar{x} \mapsto \bar{x} + (\lambda_1 - 1)\frac{(\bar{x},\bar{e}_1)}{(\bar{e}_1,\bar{e}_1)}\bar{e}_1 + (\lambda_2 - 1)\frac{(\bar{x},\bar{e}_2)}{(\bar{e}_2,\bar{e}_2)}\bar{e}_2 + (\lambda_1 - 1)(\lambda_2 - 1)\frac{(\bar{x},\bar{e}_1)(\bar{e}_1,\bar{e}_2)}{(\bar{e}_1,\bar{e}_1)(\bar{e}_2,\bar{e}_2)}\bar{e}_2.$$

Similarly, one calculates that $\theta_{\bar{e}_1,\lambda_1} \circ \theta_{\bar{e}_2,\lambda_2}$ is given by:

$$\bar{x} \mapsto \bar{x} + (\lambda_1 - 1)\frac{(\bar{x},\bar{e}_1)}{(\bar{e}_1,\bar{e}_1)}\bar{e}_1 + (\lambda_2 - 1)\frac{(\bar{x},\bar{e}_2)}{(\bar{e}_2,\bar{e}_2)}\bar{e}_2 + (\lambda_1 - 1)(\lambda_2 - 1)\frac{(\bar{x},\bar{e}_2)(\bar{e}_2,\bar{e}_1)}{(\bar{e}_1,\bar{e}_1)(\bar{e}_2,\bar{e}_2)}\bar{e}_1.$$

Hence $\theta_{\bar{e}_1,\lambda_1}$ and $\theta_{\bar{e}_2,\lambda_2}$ commute if and only if either $\bar{e}_1 \parallel \bar{e}_2$ or $(\bar{e}_1,\bar{e}_2) = 0$. The theorem now readily follows. □

6.2 The class $\mathbb{H}_n$

This class of near polygons is due to Brouwer, Cohen, Hall and Wilbrink [12]. Let A be a set of size $2n+2$, $n \geq 0$. Let $\mathbb{H}_n = (\mathcal{P}, \mathcal{L}, \mathrm{I})$ be the following incidence structure:

- $\mathcal{P}$ is the set of all partitions of A in $n+1$ sets of size 2;
- $\mathcal{L}$ is the set of all partitions of A in $n-1$ sets of size 2 and one set of size 4;
- a point $p \in \mathcal{P}$ is incident with a line $L \in \mathcal{L}$ if and only if the partition determined by p is a refinement of the partition determined by L.

Every line of $\mathbb{H}_n$ is incident with three points and every point is incident with $\binom{n+1}{2}$ lines. The incidence structure $\mathbb{H}_0$ is a point, $\mathbb{H}_1$ is a line of size 3 and $\mathbb{H}_2$ is isomorphic to the generalized quadrangle $W(2)$. (Recall Sylvester's model for $W(2)$.)

Definition. For all points x and y of $\mathbb{H}_n$ a graph $\Gamma_{x,y}$ can be defined whose vertices are the elements of A. Two vertices v_1 and v_2 are adjacent if and only if $\{v_1, v_2\} \in x$ or $\{v_1, v_2\} \in y$. Clearly, $\Gamma_{x,y}$ is the union of disjoint cycles.

Theorem 6.12. *The distance $d(x,y)$ between x and y equals $n+1-K_{x,y}$, where $K_{x,y}$ is the number of connected components of the graph $\Gamma_{x,y}$.*

Proof. Put $k := n+1-K_{x,y}$. Obviously, the theorem holds if k is equal to 0 or 1. If z_1 and z_2 are collinear points of $\mathbb{H}_n$, then $|K_{x,z_1} - K_{x,z_2}| \leq 1$. As a consequence, $k = K_{x,x} - K_{x,y} \leq \mathrm{d}(x,y)$. Now, we can find points $z_0, z_1, \ldots, z_k$ such that: (i) $z_0 = y$, (ii) $z_i \sim z_{i-1}$ and $K_{x,z_i} = K_{x,z_{i-1}} + 1$ for every $i \in \{1, \ldots, k\}$. Since $K(x, z_k) = n+1$, $z_k = x$. So, $k \geq \mathrm{d}(x,y)$. This proves the theorem. □

Theorem 6.13. *The incidence structure $\mathbb{H}_n$ is a near $2n$-gon.*

Proof. Let x be a point and let L be a line of $\mathbb{H}_n$. Let $\Gamma_{x,L}$ be the graph with vertex set A. Two vertices v_1 and v_2 are adjacent if and only if $\{v_1, v_2\} \in x$ or $\{v_1, v_2\} \in L$. $\Gamma_{x,y}$ is the union of disjoint cycles and two paths which are not closed. There is a unique way to complete the two "open paths" to two closed paths by adding two edges $\{v, v'\}$ and $\{w, w'\}$. If we replace the set $\{v, v', w, w'\}$ in the partition L by the two sets $\{v, v'\}$ and $\{w, w'\}$, then we obtain a point x' in $\mathbb{H}_n$ which is the unique point of L nearest to x by Theorem 6.12. So, $\mathbb{H}_n$ is a near polygon. By Theorem 6.12, $\mathrm{diam}(\mathbb{H}_n) \leq n$. Obviously, there exist points x and y such that $K_{x,y} = 1$, or equivalently, such that $\mathrm{d}(x,y) = n$. This proves that $\mathbb{H}_n$ is a near $2n$-gon. □

Theorem 6.14. *Every two points at distance 2 from each other have either two or three common neighbours. Hence, $\mathcal{S}$ is a dense near polygon and every quad is either a grid-quad or a $W(2)$-quad.*

Proof. Let x and y be two points of $\mathbb{H}_n$ at distance 2 from each other. Then $\Gamma_{x,y}$ has $n-1$ components. There are two possibilities:

- $\Gamma_{x,y}$ has $n-2$ components of size 2 and one component of size 6. In this case x and y have precisely three common neighbours.
- $\Gamma_{x,y}$ has $n-3$ components of size 2 and two components of size 4. In this case x and y have precisely two common neighbours. □

Definition. Let $\{A_1, A_2, \ldots, A_k\}$ denote a partition of A such that $|A_i|$ is even for every $i \in \{1, \ldots, k\}$. Put $|A_i| = 2n_i$. If P_i, $i \in \{1, \ldots, k\}$, is a partition of A_i in n_i sets of size 2, then $P_1 \cup P_2 \cup \cdots \cup P_k$ is a point of $\mathbb{H}_n$. Let $H(A_1, \ldots, A_k)$ denote the sets of all points arising in this way.

Theorem 6.15. *$H(A_1, \ldots, A_k)$ is a convex sub-$2(n+1-k)$-gon isomorphic to $\mathbb{H}_{n_1-1} \times \mathbb{H}_{n_2-1} \times \cdots \times \mathbb{H}_{n_k-1}$.*

Proof. (a) By Theorem 6.12, the maximal distance between two points of $H(A_1, \ldots, A_k)$ is equal to $n+1-k$.

(b) Suppose that x and y are two different collinear points of $H(A_1, \ldots, A_k)$, then the line L through x and y is a partition of A in $n-1$ sets of size 2 and a set $\{a_1, a_2, a_3, a_4\}$ of size 4. There exists an $i \in \{1, \ldots, k\}$ such that $\{a_1, a_2, a_3, a_4\} \subseteq A_i$. Hence, also the third point of the line xy will belong to $H(A_1, \ldots, A_k)$. This proves that $H(A_1, \ldots, A_k)$ is a subspace.

(c) Let x and y be two points of $H(A_1, \ldots, A_k)$ and let z be a point of $\mathbb{H}_n$ such that $z \sim y$ and $\mathrm{d}(x, z) = \mathrm{d}(x, y) - 1$. Let $\{a_1, a_2, a_3, a_4\}$ denote the unique set of size 4 contained in the line (i.e. partition) yz. Since $\Gamma_{x,z}$ has one component more than $\Gamma_{x,y}$, there exists an $i \in \{1, \ldots, k\}$ such that $\{a_1, a_2, a_3, a_4\} \in A_i$. Hence, z belongs to $H(A_1, \ldots, A_k)$. This proves that $H(A_1, \ldots, A_k)$ is convex.

(d) The partitions of A_i, $i \in \{1, \ldots, k\}$, determine a convex sub-$2(n_i-1)$-gon $\mathcal{A}_i$ isomorphic to $\mathbb{H}_{n_i-1}$. If $P = P_1 \cup \cdots \cup P_k$ is a point of $H(A_1, \ldots, A_k)$, with P_i a partition of A_i, then we define $\theta(P) = (P_1, P_2, \ldots, P_k)$. Obviously, θ is an isomorphism between $H(A_1, \ldots, A_k)$ and $\mathcal{A}_1 \times \cdots \times \mathcal{A}_k$. This proves the theorem. □

Theorem 6.16. *Let x and y be two points of $\mathbb{H}_n$ and let $C_1, \ldots, C_k$ denote the connected components of $\Gamma_{x,y}$. Then $\mathcal{C}(x, y)$ coincides with $H(C_1, \ldots, C_k)$.*

Proof. This follows from the fact that $\mathcal{C}(x, y)$ and $H(C_1, \ldots, C_k)$ are two convex sub-$2(n+1-k)$-gons through x and y. □

Corollary 6.17. *There exists a bijective correspondence between the partitions of A in sets of even size and the convex subpolygons of $\mathbb{H}_n$. If P is a partition of A in sets of even size, then the convex subpolygon of $\mathbb{H}_n$ corresponding with P consists of all partitions of A in $n+1$ sets of size 2 which are a refinement of the partition P.*

Theorem 6.18. *Let $\mathcal{M}_n$ denote the linear space whose points, respectively lines, are the subsets of size 2, respectively size 3, of $\{1, 2, \ldots, n+1\}$ with containment as incidence relation. Then every local space of $\mathbb{H}_n$ is isomorphic to $\mathcal{L}_{\mathbb{H}_n}$, the unique linear space obtained from $\mathcal{M}_n$ by adding lines of size 2.*

Proof. Let x denote an arbitrary point of $\mathbb{H}_n$. Then x is a partition of A in $n+1$ sets $A_1, \ldots, A_{n+1}$ of size 2. For every line L through x, we define $\theta(L) := \{i, j\}$ with i and j such that $A_i \cup A_j$ is the unique set of size 4 contained in the partition L. If Q is a $W(2)$-quad through x, then we define $\theta(x) := \{i, j, k\}$ with $A_i \cup A_j \cup A_k$ the unique set of size 6 contained in the partition associated with Q (see Theorem 6.15 and Corollary 6.17). Obviously, θ determines an isomorphism between $\mathcal{L}(\mathbb{H}_n, x)$ and $\mathcal{L}_{\mathbb{H}_n}$. $\square$

Theorem 6.19. *If A_1 is a set of size $2n$ and if A_2 is a set of size 2, then $H(A_1, A_2)$ is a big convex subpolygon. Conversely, every big convex subpolygon is obtained in this way.*

Proof. Put $A_2 = \{a, b\}$. Let x be a point of $\mathbb{H}_n$ not contained in $H(A_1, A_2)$. Let a' and b' denote the unique elements of A_1 such that $\{a, a'\}, \{b, b'\} \in x$. Then $x \setminus (\{\{a, a'\}, \{b, b'\}\}) \cup (\{\{a, b\}, \{a', b'\}\})$ is a point of $H(A_1, A_2)$ collinear with x. This proves that $H(A_1, A_2)$ is big. Consider now a convex sub-$(2n+2)$-gon $H(B_1, B_2)$ with $|B_1|, |B_2| \geq 4$. Let x be a point of $\mathbb{H}_n$ containing four sets of the form $\{b_1, b_2\}$ with $b_1 \in B_1$ and $b_2 \in B_2$. Obviously, x cannot be collinear with a point of $H(B_1, B_2)$. This proves the theorem. $\square$

Theorem 6.20. *The near polygon $\mathbb{H}_n$ has no spreads of symmetry if $n \geq 2$.*

Proof. Every line of $\mathbb{H}_n$, $n \geq 2$, is contained in a $W(2)$-quad. The proof is now completely similar to the proof of Theorem 6.3. $\square$

In Section 6.4, we will determine all valuations of $\mathbb{H}_3$.

6.3 The class $\mathbb{G}_n$

This class of near polygons is due to the author [39]. The near hexagon $\mathbb{G}_3$ was already described in [12].

6.3.1 Definition of $\mathbb{G}_n$

Let the vector space $V(2n, 4)$, $n \geq 1$, with base $\{\bar{e}_0, \ldots, \bar{e}_{2n-1}\}$ be equipped with the nonsingular hermitian form $(\bar{x}, \bar{y}) = x_0 y_0^2 + x_1 y_1^2 + \cdots + x_{2n-1} y_{2n-1}^2$ and let $H = H(2n-1, 4)$ denote the corresponding hermitian variety in $\mathrm{PG}(2n-1, 4)$. The *support* S_p of a point $p = \langle \bar{x} \rangle$ of $\mathrm{PG}(2n-1, 4)$ is the set of all $i \in \{0, \ldots, 2n-1\}$ for which $(\bar{x}, \bar{e}_i) \neq 0$. The number $|S_p|$ is called the *weight* of p. Since $\bar{x} = \sum (\bar{x}, \bar{e}_i)\, \bar{e}_i$, $|S_p|$ is equal to the number of nonzero coordinates of p. Let $X \subseteq H$ denote the set of all points of weight 2. A point of $\mathrm{PG}(2n-1, 4)$ belongs to H if and only if its weight is even. A subspace π on H is said to be *good* if it is generated by a (possibly empty) set $\mathcal{G}_\pi \subseteq H$ of points whose supports are two by two disjoint. If π is good, then $\mathcal{G}_\pi$ is uniquely determined. If $\mathcal{G}_\pi$ contains k_{2i} points of weight $2i$, $i \in \mathbb{N} \setminus \{0\}$, then π is said to be of *type* $(2^{k_2}, 4^{k_4}, \ldots)$. Let Y, respectively Y', denote the set of all good subspaces of dimension $n-1$, respectively $n-2$. Every element of Y has type (2^n). Every element of Y' has type (2^{n-1}) or $(2^{n-2}, 4^1)$.

Lemma 6.21. *If π is a good subspace on H, then there exist $\pi_1, \pi_2 \in Y$ such that $\pi = \pi_1 \cap \pi_2$.*

Proof. For every point $p = \langle \bar{x} \rangle$ of $\mathcal{G}_\pi$ we take two partitions P_p^1 and P_p^2 of S_p into $\frac{|S_p|}{2}$ subsets of size 2 in such a way that the graph $(S_p, P_p^1 \cup P_p^2)$ is a cycle of length $|S_p|$ if $|S_p| \geq 4$. If we define $A_p^k := \{\langle (\bar{x}, \bar{e}_i)\bar{e}_i + (\bar{x}, \bar{e}_j)\bar{e}_j \rangle | \{i,j\} \in P_p^k\}$, $k \in \{1,2\}$, then clearly $\langle A_p^1 \rangle \cap \langle A_p^2 \rangle = \{p\}$. If we define $A^k := \bigcup_{p \in \mathcal{G}_\pi} A_p^k$, $k \in \{1,2\}$, then $\langle A^1 \rangle \cap \langle A^2 \rangle = \langle \mathcal{G}_\pi \rangle = \pi$. Now, let N be the complement of $\bigcup_{p \in \mathcal{G}_\pi} S_p$ in $\{0, \ldots, 2n-1\}$. Clearly $|N|$ is even. If $|N| = 0$, then we put $B^1 = B^2 = \emptyset$. If $|N| \neq 0$, then we consider a partition P of N into $\frac{|N|}{2}$ sets of size 2 and an element $\alpha \in \mathbb{F}_4^* \setminus \{1\}$. We put $B^1 := \{\langle \bar{e}_i + \bar{e}_j \rangle | \{i,j\} \in P\}$ and $B^2 := \{\langle \bar{e}_i + \alpha\bar{e}_j \rangle | \{i,j\} \in P$ and $i < j\}$. Clearly $\langle B^1 \rangle \cap \langle B^2 \rangle = \emptyset$. If $\pi_k := \langle A^k \cup B^k \rangle$, $k \in \{1,2\}$, then $\pi_1, \pi_2 \in Y$ and $\pi_1 \cap \pi_2 = \pi$. □

Lemma 6.22. *The intersection of two good subspaces π_1 and π_2 is again a good subspace.*

Proof. Consider the following graph Γ on the vertex set $\{0, \ldots, 2n-1\}$. Two vertices i and j are adjacent if and only if there exists a $p \in \mathcal{G}_{\pi_1} \cup \mathcal{G}_{\pi_2}$ such that $\{i,j\} \subseteq S_p$. Let $C_1, \ldots, C_f$ denote the connected components of Γ. For every $i \in \{1, \ldots, f\}$, there is at most one point $p \in \pi_1 \cap \pi_2$ with $S_p = C_i$. We can always label the components of Γ such that the following holds for a certain $f' \in \{0, \ldots, f\}$:

(i) for every i with $1 \leq i \leq f'$, there exists a unique point $p_i \in \pi_1 \cap \pi_2$ with $S_{p_i} = C_i$;

(ii) for every i with $f' < i \leq f$, there exists no point $p \in \pi_1 \cap \pi_2$ with $S_p = C_i$.

It is now easily seen that $\pi_1 \cap \pi_2$ is good with $\mathcal{G}_{\pi_1 \cap \pi_2} = \{p_i \,|\, 1 \leq i \leq f'\}$. □

Lemma 6.23. *If π is a generator of H, then $n-2 \neq |\pi \cap X| \neq n-1$.*

Proof. We use induction on n. For $n \in \{1,2\}$, it is easily seen that every generator of H contains exactly n points of weight 2. Suppose therefore that $n \geq 3$ and let π be a generator containing the point $\langle \bar{a} \rangle = \langle (a_0, a_1, 0, 0, \ldots, 0) \rangle$. The points of $\pi \cap X$ different from $\langle \bar{a} \rangle$ are all contained in the space $\alpha \leftrightarrow X_0 = X_1 = 0$. The intersection $H' := H \cap \alpha$ is a nonsingular hermitian variety in α and $\pi' := \pi \cap \alpha$ is a generator of H'. By induction, $n-3 \neq |\pi' \cap X| \neq n-2$; hence $n-2 \neq |\pi \cap X| \neq n-1$. □

Let $\mathbb{G}_n$ be the incidence structure with point set Y and line set Y' (natural incidence). $\mathbb{G}_n$ is a substructure of $DH(2n-1,4)$. By Lemma 6.23, every generator through an element of Y' belongs to Y. Hence, every line of $\mathbb{G}_n$ is incident with three points. Let $\mathrm{d}(\pi_2, \pi_2)$ denote the distance between two points π_1 and π_2 of $DH(2n-1, 4)$.

Theorem 6.24. *Let $\pi_1, \pi_2 \in Y$. The distance between π_1 and π_2 in $\mathbb{G}_n$ is equal to $d(\pi_1, \pi_2)$.*

Proof. The proof is by induction. If $\mathrm{d}(\pi_1, \pi_2) = 1$, then $\pi_1 \cap \pi_2$ is a good subspace of dimension $n-2$ and hence belongs to Y'. As a consequence also the $\mathbb{G}_n$-distance between π_1 and π_2 is equal to 1. Suppose therefore that $\mathrm{d}(\pi_1, \pi_2) \geq 2$. Take an $x \in X \cap (\pi_1 \setminus (\pi_1 \cap \pi_2))$ and let π_3 be the unique generator through x intersecting π_2 in an $(n-2)$-dimensional subspace. Since there are at least $n-2$ elements in $X \cap \pi_2$ H-collinear with x, $|X \cap \pi_3| \geq n-1$. By Lemma 6.23, $\pi_3 \in Y$. Since $\mathrm{d}(\pi_1, \pi_3) = \mathrm{d}(\pi_1, \pi_2) - 1$, the distance between π_1 and π_3 in $\mathbb{G}_n$ is equal to $\mathrm{d}(\pi_1, \pi_2) - 1$. Since π_2 and π_3 are collinear in $\mathbb{G}_n$, the distance between π_1 and π_2 in $\mathbb{G}_n$ is at most $\mathrm{d}(\pi_1, \pi_2)$. Since $\mathbb{G}_n$ is embedded in $DH(2n-1, 4)$, this distance is at least $\mathrm{d}(\pi_1, \pi_2)$. This proves the theorem. □

Corollary 6.25. *$\mathbb{G}_n$ is a sub-$2n$-gon of $DH(2n-1, 4)$.*

Proof. Let x be a point and L a line of $\mathbb{G}_n$, then x and L are also objects of $DH(2n-1, 4)$. In the near polygon $DH(2n-1, 4)$, L contains a unique point nearest to x. By Theorem 6.24, this property also holds in $\mathbb{G}_n$. Hence $\mathbb{G}_n$ is also a near polygon. Since $\mathrm{d}(\pi_1, \pi_2) = n - 1 - \dim(\pi_1 \cap \pi_2)$ for all $\pi_1, \pi_2 \in Y$ and since there exist $\pi_1, \pi_2 \in Y$ such that $\pi_1 \cap \pi_2 = \emptyset$, see Lemma 6.21, $\mathbb{G}_n$ is a near $2n$-gon. □

The near polygon $\mathbb{G}_1$ is the unique line of size 3. The points, respectively lines, of $\mathbb{G}_2$ are all the maximal, respectively next-to maximal, subspaces of $H(3, 4)$. Hence $\mathbb{G}_2 \cong DH(3, 4) \cong Q(5, 2)$. We define $\mathbb{G}_0$ as the unique near 0-gon.

6.3.2 Subpolygons of $\mathbb{G}_n$

Theorem 6.26. *The near polygon $\mathbb{G}_n$ is dense. For every two points π_1 and π_2 of $\mathbb{G}_n$, $\mathcal{C}(\pi_1, \pi_2)$ is the unique convex sub-$[2 \cdot d(\pi_1, \pi_2)]$-gon through π_1 and π_2. Moreover, $\mathcal{C}(\pi_1, \pi_2)$ consists of all elements of Y through $\pi_1 \cap \pi_2$.*

Proof. We noticed earlier that every line of $\mathbb{G}_n$ is incident with three points. Now, let $\pi_1, \pi_2 \in Y$ such that $\mathrm{d}(\pi_1, \pi_2) = 2$, or equivalently $\dim(\pi_1 \cap \pi_2) = n-3$. Choose an $x_3 \in X \cap (\pi_2 \setminus (\pi_1 \cap \pi_2))$ and an $x_4 \in X \cap \pi_1$ not H-collinear with x_3. Let π_i, $i \in \{3, 4\}$, denote the unique generator through x_i intersecting π_{i-2} in an $(n-2)$-dimensional subspace. By the proof of Theorem 6.24, we know that π_3 and π_4 are common neighbours of π_1 and π_2. Hence $\mathbb{G}_n$ is dense. By Theorem 2.3, $\mathcal{C}(\pi_1, \pi_2)$ is the unique convex sub-$[2 \cdot \mathrm{d}(\pi_1, \pi_2)]$-gon through π_1 and π_2. Now, let F denote the set of all generators of Y through $\pi_1 \cap \pi_2$. Clearly F is a subspace of $\mathbb{G}_n$. If γ denotes a shortest path in $\mathbb{G}_n$ between two points of F, then by Theorem 6.24, γ is also a shortest path in $DH(2n-1, 4)$ and hence every point of it contains

$\pi_1 \cap \pi_2$. As a consequence every point on γ is contained in F and F is convex. If π and π' are two arbitrary elements of F, then $\pi \cap \pi'$ contains $\pi_1 \cap \pi_2$ and hence $\mathrm{d}(\pi, \pi') = n - 1 - \dim(\pi \cap \pi') \leq n - 1 - \dim(\pi_1 \cap \pi_2) = \mathrm{d}(\pi_1, \pi_2)$. As a consequence the diameter of F is at most $\mathrm{d}(\pi_1, \pi_2)$. Since F contains π_1 and π_2, the diameter is precisely $\mathrm{d}(\pi_1, \pi_2)$. Since F is a convex sub-$[2 \cdot \mathrm{d}(\pi_1, \pi_2)]$-gon through π_1 and π_2, it coincides with $\mathcal{C}(\pi_1, \pi_2)$. □

For every convex subspace F of $\mathbb{G}_n$, let π_F denote the intersection of all points of F regarded as generators of H. Since there exist elements $\pi_1, \pi_2 \in Y$ such that $\pi_1 \cap \pi_2 = \emptyset$, $\pi_{\mathbb{G}_n} = \emptyset$.

Theorem 6.27. (a) *There exists a one-to-one correspondence between the convex subspaces of $\mathbb{G}_n$ and the good subspaces on H.*

(b) *If F_1 and F_2 are two convex subpolygons, then $F_1 \subseteq F_2$ if and only if $\pi_{F_2} \subseteq \pi_{F_1}$.*

Proof. Let F denote an arbitrary convex subpolygon of $\mathbb{G}_n$. If π_1 and π_2 denote two points of F at maximal distance from each other, then $F = \mathcal{C}(\pi_1, \pi_2)$. By Theorem 6.26, $\pi_F = \pi_1 \cap \pi_2$. Hence π_F is good by Lemma 6.22. Conversely, suppose that π is a good subspace on H. If $\pi = \pi_F$, then F necessarily consists of all elements of Y through π. Hence, the equation $\pi_F = \pi$ has at most one solution for F. It suffices to show that this equation has at least one solution. By Lemma 6.21, there exist elements $\pi_1, \pi_2 \in Y$ such that $\pi = \pi_1 \cap \pi_2$. If we put F equal to $\mathcal{C}(\pi_1, \pi_2)$, then by Theorem 6.26, $\pi_F = \pi_1 \cap \pi_2 = \pi$. This proves part (a). Part (b) follows from the fact that the points of a convex subpolygon F are precisely the generators of Y through π_F. □

Corollary 6.28. *Let F_1 and F_2 be two convex subpolygons of $\mathbb{G}_n$ and let $F_3 = \mathcal{C}(F_1, F_2)$. Then $\pi_{F_3} = \pi_{F_1} \cap \pi_{F_2}$.*

Proof. Since F_3 is the smallest convex subpolygon through F_1 and F_2, π_{F_3} is the biggest good subspace contained in π_{F_1} and π_{F_2}. The result now easily follows from Lemma 6.22. □

Theorem 6.29. *Let p denote an arbitrary point of weight $2n$ in $\mathrm{PG}(2n-1, 4)$, then $p \in H$ and the set of all generators of Y through p determines a convex sub-$2(n-1)$-gon isomorphic to $\mathbb{H}_{n-1}$.*

Proof. Put $p = \langle \alpha_0 \bar{e}_0 + \cdots + \alpha_{2n-1} \bar{e}_{2n-1} \rangle$. The set $\{p\}$ is a good subspace of H and hence, by Theorem 6.27, the set of all generators of Y through p determines a convex sub-$2(n-1)$-gon $\mathcal{B}$. The set $\{0, \ldots, 2n-1\}$ has size $2n$ and hence, by Section 6.2, a near $2(n-1)$-gon $\mathcal{A} \cong \mathbb{H}_{n-1}$ can be constructed from this set. For every point P of $\mathcal{A}$, i.e. for every partition P of $\{0, \ldots, 2n-1\}$ into n sets of size 2, we put $\phi(P) := \langle \{ \langle \alpha_i \bar{e}_i + \alpha_j \bar{e}_j \rangle | \{i, j\} \in P \} \rangle$. Clearly $\phi(P)$ is a generator of Y through p. Conversely, every generator of Y through p is of the form $\phi(P)$ for some point P of $\mathcal{A}$. We will now show that ϕ determines an isomorphism between the collinearity graphs of $\mathcal{A}$ and $\mathcal{B}$. If P_1 and P_2 are two collinear points of $\mathcal{A}$, then

$\phi(P_1) \cap \phi(P_2)$ is a good subspace of type $(2^{n-2}, 4^1)$; hence $\phi(P_1)$ and $\phi(P_2)$ are collinear in $\mathcal{B}$. Conversely, suppose that $\phi(P_1)$ and $\phi(P_2)$ are collinear in $\mathcal{B}$, then $\phi(P_1) \cap \phi(P_2)$ is a good subspace of type (2^{n-1}) or $(2^{n-2}, 4^1)$. If $\phi(P_1) \cap \phi(P_2)$ has type (2^{n-1}), then $|P_1 \cap P_2| \geq n-1$ and hence $P_1 = P_2$, a contradiction. As a consequence $\phi(P_1) \cap \phi(P_2)$ has type $(2^{n-2}, 4^1)$ and P_1 and P_2 are collinear in $\mathcal{A}$. Since the collinearity graphs of $\mathcal{A}$ and $\mathcal{B}$ are isomorphic, $\mathcal{A}$ and $\mathcal{B}$ themselves are isomorphic. □

Theorem 6.30. *The convex sub-$(n-k)$-gons, $k \in \{0, \ldots, n\}$, of $\mathbb{G}_n$ are of the form $\mathbb{H}_{n_1-1} \times \cdots \times \mathbb{H}_{n_k-1} \times \mathbb{G}_{n_{k+1}}$ with $n_1, \ldots, n_k \geq 1$, $n_{k+1} \geq 0$ and $n_1 + \cdots + n_{k+1} = n$.*

Proof. Let F denote an arbitrary convex sub-$(n-k)$-gon, $k \in \{0, \ldots, n\}$, and put $\mathcal{G}_{\pi_F} = \{p_1, \ldots, p_k\}$. Let S_i, $i \in \{1, \ldots, k\}$, denote the support of p_i, and let $S_{k+1} = \{0, \ldots, 2n-1\} \setminus (S_1 \cup \cdots \cup S_k)$. For every $i \in \{1, \ldots, k+1\}$, we put $|S_i| = 2n_i$ and $\alpha_i := \langle \bar{e}_j | j \in S_i \rangle$. Clearly, $n_1, \ldots, n_k \geq 1$, $n_{k+1} \geq 0$ and $n_1 + \cdots + n_{k+1} = n$. Also $\alpha_i \cap H$ is a nonsingular hermitian variety of type $H(2n_i - 1, 4)$. If π is an arbitrary point of F, or equivalently, an arbitrary generator of Y through π_F, then $\pi = \langle \pi \cap \alpha_1, \ldots, \pi \cap \alpha_k, \pi \cap \alpha_{k+1} \rangle$. Moreover, $\pi \cap \alpha_i$ is a generator of $\alpha_i \cap H$ containing n_i points of weight 2, and $p_i \in \pi \cap \alpha_i$ if $i \neq k+1$. Conversely, if β_i, $i \in \{1, \ldots, k+1\}$, is a generator of $\alpha_i \cap H$ containing n_i vertices of weight 2 such that $p_i \in \beta_i$ if $i \leq k$, then $\langle \beta_1, \ldots, \beta_{k+1} \rangle$ is a generator of F through π_F. Hence, by Theorem 6.29, the map $\pi \rightarrow (\pi \cap \alpha_1, \ldots, \pi \cap \alpha_k, \pi \cap \alpha_{k+1})$ determines a bijection between the point sets of the near polygons F and $\mathbb{H}_{n_1-1} \times \cdots \times \mathbb{H}_{n_k-1} \times \mathbb{G}_{n_{k+1}}$. Now, two points π_1 and π_2 of F are collinear if and only if $\dim(\pi_1 \cap \pi_2) = n-2$. This happens if and only if there exists a $j \in \{1, \ldots, k+1\}$ such that $\dim(\pi_1 \cap \pi_2 \cap \alpha_j) = n_j - 2$ and $\dim(\pi_1 \cap \pi_2 \cap \alpha_i) = n_i - 1$ for every $i \in \{1, \ldots, k+1\} \setminus \{j\}$. These conditions are equivalent with $\dim((\pi_1 \cap \alpha_j) \cap (\pi_2 \cap \alpha_j)) = n_j - 2$ and $\pi_1 \cap \alpha_i = \pi_2 \cap \alpha_i$. Hence π_1 and π_2 are collinear in F if and only if $(\pi_1 \cap \alpha_1, \ldots, \pi_1 \cap \alpha_{k+1})$ and $(\pi_2 \cap \alpha_1, \ldots, \pi_2 \cap \alpha_{k+1})$ are collinear in $\mathbb{H}_{n_1-1} \times \cdots \times \mathbb{H}_{n_k-1} \times \mathbb{G}_{n_{k+1}}$. Hence, the collinearity graphs of F and $\mathbb{H}_{n_1-1} \times \cdots \times \mathbb{H}_{n_k-1} \times \mathbb{G}_{n_{k+1}}$ are isomorphic. So, the near polygons themselves are also isomorphic. □

6.3.3 Lines and quads in $\mathbb{G}_n$

Let $n \geq 3$. If L is a line of $\mathbb{G}_n$, then there are two possibilities for π_L $(= L)$:

(a) π_L has type (2^{n-1});

(b) π_L has type $(2^{n-2}, 4^1)$.

If Q is a quad of $\mathbb{G}_n$, then there are four possibilities for π_Q.

(i) π_Q has type (2^{n-2}).
By Lemma 6.23, each of the 27 generators through π_Q belongs to Y, proving that Q is a $Q(5,2)$-quad. The quad Q has 18 lines of type (a) and 27 lines of type (b). The 18 lines of type (a) define three grids which partition the point set of Q.

(ii) π_Q has type $(2^{n-3}, 6^1)$.
From the 27 generators through π_Q, 15 are contained in Y, proving that Q is a $W(2)$-quad. Clearly Q contains only lines of type (b).

(iii) π_Q has type $(2^{n-3}, 4^1)$.
From the 27 generators through π_Q, nine are contained in Y, proving that Q is a grid. The quad Q contains three lines of type (a) and three lines of type (b). Three lines of the same type partition the point set of Q.

(iv) π_Q has type $(2^{n-4}, 4^2)$.
This type of quad only exists if $n \geq 4$. From the 27 generators through π_Q, nine are contained in Y, proving that Q is a grid. All six lines of Q have type (b).

By Lemma 6.27, it then easily follows:

Theorem 6.31. *Consider the near polygon $\mathbb{G}_n$ with $n \geq 3$. Then*

- *each point is contained in n lines of type (a) and $3\frac{n(n-1)}{2}$ lines of type (b);*
- *each line of type (a) is contained in exactly $n-1$ $Q(5,2)$-quads, 0 $W(2)$-quads and $3\frac{(n-1)(n-2)}{2}$ grid-quads;*
- *each line of type (b) is contained in a unique $Q(5,2)$-quad, $3(n-2)$ $W(2)$-quads and $\frac{(n-2)(3n-7)}{2}$ grid-quads;*
- *each line is contained in exactly $\frac{(n-1)(3n-4)}{2}$ quads.*

In the sequel lines of type (a) in $\mathbb{G}_n$, $n \geq 3$, will be called *special*, while lines of type (b) are called *ordinary*. Clearly, a line is special if and only if it is not contained in a $W(2)$-quad. For every permutation σ of $\{0, \ldots, 2n-1\}$ and for every $\lambda_0, \ldots, \lambda_{2n-1} \in \mathbb{F}_4^*$, the linear transformation of $V(2n,4)$ defined by $\bar{e}_i \mapsto \lambda_i \bar{e}_{\sigma(i)}$, $i \in \{0, \ldots, 2n-1\}$, determines an automorphism of $\mathbb{G}_n$. Using these automorphisms it is easily seen that any two lines of the same type are in the same $\mathrm{Aut}(\mathbb{G}_n)$-orbit. Similarly, any two quads of the same type are contained in the same $\mathrm{Aut}(\mathbb{G}_n)$-orbit. Since a special line can never be mapped to an ordinary line, $\mathrm{Aut}(\mathbb{G}_n)$ has two orbits on the set of lines and three or four orbits on the set of quads depending on whether $n = 3$ or $n \geq 4$. In Section 6.3.5 we will determine $\mathrm{Aut}(\mathbb{G}_n)$.

Remark. The above remarks on the orbits of $\mathrm{Aut}(\mathbb{G}_n)$, $n \geq 3$, do not hold for $\mathbb{G}_2$. Since $\mathbb{G}_2 \cong Q(5,2)$ all lines are in the same orbit.

6.3.4 Some properties of $\mathbb{G}_n$

Theorem 6.32. *The near $2n$-gon $\mathbb{G}_n$, $n \geq 1$, has order $(s,t) = (2, \frac{3n^2-n-2}{2})$ and $v = \frac{3^n \cdot (2n)!}{2^n \cdot n!}$ points.*

Proof. Clearly, the theorem holds if $n \in \{1, 2\}$. So suppose that $n \geq 3$. $H(2n-1, 4)$ has exactly $\frac{3^n \cdot (2n)!}{2^n \cdot n!}$ good subspaces of type (2^n). We noticed earlier that every line is incident with exactly $s + 1 = 3$ points, and by Theorem 6.31, it follows that $t + 1 = n + 3\frac{n(n-1)}{2}$. □

Theorem 6.33. *Let F be a convex subpolygon of $\mathbb{G}_n$ isomorphic to $\mathbb{G}_k$, $k \geq 2$, and let x denote an arbitrary point of F. Then π_F has type (2^{n-k}) and precisely k from the n special lines through x are contained in F.*

Proof. Recall that no near polygon of type $\mathbb{H}_l$, $l \geq 0$, has a $Q(5,2)$-quad. If $F \cong \mathbb{H}_l \times \mathcal{A}$ for some $l \geq 1$ and some dense near $2(k-l)$-gon $\mathcal{A}$, then F has a line that is not contained in a $Q(5,2)$-quad, contradicting Theorem 6.31. By the proof of Theorem 6.30, it then follows that that π_F has type (2^{n-k}). Theorem 6.27 allows us to count the number of special lines through x which are also contained in F. It is easily seen that this number equals k. □

Theorem 6.34. *If $L_1, \ldots, L_k$ are different special lines of $\mathbb{G}_n$, $n \geq 3$, through a given point x, then $\mathcal{C}(L_1, \ldots, L_k) \cong \mathbb{G}_k$.*

Proof. Put $F = \mathcal{C}(L_1, \ldots, L_k)$. By Corollary 6.28, $\pi_F = \pi_{L_1} \cap \cdots \cap \pi_{L_k}$. Every π_{L_i}, $i \in \{1, \ldots, k\}$, is a good subspace of type (2^{n-1}) contained in the good subspace of type (2^n) associated with x. Hence $\pi_F = \pi_{L_1} \cap \cdots \cap \pi_{L_k}$ is a good subspace of type (2^{n-k}). By Theorem 6.30, $F \cong \underbrace{\mathbb{H}_0 \times \cdots \times \mathbb{H}_0}_{n-k} \times \mathbb{G}_k \cong \mathbb{G}_k$. □

Theorem 6.35. *Let F be a convex sub-$2(n-1)$-gon of $\mathbb{G}_n$, $n \geq 3$.*

(a) *If $F \cong \mathbb{G}_{n-1}$, then F is big in $\mathbb{G}_n$.*

(b) *If F is big in $\mathbb{G}_n$, then $F \cong \mathbb{G}_{n-1}$ and π_F has type (2^1).*

Proof. (a) If $F \cong \mathbb{G}_{n-1}$, then the total number of points at distance at most 1 from F is equal to $|F| \cdot (1 + 2(t - t_F))$ which is exactly the total number of points in $\mathbb{G}_n$. Hence F is big in $\mathbb{G}_n$.

(b) Take a line L intersecting F in a point, then L is contained in precisely $\frac{(n-1)(3n-4)}{2}$ quads, see Theorem 6.31. Since F is big, each of these $\frac{(n-1)(3n-4)}{2}$ quads meets F in a line by Theorem 1.7. Hence $t_F + 1 = \frac{(n-1)(3n-4)}{2}$. Since F is a convex sub-$2(n-1)$-gon, π_F has type $((2k)^1)$ for a certain $k \in \{1, \ldots, n\}$. By Theorem 6.30, $F \cong \mathbb{H}_{k-1} \times \mathbb{G}_{n-k}$. Hence $\frac{(n-1)(3n-4)}{2} = t_F + 1 = \frac{k(k-1)}{2} + \frac{(n-k)(3n-3k-1)}{2}$ or $(k-1)(6n-4k-4) = 0$. Now $6n-4k-4 = 4(n-k)+2n-4 > 0$ since $n \geq 3$. Hence $k = 1$, $F \cong \mathbb{G}_{n-1}$ and π_F has type (2^1). □

6.3.5 Determination of $\mathrm{Aut}(\mathbb{G}_n)$, $n \geq 3$

Let $n \geq 3$ and let B denote the set of all big convex sub-$2(n-1)$-gons of $\mathbb{G}_n$ isomorphic to $\mathbb{G}_{n-1}$, or equivalently, the set of all convex subpolygons F for which

π_F has type (2^1). Consider the following relation R on the elements of B: $(F_1, F_2) \in R \Leftrightarrow (F_1 = F_2)$ or $(F_1 \cap F_2 = \emptyset$ and every line meeting F_1 and F_2 is special).

Lemma 6.36. *The relation R is an equivalence relation and each equivalence class contains exactly three elements.*

Proof. For every element F of B, let C_F denote the set of all elements $F' \in B$ satisfying $(F, F') \in R$. If F_1 and F_2 are two elements of B such that $\pi_{F_1} = \langle \bar{e}_i + \alpha_1 \bar{e}_j \rangle$ and $\pi_{F_2} = \langle \bar{e}_i + \alpha_2 \bar{e}_j \rangle$, then one readily verifies that $(F_1, F_2) \in R$. Hence $|C_F| \geq 3$ for every $F \in B$. It now suffices to prove that $|C_F| \leq 3$. Let L denote an arbitrary special line intersecting F in a point. If F' is an element of C_F, then F' intersects L in a point. Now, each point x on L is contained in at most one element of C_F, namely the element of B generated by the $n-1$ special lines through x different from L. Hence $|C_F| \leq 3$. This proves our lemma. □

Clearly the equivalence classes are in bijective correspondence with the pairs $\{i, j\} \subseteq \{0, \ldots, 2n-1\}$. Consider now the graph Γ whose vertices are the equivalence classes, with two classes C_1 and C_2 adjacent if and only if $F_1 \cap F_2 = \emptyset$ for every $F_1 \in C_1$ and every $F_2 \in C_2$. Clearly two vertices are adjacent if and only if the corresponding pairs have one element in common. Γ is a so-called triangular graph.

If F_1 and F_2 are two elements of B satisfying $\pi_{F_1} = \langle \bar{e}_0 + r\bar{e}_1 \rangle$ and $\pi_{F_2} = \langle \bar{e}_0 + s\bar{e}_1 \rangle$, $r \neq s$, then the reflection $F_3 = \mathcal{R}_{F_2}(F_1)$ of F_1 about F_2 is the unique element of C_{F_1} different from F_1 and F_2; hence $\pi_{F_3} = \langle \bar{e}_0 + (r+s)\bar{e}_1 \rangle$.

Lemma 6.37. *If F_1 and F_2 are two elements of B satisfying $\pi_{F_1} = \langle \bar{e}_0 + r\bar{e}_1 \rangle$ and $\pi_{F_2} = \langle \bar{e}_0 + s\bar{e}_2 \rangle$, then $F_3 := \mathcal{R}_{F_2}(F_1)$ satisfies $\pi_{F_3} = \langle \bar{e}_1 + r^{-1}s\bar{e}_2 \rangle$.*

Proof. Every point p of F_1 is of the form $\langle \bar{e}_0 + r\bar{e}_1, \bar{e}_2 + t\bar{e}_i, \bar{v}_3, \ldots, \bar{v}_n \rangle$ for some $i \in \{3, \ldots, 2n-1\}$, some $t \in \mathbb{F}_4^*$ and some vectors $\bar{v}_j$, $j \in \{3, \ldots, n\}$, of weight 2. The unique line L through p intersecting F_2 is then equal to $\langle \bar{e}_0 + r\bar{e}_1 + s\bar{e}_2 + st\bar{e}_i, \bar{v}_3, \ldots, \bar{v}_n \rangle$. The point $\langle \bar{e}_0 + st\bar{e}_i, \bar{e}_1 + r^{-1}s\bar{e}_2, \bar{v}_3, \ldots, \bar{v}_n \rangle$ of L is not contained in $F_1 \cup F_2$ and hence belongs to F_3. Considering all possibilities for i, t and $\bar{v}_j$, $j \in \{3, \ldots, 2n-1\}$, we easily see that $\pi_{F_3} = \langle \bar{e}_1 + r^{-1}s\bar{e}_2 \rangle$. □

Theorem 6.38. *For every permutation ϕ of $\{0, \ldots, 2n-1\}$, every automorphism θ of $\mathbb{F}_4$, and all $\lambda_0, \ldots, \lambda_{2n-1} \in \mathbb{F}_4^*$, the semilinear map $V(2n,4) \to V(2n,4) : \sum \alpha_i \bar{e}_i \mapsto \sum \lambda_i \alpha_i^\theta \bar{e}_{\phi(i)}$ induces an automorphism of $\mathbb{G}_n$. Conversely, every automorphism of $\mathbb{G}_n$, $n \geq 3$, is obtained in this way.*

Proof. Clearly every semilinear map $V(2n,4) \to V(2n,4) : \sum \alpha_i \bar{e}_i \mapsto \sum \lambda_i \alpha_i^\theta \bar{e}_{\phi(i)}$ induces an automorphism of $\mathbb{G}_n$. We will now prove that every $\mu \in \mathrm{Aut}(\mathbb{G}_n)$ is derived from a semilinear map. The action of μ on the set B determines an action on the vertices of Γ. Clearly, that action permutes the $2n$ maximal cliques of size $2n-1$ in Γ. Thus, there exists a permutation ϕ of $\{0, \ldots, 2n-1\}$ such that, if C is the equivalence class corresponding to the pair $\{i, j\}$, then $\mu(C)$ is the class corresponding to $\{\phi(i), \phi(j)\}$. Now, fix $i, j \in \{0, \ldots, 2n-1\}$ with $i \neq j$.

For all $r \in \mathbb{F}_4^*$, μ maps the element $\langle \bar{e}_i + r\bar{e}_j \rangle$ of B to an element of the form $\langle \bar{e}_{\phi(i)} + r'\bar{e}_{\phi(j)} \rangle$ (notice that we identify each element $F \in B$ with π_F); hence there exists an $\epsilon_{ij} \in \{1, 2\}$ and a $\lambda_{ij} \in \mathbb{F}_4^*$ such that $\mu(\langle \bar{e}_i + r\bar{e}_j \rangle) = \langle \bar{e}_{\phi(i)} + \lambda_{ij} r^{\epsilon_{ij}} \bar{e}_{\phi(j)} \rangle$ for all $r \in GF(4)^*$. Clearly, $\lambda_{ji} = \lambda_{ij}^{-1}$ and $\epsilon_{ji} = \epsilon_{ij}$ for all $i, j \in \{0, \ldots, 2n-1\}$ with $i \neq j$. Put λ_{ii} equal to 1 for all $i \in \{0, \ldots, 2n-1\}$. Now take mutually distinct $i, j, k \in \{0, \ldots, 2n-1\}$. For all $r, s \in \mathbb{F}_4^*$, the reflection of $\langle \bar{e}_i + r\bar{e}_j \rangle$ around $\langle \bar{e}_i + s\bar{e}_k \rangle$ equals $\langle \bar{e}_j + r^{-1}s\bar{e}_k \rangle$. Since $\mu \in \mathrm{Aut}(\mathbb{G}_n)$, the reflection of $\langle \bar{e}_{\phi(i)} + \lambda_{ij} r^{\epsilon_{ij}} \bar{e}_{\phi(j)} \rangle$ around $\langle \bar{e}_{\phi(i)} + \lambda_{ik} s^{\epsilon_{ik}} \bar{e}_{\phi(k)} \rangle$ equals $\langle \bar{e}_{\phi(j)} + \lambda_{jk} (r^{-1}s)^{\epsilon_{jk}} \bar{e}_{\phi(k)} \rangle$, or equivalently, $\lambda_{ij}^{-1} r^{-\epsilon_{ij}} \lambda_{ik} s^{\epsilon_{ik}} = \lambda_{jk} r^{-\epsilon_{jk}} s^{\epsilon_{jk}}$. Since this holds for all $r, s \in \mathbb{F}_4^*$, $\lambda_{ij}\lambda_{jk} = \lambda_{ik}$, $\epsilon_{ij} = \epsilon_{jk}$ and $\epsilon_{ik} = \epsilon_{jk}$. It now easily follows that $\epsilon_{ij} = \epsilon_{01} = \epsilon$ and $\lambda_{ij} = \lambda_{0i}^{-1}\lambda_{0j}$ for all $i, j \in \{0, \ldots, 2n-1\}$ with $i \neq j$. For all $r \in \mathbb{F}_4^*$ and all $j, k \in \{0, \ldots, 2n-1\}$ with $j \neq k$, $\mu(\langle \bar{e}_j + r\bar{e}_k \rangle) = \langle \bar{e}_{\phi(j)} + \lambda_{0j}^{-1} \lambda_{0k} r^{\epsilon} \bar{e}_{\phi(k)} \rangle = \langle \lambda_{0j} \bar{e}_{\phi(j)} + \lambda_{0k} r^{\epsilon} \bar{e}_{\phi(k)} \rangle$. The action of μ on the elements of B completely determines the action of μ on the points of $\mathbb{G}_n$. For, if p is a point of $\mathbb{G}_n$, then $\mu(p) = \bigcap \mu(F)$ where F ranges over all the n elements of B through p. Hence μ is induced by the semi-linear map $\sum \alpha_i \bar{e}_i \mapsto \sum \lambda_{0i} \alpha_i^{\epsilon} \bar{e}_{\phi(i)}$. □

Remark. We have $|\mathrm{Aut}(\mathbb{G}_n)| = 2 \cdot 3^{2n-1} \cdot (2n)!$. The condition $n \geq 3$ in Theorem 6.38 is necessary. For $n = 2$, the natural distinction between lines of type (a) and lines of type (b) disappears. Since $\mathbb{G}_2 \cong Q(5, 2)$, $|\mathrm{Aut}(\mathbb{G}_2)| = |P\Gamma U(4, 4)| = 103680$, while $2 \cdot 3^3 \cdot 4! = 1296$.

6.3.6 Spreads in $\mathbb{G}_n$

For every $i, j \in \{0, \ldots, 2n-1\}$ with $i \neq j$, let $A_{i,j}$ denote the set of all good subspaces α on $H = H(2n-1, 4)$ that satisfy the following properties:

- α has type (2^{n-1});
- $\langle \langle \bar{e}_i + r\bar{e}_j \rangle, \alpha \rangle$ is a generator of H for every $r \in \mathbb{F}_4^*$.

Clearly, $\bigcup_{0 \leq i < j \leq 2n-1} A_{i,j}$ is the set of all special lines of $\mathbb{G}_n$. For every $i \in \{0, \ldots, 2n-1\}$, we put $B_i := \bigcup_{j \neq i} A_{i,j}$. Obviously B_i consists of all good subspaces of type (2^{n-1}) contained in $\langle \bar{e}_i \rangle^{\zeta} \cap H$. Here ζ denotes the hermitian polarity associated with H.

Lemma 6.39. *Let $n \geq 2$. For every $i \in \{0, \ldots, 2n-1\}$, B_i is a spread of symmetry of $\mathbb{G}_n$. As a consequence, B_i is also an admissible spread.*

Proof. If π is a point of $\mathbb{G}_n$, i.e. a good subspace of type (2^n), then π contains a unique point of the form $\langle \bar{e}_i + r\bar{e}_j \rangle$. Clearly $\langle (X \cap \pi) \setminus \{\langle \bar{e}_i + r\bar{e}_j \rangle\} \rangle$ is the unique line of B_i incident with π. This proves that B_i is a spread. For every $\lambda \in \mathbb{F}_4^*$, the linear map $\bar{e}_i \mapsto \lambda \bar{e}_i$, $\bar{e}_j \mapsto \bar{e}_j$ for all $j \neq i$, induces an automorphism $\theta_{\lambda,i}$ of $\mathbb{G}_n$ which fixes each line of S. Clearly, $\{\theta_{\lambda,i} \mid \lambda \in \mathbb{F}_4^*\}$ acts regularly on every line of B_i, proving that B_i is a spread of symmetry. □

Lemma 6.40. *An admissible spread S of $\mathbb{G}_n$, $n \geq 3$, contains only special lines.*

Proof. Suppose that S has an ordinary line L and let x denote an arbitrary point of L. By Theorems 6.31 and 6.33, there exists a unique pair $\{L_1, L_2\}$ of special lines through x such that $L \in \mathcal{C}(L_1, L_2)$. Let L_3 denote a special line through x different from L_1 and L_2 and let $\mathcal{H}$ denote the hex $\mathcal{C}(L_1, L_2, L_3)$. By Theorem 6.34, $\mathcal{H} \cong \mathbb{G}_3$. By Theorem 4.5, the spread S induces an admissible spread S' in $\mathcal{H}$. By Theorem 6.31, there exist two $W(2)$-quads Q_1 and Q_2 in $\mathcal{H}$ through the line L. Let S_i, $i \in \{1, 2\}$, denote the spread of Q_i induced by S'. Let L' be an element of S_2 different from L, let Q_3 denote a $Q(5, 2)$-quad of $\mathcal{H}$ through L' and let S_3 denote the spread of Q_3 induced by S'. Now, Q_1 and Q_3 are disjoint, and since Q_3 is big in $\mathcal{H}$, every point of Q_1 has distance one to a unique point of Q_3. As a consequence Q_1 projects to a subGQ Q_4 of Q_3 isomorphic to $W(2)$. If $y \in Q_4$ then y is collinear with a unique point y' of Q_1 and y' is contained in a unique line M of S_1. The unique line of S_3 through y is contained in the quads $\mathcal{C}(M, y)$ and Q_3 and hence coincides with the line $\mathcal{C}(M, y) \cap Q_3$ which is precisely the projection of M on Q_3. As a consequence the spread S_1 projects to a spread S_4 of Q_4 and $S_4 \subseteq S_3$. Let z be a point of $Q_3 \setminus Q_4$. Through z there is a line of S_3 and five lines intersecting an element of S_4. Hence, the point z of Q_3 is contained in at least six lines, contradicting $Q_3 \cong Q(5, 2)$. □

Lemma 6.41. *Let S be a spread of $\mathbb{G}_n$, $n \geq 3$, satisfying*

(a) *every line of S is special,*

(b) *if a grid-quad contains one line of S, then it contains exactly three lines of S.*

Then $S = B_i$ for a certain $i \in \{0, \ldots, 2n-1\}$.

Proof. Suppose that S contains a special line K of the set $A_{2n-2,2n-1}$, e.g. let $K = \langle\langle \alpha_0 \bar{e}_0 + \alpha_1 \bar{e}_1 \rangle, \langle \alpha_2 \bar{e}_2 + \alpha_3 \bar{e}_3 \rangle, \ldots, \langle \alpha_{2n-4} \bar{e}_{2n-4} + \alpha_{2n-3} \bar{e}_{2n-3} \rangle\rangle$ for certain $\alpha_0, \ldots, \alpha_{2n-3} \in \mathbb{F}_4^*$. Now, for every $\lambda \in \mathbb{F}_4^*$, the grid-quad Q for which $\pi_Q = \langle\langle \alpha_0 \bar{e}_0 + \alpha_1 \bar{e}_1 + \lambda\alpha_2 \bar{e}_2 + \lambda\alpha_3 \bar{e}_3 \rangle, \ldots, \langle \alpha_{2n-4} \bar{e}_{2n-4} + \alpha_{2n-3} \bar{e}_{2n-3} \rangle\rangle$ contains K. Hence, the two other lines in Q disjoint from K are also contained in S, or equivalently, $\langle\langle \alpha_0 \bar{e}_0 + \lambda\alpha_2 \bar{e}_2 \rangle, \langle \alpha_1 \bar{e}_1 + \lambda\alpha_3 \bar{e}_3 \rangle, \ldots, \langle \alpha_{2n-4} \bar{e}_{2n-4} + \alpha_{2n-3} \bar{e}_{2n-3} \rangle\rangle \in S$ and $\langle\langle \alpha_0 \bar{e}_0 + \lambda\alpha_3 \bar{e}_3 \rangle, \langle \alpha_1 \bar{e}_1 + \lambda\alpha_2 \bar{e}_2 \rangle, \ldots, \langle \alpha_{2n-4} \bar{e}_{2n-4} + \alpha_{2n-3} \bar{e}_{2n-3} \rangle\rangle \in S$. Applying this several times, we see that every line of $A_{2n-2,2n-1}$ belongs to S. Hence S is a union of sets of the form $A_{i,j}$. Since $S = \frac{|Y|}{3}$, S is the union of $2n-1$ sets of the form $A_{i,j}$. For all $i, j, k, l \in \{0, \ldots, 2n-1\}$ with $i \neq j$, $k \neq l$ and $\{i, j\} \cap \{k, l\} = \emptyset$, $A_{i,j} \cup A_{k,l}$ always contains two intersecting lines. The lemma now easily follows. □

Corollary 6.42. *The spreads B_i, $i \in \{0, \ldots, 2n-1\}$, are the only admissible spreads in $\mathbb{G}_n$, $n \geq 3$.*

Proof. This follows immediately from Theorem 4.5 and Lemmas 6.39, 6.40, 6.41. □

Theorem 6.43. *For all $i, j \in \{0, 1, \ldots, 2n-1\}$, B_i is compatible with B_j.*

Proof. Let G_i, $i \in \{0, \ldots, 2n-1\}$, denote the group of automorphisms of $\mathbb{G}_n$ fixing each line of S_i. By Theorem 4.43, $G_i = \{\theta_{i,\lambda} \,|\, \lambda \in \mathbb{F}_4^*\}$, with $\theta_{i,\lambda}$ as defined in the proof of Lemma 6.39. It is now easy to see that $[G_i, G_j] = 0$ for all $i, j \in \{0, \ldots, 2n-1\}$. □

So, the near polygon $\mathbb{G}_n$, $n \geq 3$, has $2n$ spreads of symmetry and all these spreads are mutually compatible. The generalized quadrangle $\mathbb{G}_2 \cong Q(5,2) \cong DH(3,4)$ has more than $2n = 4$ spreads of symmetry and not every pair of such spreads is compatible, as we have seen in Theorem 6.11.

6.3.7 Valuations of $\mathbb{G}_3$

Definition. Let $\overline{W(2)}$ denote the linear space derived from the generalized quadrangle $W(2)$ by adding its six ovoids as extra lines.

Lemma 6.44. *A linear space $\mathcal{L}$ is isomorphic to $\overline{W(2)}$ if and only if each point of $\mathcal{L}$ is incident with exactly three lines of size* 3 *and two lines of size* 5.

Proof. One calculates that $\mathcal{L}$ has fifteen points, fifteen lines of size 3 and six lines of size 5. If L is a line of size 5 then precisely 20 other lines meet L in a point. Hence no line is disjoint from L. Let $\mathcal{L}'$ be the partial linear space obtained from $\mathcal{L}$ by removing all lines of size 5. We will show that $\mathcal{L}' \cong W(2)$. Obviously, $\mathcal{L}'$ has order $(2,2)$. Let (y, L) be a non-incident point-line pair in $\mathcal{L}'$. Because both lines of size 5 through x intersect L in a point, exactly one line of size 3 through x intersects L in a point. It follows that $\mathcal{L}'$ is the generalized quadrangle of order 2. Since every line of size 5 in $\mathcal{L}$ intersects every line of size 3 in exactly one point, every line of size 5 determines an ovoid in $\mathcal{L}$. This proves the lemma. □

We have shown above that there exists a distance-preserving embedding of $\mathbb{G}_3$ in $DH(5,4)$. So, every (classical) valuation of $DH(5,4)$ induces a valuation of $\mathbb{G}_3$. Since $\mathbb{G}_3$ has $Q(5,2)$-quads, there are no ovoids in $\mathbb{G}_3$.

Theorem 6.45. *If f is a nonclassical valuation of $\mathbb{G}_3$, then $G_f \cong \overline{W(2)}$ and f is induced by a (classical) valuation of $DH(5,4)$.*

Proof. Since f is not classical, $f(u) \in \{0,1,2\}$ for every point u of $\mathbb{E}_3$ by Theorem 5.9. Since $Q(5,2)$ has no ovoids, every $Q(5,2)$-quad contains a unique point of O_f. Since there are 45 $Q(5,2)$-quads and since every point is contained in three $Q(5,2)$-quads, $|O_f| = \frac{45}{3} = 15$. Now, choose an arbitrary point x of O_f and suppose that x is contained in α_x special grid-quads and β_x special $W(2)$-quads. Since f is not ovoidal, G_f is a linear space and we find that

$$2\alpha_x + 4\beta_x = 14.$$

Since no special quads intersect in a line,

$$2\alpha_x + 3\beta_x \leq 12.$$

Hence, $\beta_x \geq 2$. From the structure of the local spaces, see Theorem 6.31, it follows that the possibility $(\alpha_x, \beta_x) = (1, 3)$ is impossible. Hence, $(\alpha_x, \beta_x) = (3, 2)$. By Lemma 6.44, $G_f \cong \overline{W(2)}$.

Let Q be a special quad such that $x \notin Q$. Let Q' denote the unique $Q(5,2)$-quad of $DH(5,4)$ containing Q, let x' denote the unique point of Q' collinear with x, let g' denote the classical valuation of $DH(5,4)$ determined by $g'(x') = 0$ and let g denote the valuation of $\mathbb{G}_3$ induced by g'. Since $|O_f| = |O_g| = 15$, $G_f \cong G_g \cong \overline{W(2)}$ and $\{x\} \cup (\Gamma_2(x) \cap Q) \subseteq O_f \cap O_g$, $O_f = O_g$. Since every point y of $\mathbb{G}_3$ is contained in a $Q(5,2)$-quad, $\mathrm{d}(y, O_f) \leq 2$ and $f(y) = \mathrm{d}(y, O_f) = \mathrm{d}(y, O_g) = g(y)$. This proves the theorem. □

6.4 The class $\mathbb{I}_n$

This class of near polygons is due to Brouwer, Cohen, Hall and Wilbrink [12]. Consider a nonsingular quadric $Q(2n,2)$, $n \geq 2$, in $\mathrm{PG}(2n,2)$ and a hyperplane Π of $\mathrm{PG}(2n,2)$ intersecting $Q(2n,2)$ in a nonsingular hyperbolic quadric $Q^+(2n-1,2)$.

Theorem 6.46. *The generators of $Q(2n,2)$ not contained in $Q^+(2n-1,2)$ form a subspace X of $DQ(2n,2)$.*

Proof. Let π_1 and π_2 be two generators of $Q(2n,2)$ which are not contained in $Q^+(2n-1,2)$ intersecting each other in an $(n-2)$-dimensional subspace. The subspace $\Pi \cap \pi_1$ is contained in three generators of $Q(2n,2)$. One of these generators is π_1 and the other two are contained in $Q^+(2n-1,2)$. Hence, $\pi_1 \cap \pi_2 \neq \Pi \cap \pi_1$. Hence none of the generators through $\pi_1 \cap \pi_2$ is contained in $Q^+(2n-1,2)$. This proves the theorem. □

Let $\mathbb{I}_n$ denote the following incidence structure:

- the points of $\mathbb{I}_n$ are the maximal subspaces of $Q(2n,2)$ not contained in $Q^+(2n-1,2)$;
- the lines of $\mathbb{I}_n$ are the next-to-maximal subspaces of $Q(2n,2)$ not contained in $Q^+(2n-1,2)$;
- incidence is reverse containment.

The incidence structure $\mathbb{I}_n$ is a subgeometry of $DQ(2n,2)$. Every line of $\mathbb{I}_n$ is incident with three points. Let $\mathrm{d}(\pi_1, \pi_2)$ denote the distance between two points π_1 and π_2 of $DQ(2n,2)$.

Theorem 6.47. *The distance between two points π_1 and π_2 of $\mathbb{I}_n$ in the geometry $\mathbb{I}_n$ is equal to $d(\pi_1, \pi_2)$.*

Proof. We will prove this by induction on the distance $\mathrm{d}(\pi_1, \pi_2)$. Obviously, the theorem holds if $\mathrm{d}(\pi_1, \pi_2) \leq 1$. So, suppose $\mathrm{d}(\pi_1, \pi_2) \geq 2$. Obviously, the distance between π_1 and π_2 in $\mathbb{I}_n$ is at least $\mathrm{d}(\pi_1, \pi_2)$. Let x be a point of π_1 not contained in $\pi_2 \cup \Pi$ and let π_3 be the unique generator through x intersecting π_2 in an $(n-2)$-dimensional subspace. Then $\mathrm{d}(\pi_1, \pi_3) = \mathrm{d}(\pi_1, \pi_2) - 1$. The generator π_3 is a point of $\mathbb{I}_n$ and the distance between π_1 and π_3 in $\mathbb{I}_n$ is equal to $\mathrm{d}(\pi_1, \pi_3)$. Hence, the distance between π_1 and π_2 in $\mathbb{I}_3$ is at most (and hence precisely) $\mathrm{d}(\pi_1, \pi_2)$. This proves the theorem. □

Theorem 6.48. *$\mathbb{I}_n$ is a dense near $2n$-gon.*

Proof. Let (π, L) be a point-line pair of $\mathbb{I}_n$. Since π and L are also objects of the near polygon $DQ(2n, 2)$ and since distances in $\mathbb{I}_n$ are inherited from distances in $DQ(2n, 2)$, L contains a unique point nearest to π (in $\mathbb{I}_n$). This proves that $\mathbb{I}_n$ is a near polygon. By Theorem 6.47, the distance between two points of $\mathbb{I}_n$ is at most n. If π_1 and π_2 are two disjoint generators of $Q(2n, 2)$ not contained in $Q^+(2n-1, 2)$, then $\mathrm{d}(\pi_1, \pi_2) = n$, proving that $\mathbb{I}_n$ is a slim near $2n$-gon. Now, take two points π_1 and π_2 at distance 2 from each other. So, $\dim(\pi_1 \cap \pi_2) = n - 2$. Let x be a point of π_1 not contained in $\pi_2 \cup \Pi$, let π_3 be the unique generator through x intersecting π_2 in an $(n-2)$-dimensional subspace, let y be a point of π_2 not contained in $\pi_3 \cup \Pi$ and let π_4 be the unique generator through y intersecting π_1 in an $(n-2)$-dimensional subspace. Then π_3 and π_4 are two common neighbours of π_1 and π_2. This proves that $\mathbb{I}_n$ is dense. □

Theorem 6.49. *Let α be an $(n-1-i)$-dimensional subspace of $Q(2n, 2)$ which is not contained in $Q^+(2n-1, 2)$ if $i \in \{0, 1\}$. Then the set F_α of all generators through α not contained in $Q^+(2n-1, 2)$ defines a convex subpolygon of $\mathbb{I}_n$. If $i \geq 2$ and α is not contained in $Q^+(2n-1, 2)$, then $F_\alpha \cong DQ(2i, 2)$. If $i \geq 2$ and α is contained in $Q^+(2n-1, 2)$, then $F_\alpha \cong \mathbb{I}_i$.*

Proof. Let π_1 and π_2 be two different collinear points of $\mathbb{I}_n$ belonging to F_α and let π_3 denote the unique third point on the line through π_1 and π_2. Since $\alpha \subseteq \pi_1 \cap \pi_2$, $\alpha \subseteq \pi_3$. Hence, π_3 is also a point of F_α and F_α is a subspace.

Let π_1 and π_2 be two arbitrary points of F_α. Every point on a shortest path in $DQ(2n, 2)$ between π_1 and π_2 contains α. Hence, every point on a shortest path in $\mathbb{I}_n$ between π_1 and π_2 contains α. This proves that F_α is convex.

If $i \geq 2$, then the subspaces through α define a polar space of type $Q(2i, 2)$. If α is contained in $Q^+(2n-1, 2)$, then the subspaces through α contained in $Q^+(2n-1, 2)$ define a polar space of type $Q^+(2i-1, 2)$. The theorem now readily follows. □

Theorem 6.50. *If π_1 and π_2 are two points of $\mathbb{I}_n$, then $F_{\pi_1 \cap \pi_2}$ is the unique convex sub-$[2 \cdot d(\pi_1, \pi_2)]$-gon of $\mathbb{I}_n$ through π_1 and π_2.*

Proof. This follows from the fact that $\mathcal{C}(\pi_1, \pi_2)$ and $F_{\pi_1 \cap \pi_2}$ are two convex sub-$[2 \cdot \mathrm{d}(\pi_1, \pi_2)]$-gons through π_1 and π_2. □

Theorem 6.51. *Every local space of $\mathbb{I}_n$ is isomorphic to $\mathrm{PG}(n-1, 2)'$, the linear space derived from the point-line system of $\mathrm{PG}(n-1, 2)$ by deleting a point.*

Proof. A point π of $\mathbb{I}_n$ is an $(n-1)$-dimensional space of $Q(2n,2)$. The lines through π correspond with the hyperplanes of π different from $\pi \cap \Pi$ and the quads through π correspond with the $(n-3)$-dimensional subspaces of π. This proves the theorem. □

Theorem 6.52. *Let $n \geq 3$ and let F be a convex sub-$2(n-1)$-gon of $\mathbb{I}_n$. If $F \cong DQ(2n-2,2)$, then F is big in $\mathbb{I}_n$. If $F \cong \mathbb{I}_{n-1}$, then F is not big in $\mathbb{I}_n$.*

Proof. Suppose $F \cong DQ(2n-2,2)$. Then F consists of all generators through a point p of $Q(2n,2)$ not contained in Π. Let π denote an arbitrary point of $\mathbb{I}_n$ not contained in F. Let π' denote the unique generator through p intersecting π in an $(n-2)$-dimensional subspace. Then π' is the unique point of F collinear with π. This proves that F is big.

Suppose that $F \cong \mathbb{I}_{n-1}$. Then F consists of all generators through a point p of $Q^+(2n-1,2)$ which are not contained in $Q^+(2n-1,2)$. Let α be a generator of $Q^+(2n-1,2)$ through p, let β be an $(n-2)$-dimensional subspace of α not containing p and let γ denote the unique generator through β which is not contained in $Q^+(2n-1,2)$. Obviously, $\mathrm{d}(\gamma, F) \geq 2$, proving that F is not big. □

Theorem 6.53. *The near polygon $\mathbb{I}_n$ does not contain spreads of symmetry if $n \geq 3$.*

Proof. Similarly as in the proof of Theorem 6.3, this follows from the fact that every line of $\mathbb{I}_n$ is contained in a $W(2)$-quad. □

We will now determine all valuations of $\mathbb{I}_3$. The above-mentioned embedding of $\mathbb{I}_3$ in $DQ(6,2)$ is distance-preserving. Hence by Theorem 5.7, every valuation f of $DQ(6,2)$ induces a valuation $\tilde{f}$ of $\mathbb{I}_3$. In this way, we obtain at least four classes of valuations of $\mathbb{I}_3$.

- If f is a classical valuation of $DQ(6,2)$ such that the unique point x with value 0 belongs to $\mathbb{I}_3$, then $\tilde{f}$ is also a classical valuation of $\mathbb{I}_3$.
- If f is a classical valuation of $DQ(6,2)$ such that the unique point x with value 0 does not belong to $\mathbb{I}_3$, then $\tilde{f}$ is not a classical valuation. Every line through x corresponds with a line of $Q(6,2)$ contained in $Q^+(5,2)$. Through this line there are three generators, two are also generators of $Q^+(5,2)$ and one is not contained in $Q^+(5,2)$. This proves that every line through x meets $\mathbb{I}_3$ in a unique point. So, $|O_{\tilde{f}}| = 7$. Also, $\tilde{f}$ admits only special grid-quads. Hence, $G_{\tilde{f}} \cong \mathrm{PG}(2,2)$.
- If f is the extension of an ovoidal valuation in a $W(2)$-quad of $DQ(6,2)$ which is also contained in $\mathbb{I}_3$, then $\tilde{f}$ is the extension of an ovoidal valuation of Q (in the near hexagon $\mathbb{I}_3$). We have $O_{\tilde{f}} = O_f$.
- If f is the extension of an ovoidal valuation of a $W(2)$-quad Q of $DQ(6,2)$ which is not contained in $\mathbb{I}_3$, then $O_{\tilde{f}}$ is an ovoid in the grid-quad $Q \cap \mathbb{I}_3$.

Theorem 6.54. *For every point x of $\mathbb{I}_3$, there exists a path of length 7 in $\Gamma_3(x)$. As a consequence, $\mathbb{I}_3$ has no semi-classical valuations.*

Proof. Let $x_1, x_2 \in \Gamma_1(x)$ such that $\mathcal{C}(x_1, x_2) \cong W(2)$. Let L_1 and L_2 be lines of $\mathcal{C}(x_1, x_2)$ such that $L_1 \cap L_2 = \emptyset$ and $x_i \in L_i$, $L_i \neq xx_i$ for every $i \in \{1, 2\}$. Let Q_i, $i \in \{1, 2\}$, denote the unique $W(2)$-quad through L_i different from $\mathcal{C}(x_1, x_2)$. Then Q_1 and Q_2 are disjoint big quads of $\mathbb{I}_3$ and we can follow a similar reasoning as in the proof of Theorem 6.4. □

By Lemma 6.6, $\mathbb{I}_3$ does not have ovoids and hence also no ovoidal valuations.

Theorem 6.55. *Let f be a nonclassical valuation of $\mathbb{I}_3$. Then f is induced by a valuation of $DQ(6, 2)$.*

Proof. If $|O_f| = 1$, then f is a classical valuation by Theorems 5.33 and 6.54. So, suppose that $|O_f| \geq 2$. Since f is not ovoidal, every two points of O_f lie at distance 2 from each other by Theorem 5.34. We distinguish the following possibilities.

- There exists a special $W(2)$-quad Q. For every point outside Q, there exists a point of $O_f \cap Q$ at distance 3 from that point. Hence, $O_f = O_f \cap Q$. For every point x of $\mathbb{I}_3$, we have $f(x) = \mathrm{d}(x, O_f) = \mathrm{d}(x, \pi_Q(x)) + \mathrm{d}(\pi_Q(x), O_f)$ by Theorem 5.12. So, f is the extension of an ovoidal valuation in a $W(2)$-quad.

- $G_f \cong \mathbb{L}_3$. Let Q denote the unique special grid-quad. By the discussion before Theorem 6.54, there exists a unique valuation f' such that (i) $O_{f'} = O_f$ and (ii) f' is induced by a valuation of $DQ(6, 2)$. Let m_i, $i \in \{0, 1, 2\}$, denote the total number of points with f-value i. By Theorem 5.13, $m_0 - \frac{m_1}{2} + \frac{m_2}{4} = 0$. It follows from $m_0 = 3$ and $m_0 + m_1 + m_2 = 105$ that $m_0 = 3$, $m_1 = 38$ and $m_2 = 64$. By Theorem 5.12, f is completely determined by O_f: the 64 points with value 2 are on the one hand the 48 points in $\Gamma_1(Q)$ which are not collinear with a point of O_f and the 16 points of $\Gamma_2(Q)$ which lie at distance 2 from at least one point of O_f. Since $O_f = O_{f'}$, $f = f'$ and f is induced by a valuation of $DQ(6, 2)$.

- G_f has only lines of size 3 and contains a point x and a line L such that $x \notin L$. There exist no points with value 3 by Theorem 5.9. So, every $W(2)$-quad contains a unique point of O_f. It follows that $|O_f| = 7$ and $G_f \cong \mathrm{PG}(2, 2)$. Let Q denote the grid-quad corresponding with L. Let Q' denote the unique $W(2)$-quad of $DQ(6, 2)$ containing Q, let x' denote the unique point of Q' collinear with x, let g' denote the classical valuation of $DQ(6, 2)$ determined by $g'(x') = 0$ and let g denote the valuation of $\mathbb{I}_3$ induced by g. Since $|O_f| = |O_g| = 7$, $G_f \cong G_g \cong \mathrm{PG}(2, 2)$ and $\{x\} \cup (\Gamma_2(x) \cap Q) \subseteq O_f \cap O_g$, we must have $O_f = O_g$. Since every point of $\mathbb{I}_3$ is contained in a $W(2)$-quad which contains a unique point of O_f, $\mathrm{d}(x, O_f) \leq 2$ and $f(x) = \mathrm{d}(x, O_f) = \mathrm{d}(x, O_g) = g(x)$. This proves the theorem. □

6.5 The near hexagon $\mathbb{E}_1$

The near hexagon $\mathbb{E}_1$ is due to Shult and Yanushka [91].

6.5.1 Description of $\mathbb{E}_1$ in terms of the extended ternary Golay code

Let $\mathbb{F}_3^{12}$ denote the 12-th dimensional vector space over the finite field $\mathbb{F}_3$. Let $B = \{e_1, \ldots, e_{12}\}$ denote a base of $\mathbb{F}_3^{12}$. Let M denote the following matrix with entries M_{ij}, $1 \leq i \leq 6$ and $1 \leq j \leq 12$, belonging to $\mathbb{F}_3$.

$$\begin{bmatrix} 1 & 0 & 0 & 0 & 0 & 0 & 1 & 1 & 1 & 1 & 1 & 0 \\ 0 & 1 & 0 & 0 & 0 & 0 & 0 & 1 & -1 & -1 & 1 & -1 \\ 0 & 0 & 1 & 0 & 0 & 0 & 1 & 0 & 1 & -1 & -1 & -1 \\ 0 & 0 & 0 & 1 & 0 & 0 & -1 & 1 & 0 & 1 & -1 & -1 \\ 0 & 0 & 0 & 0 & 1 & 0 & -1 & -1 & 1 & 0 & 1 & -1 \\ 0 & 0 & 0 & 0 & 0 & 1 & 1 & -1 & -1 & 1 & 0 & -1 \end{bmatrix}.$$

Put $v_i := \sum_{j=1}^{12} M_{ij} e_j$ for every $i \in \{1, \ldots, 6\}$. We call the subspace C of $\mathbb{F}_3^{12}$ generated by the vectors v_i, $i \in \{1, \ldots, 6\}$, the *extended ternary Golay code* ([22]).

For all vectors $x = \sum x_i e_i$ and $y = \sum y_i e_i$, we define

$$\langle x, y \rangle := \sum x_i y_i.$$

If $\langle x, y \rangle = 0$, then we say that x is *orthogonal* with y.

Lemma 6.56. *Every two vectors of C are orthogonal. If x is a vector of $\mathbb{F}_3^{12}$ which is orthogonal with every vector of C, then $x \in C$.*

Proof. One easily verifies that $\langle v_i, v_j \rangle = 0$ for all $i, j \in \{1, \ldots, 6\}$. It then follows that every vector of C is orthogonal with every vector of C. The six equations $\langle x, v_i \rangle = 0$, $i \in \{1, \ldots, 6\}$, determine a 6-dimensional subspace of $\mathbb{F}_3^{12}$ which contains C and hence is equal to C. □

Lemma 6.57. *Every vector of C has weight* 0, 6, 9 *or* 12.

Proof. Let x be an arbitrary nonzero vector of C. Since x is orthogonal with itself, its weight is equal to either 0, 3, 6, 9 or 12. One easily sees that the following holds.

- Every vector of the form αv_i with $\alpha \neq 0$ and $i \in \{1, \ldots, 6\}$ has weight 6.
- Every vector of the form $\alpha v_i + \beta v_j$ with $\alpha\beta \neq 0$ and $1 \leq i < j \leq 6$ has weight at least 4 and hence weight at least 6.
- Every vector of the form $\alpha v_i + \beta v_j + \gamma v_k$ with $\alpha\beta\gamma \neq 0$ and $1 \leq i < j < k \leq 6$ has weight at least 4 and hence weight at least 6.
- Every vector of the form $\alpha_1 v_1 + \cdots + \alpha_6 v_6$ where at least four of the α_i's are different from 0 has weight at least 4 and hence weight at least 6.

The lemma now easily follows. □

Remark. If all nonzero vectors of a 6-dimensional subspace of $\mathbb{F}_3^{12}$ have weight at least 6, then by [66] and [83], this subspace is isomorphic to the extended ternary Golay code.

Lemma 6.58. *Every five columns of M are linearly independent.*

Proof. Suppose the contrary. Let $w = \sum w_i e_i$ be a nonzero vector of weight at most 5 such that $M \cdot [w_1, \ldots, w_{12}]^T = 0$. By Lemma 6.56, $w \in C$. But this contradicts Lemma 6.57. □

Now, let $\mathbb{E}_1$ be the following incidence structure:

- the points of $\mathbb{E}_1$ are all the cosets $v + C$, $v \in \mathbb{F}_3^{12}$;
- the lines of $\mathbb{E}_1$ are all the sets of the form $\{v + C, v + e_i + C, v - e_i + C\}$;
- incidence is containment.

The incidence structure $\mathbb{E}_1$ has 729 points. Every line of $\mathbb{E}_1$ is incident with three points and every point is incident with 12 lines. So, $\mathbb{E}_1$ has 2916 lines.

Theorem 6.59. *$\mathbb{E}_1$ is a regular near hexagon with parameters $(s, t_2, t) = (2, 1, 11)$. Every local space of $\mathbb{E}_1$ is isomorphic to the complete graph K_{12} on 12 vertices.*

Proof. If $v + C$ is a point of $\mathbb{E}_1$, then the points of $\mathbb{E}_1$ at distance 1, 2, respectively 3, from $v + C$ are the points of the form $v \pm e_i + C$, $v \pm e_i \pm e_j + C$, respectively $v \pm e_i \pm e_j \pm e_k + C$. By Lemma 6.58, $|\Gamma_1(v + C)| = 24$ and $|\Gamma_2(v + C)| = 264$. Moreover, every two points at distance 2 have precisely two common neighbours. If x is a point of $\mathbb{E}_1$ and if L is a line of $\mathbb{E}_1$ such that $\mathrm{d}(x, L) \leq 2$, then by Lemma 6.58, L contains a unique point nearest to x. Hence, for a given point x there are 12 lines through x, $24 \cdot 11 = 264$ lines at distance 1 from x and $264 \cdot 10$ lines at distance 2 from x. In this way we obtain $12 + 264 + 2640 = 2916$ lines. Since these are all the lines, $\mathbb{E}_1$ is a near hexagon. The theorem now readily follows. □

Theorem 6.60. *Every regular near hexagon with parameters $(s, t_2, t) = (2, 1, 11)$ is isomorphic to $\mathbb{E}_1$.*

Proof. This was proved in [7] using the results of [91]. □

6.5.2 Description of $\mathbb{E}_1$ in terms of the Coxeter cap

The 12 columns of the matrix M define a set $\mathcal{K}$ of 12 points in $\mathrm{PG}(5, 3)$. This set of 12 points, which was discovered by Coxeter [28], has several nice properties. For every point x of $\mathrm{PG}(5, 3)$, define the *generating index* $i_{\mathcal{K}}(x)$ of x as the minimal number of points of $\mathcal{K}$ which are necessary to generate a subspace containing x.

Lemma 6.61. (a) *Every point has index at most* 3.

(b) *If L is a line of $\mathrm{PG}(5, 3)$ through a point x of $\mathcal{K}$, then $L \setminus \{x\}$ contains a unique point with smallest index.*

Proof. Let x be a given point of PG(5, 3). By Lemma 6.58, there are $|\mathcal{K}\setminus\{x\}| = 11$ lines through x containing besides x a necessarily unique point with index 1, and $2 \cdot \binom{|\mathcal{K}\setminus\{x\}|}{2} = 110$ lines through x containing besides x a necessarily unique point with index 2 but no point with index 1. The lemma now follows from the fact that there are precisely 121 lines through x. □

Now, embed PG(5, 3) as a hyperplane Π_∞ in a 6-dimensional projective space Π. Then we can define the following geometry $T_5^*(\mathcal{K})$:

- the points of $T_5^*(\mathcal{K})$ are the affine points of Π, i.e. the points of Π not contained in Π_∞;
- the lines of $T_5^*(\mathcal{K})$ are the lines of Π not contained in Π_∞ and intersecting Π_∞ in a point of $\mathcal{K}$;
- incidence is derived from Π.

Theorem 6.62. (a) *Let x and y be two different points of $T_5^*(\mathcal{K})$ and let z be the intersection point of the line xy with the hyperplane Π_∞. Then $\mathrm{d}(x,y) = i_\mathcal{K}(z)$.*

(b) *$T_5^*(\mathcal{K})$ is a regular near hexagon with parameters $(s, t_2, t) = (2, 1, 11)$.*

Proof. (a) Obviously, $\mathrm{d}(x,y) = 1$ if and only if $z \in \mathcal{K}$, i.e. if and only if $i_\mathcal{K}(z) = 1$.

If $i_\mathcal{K}(z) = 2$, then there exists a unique line L_z through z containing two different points $z_1, z_2 \in \mathcal{K}$. Then the points $xz_1 \cap yz_2$ and $xz_2 \cap yz_1$ are common neighbours of x and y, proving that $\mathrm{d}(x,y) = 2$.

If $\mathrm{d}(x,y) = 2$ and if u is a common neighbour of x and y, then the points $xy \cap \Pi_\infty$, $xu \cap \Pi_\infty$ and $yu \cap \Pi_\infty$ lie on a line. So, $i_\mathcal{K}(z) = 2$. Since there exists a unique line L_z through z containing two different points $z_1, z_2 \in \mathcal{K}$ (Lemma 6.58), we see that $xz_1 \cap yz_2$ and $xz_2 \cap yz_1$ are the only common neighbours of z_1 and z_2.

If $i_\mathcal{K}(z) = 3$, then there exist points z_1, z_2 and z_3 such that $z \in \langle z_1, z_2, z_3\rangle$. Put $\{z'\} = zz_3 \cap z_1z_2$ and $\{u\} = z'x \cap z_3y$. Since $i_\mathcal{K}(z') = 2$, $\mathrm{d}(x,u) = 2$. Since $i_\mathcal{K}(z_3) = 1$, $\mathrm{d}(u,y) = 1$. This proves that $\mathrm{d}(x,y) = 3$.

(b) Let x denote an arbitrary point of $T_5^*(\mathcal{K})$ and let L be an arbitrary line of $T_5^*(\mathcal{K})$. Put $L \cap \Pi_\infty = \{z\}$. If $x \in L$, then obviously L contains a unique point nearest to x. Suppose therefore that $x \notin L$. Then the plane $\langle x, L\rangle$ intersects Π_∞ in a line L' through z. By Lemma 6.61, $L' \setminus \{z\}$ contains a unique point z' with smallest index. The point $z'x \cap L$ is the unique point of L nearest to x. This proves that $T_5^*(\mathcal{K})$ is a near polygon. By (a) and Lemma 6.61, $T_5^*(\mathcal{K})$ is a regular near hexagon with parameters $(s, t_2, t) = (2, 1, 11)$. □

Theorem 6.63. *Let L be a line of Π_∞ intersecting $\mathcal{K}$ in two points and let α be a plane through L not contained in Π_∞. Then the points of α not contained in L form a quad of $T_5^*(\mathcal{K})$. Conversely, every quad of $T_5^*(\mathcal{K})$ is obtained in this way.*

Proof. By Theorem 6.62, we know that every quad is a (3×3)-grid. Obviously, the points in α determine a subgrid of $T_5^*(\mathcal{K})$ which is necessarily a quad. Conversely, suppose that x and y are two points at distance 2 from each other. Let Q denote the unique quad through x and y and let z denote the unique point on the line xy contained in Π_∞. Through z there exists a unique line containing two different points z_1 and z_2 of $\mathcal{K}$. Now, the plane $\langle z_1, z_2, x\rangle$ determines a quad through x and y and this quad necessarily coincides with Q. □

Theorem 6.64. *Let α be a hyperplane of Π intersecting Π_∞ in a four-dimensional subspace disjoint with $\mathcal{K}$. Then the affine points of α define an ovoid of $\mathcal{K}$. Conversely, every ovoid is obtained in this way.*

Proof. Every line of Π_∞ intersects α in a unique point, proving that the affine points of α define an ovoid. Conversely, let O be an ovoid of $T_5^*(\mathcal{K})$. In order to show that O comes from a hyperplane it suffices to show that for every two different points x_1 and x_2 of O, also the third affine point x_3 on the line x_1x_2 belongs to O. We distinguish two possibilities. In both cases, let z be the intersection point of the line x_1x_2 and the hyperplane Π_∞.

- $\mathrm{d}(x_1, x_2) = 2$. Let Q denote the unique quad through x_1 and x_2. By Theorem 6.63, x_3 also belongs to Q. Now, $O\cap Q$ is the unique ovoid of Q through the points x_1 and x_2. Since $i_{\mathcal{K}}(z) = 2$, x_1, x_2 and x_3 are three points of Q at mutual distance 2. It follows that $x_3 \in O$.
- $\mathrm{d}(x_1, x_2) = 3$. Since $i_{\mathcal{K}}(z) = 3$, there exist three points z_1, z_2 and z_3 such that $z \in \langle z_1, z_2, z_3\rangle$. By Lemma 6.58, z_1, z_2 and z_3 are the only points of $\mathcal{K}$ contained in $\langle z_1, z_2, z_3\rangle$. Consider now the three-dimensional subspace $\langle z_1, z_2, z_3, x, y\rangle$ and let X denote the set of affine points contained in this subspace. By looking at the points, lines and of $T_5^*(\mathcal{K})$ which are contained in X, we see that the subspace X induces a near hexagon H isomorphic to $\mathbb{L}_3 \times \mathbb{L}_3 \times \mathbb{L}_3$. The subhexagon H is not necessarily convex. The set $O\cap H$ is the unique ovoid of H through the points x_1 and x_2. Since $i_{\mathcal{K}}(z) = 3$, x_1, x_2 and x_3 form a set of points of H at mutual distance 3. It follows that $x_3 \in O \cap H$. □

Remark. There are 12 hyperplanes in Π_∞ disjoint from $\mathcal{K}$. These hyperplanes form a Coxeter cap in the dual projective space Π_∞^*.

Theorem 6.65. *Let x be be a point of $\mathcal{K}$ and let S_x denote the set of lines of Π through x not contained in Π_∞. Then S_x is a spread of symmetry of $T_5^*(\mathcal{K})$. Conversely, every spread of symmetry of $T_5^*(\mathcal{K})$ is obtained in this way.*

Proof. Obviously, S_x is a spread of $T_5^*(\mathcal{K})$. Let G denote the group of elations with center x and axis Π_∞. Then G induces a group of automorphisms of $T_5^*(\mathcal{K})$ acting transitively on every line of S_x. Hence, S_x is a spread of symmetry of $T_5^*(\mathcal{K})$. Conversely, let S be a spread of symmetry of $T_5^*(\mathcal{K})$. For every line L of S, let x_L denote the unique point of $\mathcal{K}$ contained in L. We need to show that $x_K = x_L$ for every two lines K and L of S. By connectedness of $T_5^*(\mathcal{K})$, we may suppose that $\mathrm{d}(K,L)=1$. Since S is a spread of symmetry, it is also a regular spread. So, K and L are contained in a subgrid which is also a quad. By Theorem 6.63, it then follows that $x_K = x_L$. □

Corollary 6.66. *The near hexagon $\mathbb{E}_1$ has, up to isomorphism, only one spread of symmetry.*

Proof. By [28], the automorphism group of Π_∞ fixing the set $\mathcal{K}$ acts 5-transitively on the set of points of $\mathcal{K}$ (see also [99]). The result now readily follows. □

Theorem 6.67. *If x_1 and x_2 are two points of $\mathcal{K}$, then the spreads of symmetry S_{x_1} and S_{x_2} are compatible.*

Proof. Let G_i, $i \in \{1,2\}$, denote the group of elations of Π with center x_i and axis Π_∞. G_i induces a group $\tilde{G}_i$ of automorphisms of $T_5^*(\mathcal{K})$. By Theorem 4.43, $\tilde{G}_i$ is the whole group of automorphisms fixing each line of S_{x_i}. The theorem now follows from the fact that any element of G_1 commutes with every element of G_2. □

Remark. The spreads of symmetry of $\mathbb{E}_1$ can easily be recognized in the model of $\mathbb{E}_1$ which makes use of the extended ternary Golay code $C \subset \mathbb{F}_3^{12}$. For every $i \in \{1,\ldots,12\}$, let S_i denote the set of all lines of the form $\{v+C, v+e_i+C, v-e_i+C\}$. Obviously, S_i is a spread of $\mathbb{E}_1$. The three translations $v \mapsto v+\lambda e_i$, $\lambda \in \mathbb{F}_3$, determine a group of automorphisms of $\mathbb{E}_1$ fixing each line of S_i and acting regularly on each line of S_i. So, $S_1,\ldots,S_{12}$ are the twelve spreads of symmetry of $\mathbb{E}_1$.

Theorem 6.68. *For every point x of $\mathbb{E}_3$, there exists a closed path of length 7 in $\Gamma_3(x)$. As a consequence, $\mathbb{E}_3$ has no semiclassical valuations.*

Proof. Take a 3-dimensional subspace α of Π_∞ for which $|\alpha \cap \mathcal{K}| = 4$. By [28], the automorphism group of Π_∞ fixing the set $\mathcal{K}$ acts 5-transitively on the set of points of $\mathcal{K}$. So, we may suppose that α is generated by the first four columns of the matrix M. Let β be a plane generated by three points of $\alpha \cap \mathcal{K}$. Let $\gamma \notin \{\alpha, \langle x,\beta\rangle\}$ be a 3-dimensional subspace of $\langle x,\alpha\rangle$ through β. The points of $\gamma \setminus \Pi_\infty$ determine a subhexagon C of $T_5^*(\mathcal{K})$ which is isomorphic to $\mathbb{L}_3 \times \mathbb{L}_3 \times \mathbb{L}_3$. The points of C can be labeled with the triples (i,j,k), $i,j,k \in \{-1,0,1\}$, such that $(i,j,k) \sim (i',j',k')$ if and only if these triples agree in exactly two positions. If $(0,0,0)$ is the unique point of $\gamma \setminus \Pi_\infty$ collinear with x, then $\mathrm{d}(x,(0,0,0)) = 1$ and

$$\mathrm{d}(x,(1,0,0)) = \mathrm{d}(x,(-1,0,0)) = \mathrm{d}(x,(0,1,0)) = 2,$$

$$\mathrm{d}(x,(0,-1,0)) = \mathrm{d}(x,(0,0,1)) = \mathrm{d}(x,(0,0,-1)) = 2,$$

$$\mathrm{d}(x,(1,1,0)) = \mathrm{d}(x,(1,-1,0)) = \mathrm{d}(x,(-1,1,0)) = \mathrm{d}(x,(-1,-1,0)) = 3,$$

$$\mathrm{d}(x,(1,0,1)) = \mathrm{d}(x,(1,0,-1)) = \mathrm{d}(x,(-1,0,1)) = \mathrm{d}(x,(-1,0,-1)) = 3,$$

$$\mathrm{d}(x,(0,1,1)) = \mathrm{d}(x,(0,1,-1)) = \mathrm{d}(x,(0,-1,1)) = \mathrm{d}(x,(0,-1,-1)) = 3.$$

We may suppose that $\mathrm{d}(x,(1,1,1)) = 3$. Then

$$\mathrm{d}(x,(1,1,-1)) = \mathrm{d}(x,(-1,1,1)) = \mathrm{d}(x,(1,-1,1)) = \mathrm{d}(x,(-1,-1,-1)) = 2$$

and

$$\mathrm{d}(x,(-1,-1,1)) = \mathrm{d}(x,(-1,1,-1)) = \mathrm{d}(x,(1,-1,-1)) = 3.$$

The closed path

$$(0,-1,-1),(1,-1,-1),(1,0,-1),(1,0,1),(1,1,1),(0,1,1),(0,-1,1),(0,-1,-1)$$

has length 7 and is completely contained in $\Gamma_3(x)$. □

6.5.3 The valuations of $\mathbb{E}_1$

In this paragraph, we will show the following.

Theorem 6.69. *Every valuation of $\mathbb{E}_1$ is either classical or ovoidal.*

We take the proof of [63]. Let f be a valuation of $\mathbb{E}_1$. There are three possibilities.

- The maximal value attained by f is equal to 3. Then f is a classical valuation by Theorem 5.9.
- The maximal value attained by f is equal to 1. Then f is an ovoidal valuation by Theorem 5.9.
- The maximal value attained by f is equal to 2. We will prove that this case cannot occur.

Suppose that the maximal value attained by f is equal to 2. Let O_f denote the set of points of $\mathbb{E}_1$ with value 0. By Theorem 5.34, every two different points of O_f lie at distance 2 from each other.

Let θ denote an isomorphism between $\mathbb{E}_1$ and $T_5^*(\mathcal{K})$. So, θ is a bijection between the point sets of $\mathbb{E}_1$ and $\mathrm{AG}(6,3) := \Pi\setminus\Pi_\infty$ such that $\mathrm{d}(x,y) = i_{\mathcal{K}}[\theta(x)\theta(y)\cap\Pi_\infty]$ for all points x and y in $\mathbb{E}_1$ with $x \neq y$.

Lemma 6.70. *The set $\theta(O_f)$ is a proper subspace of* $\mathrm{AG}(6,3)$. *As a consequence,* $|O_f| \in \{1,3,9,27,81,243\}$.

Proof. Let $\theta(x_1)$ and $\theta(x_2)$ denote two arbitrary different points of $\theta(O_f)$. Then the quad $\mathcal{C}(\theta(x_1),\theta(x_2))$ of $T_5^*(\mathcal{K})$ consists of all affine points in a plane α of Π and the intersection line $\alpha\cap\Pi_\infty$ contains precisely two points of $\mathcal{K}$. The third affine point of the line $\theta(x_1)\theta(x_2)$ is not $T_5^*(\mathcal{K})$-collinear with $\theta(x_1)$ and $\theta(x_2)$ and hence equals $\theta(x_3)$, where x_3 is the unique point of $\mathcal{C}(x_1,x_2)$ not collinear with x_1 and x_2. Since $|O_f\cap\mathcal{C}(x_1,x_2)| \geq 2$, $O_f\cap\mathcal{C}(x_1,x_2)$ is an ovoid of $\mathcal{C}(x_1,x_2)$ by Theorem 5.7 and Corollary 5.10. Hence, $x_3 \in O_f$ and $\theta(x_3) \in \theta(O_f)$. This proves that $\theta(O_f)$ is a subspace of $\mathrm{AG}(6,3)$. Since no two points of O_f are collinear, $\theta(O_f) \neq \mathrm{AG}(6,3)$. It follows that $|O_f| \in \{1,3,9,27,81,243\}$. □

Lemma 6.71. $|O_f| \leq 13$.

Proof. Let $x \in O_f$ and let V denote the set of all quads of the form $\mathcal{C}(x, z)$ where $z \in O_f \setminus \{x\}$. Suppose that $|O_f| \geq 14$. Then $|V| \geq \frac{14-1}{2}$ or $|V| \geq 7$. Since there are precisely 12 lines through x, there exists a line L through x which is contained in two quads Q_1 and Q_2 of V. Let u denote an arbitrary point of $L \setminus \{x\}$, let u_1 denote the unique point of $(Q_1 \cap O_f) \setminus \{x\}$ collinear with u and let u_2 denote the unique point of $Q_2 \cap O_f$ not collinear with u. Then $\mathrm{d}(u_1, u_2) = \mathrm{d}(u_1, u) + \mathrm{d}(u, u_2) = 3$, contradicting Theorem 5.34. □

Lemma 6.72. *The case* $|O_f| = 1$ *cannot occur.*

Proof. Suppose that $O_f = \{x\}$. By Theorem 5.12, $f(x) = \mathrm{d}(x, y)$ for every point y at distance at most 2 from x. Since the maximal value attained by f is equal to 2, it follows by property (V_2) that every point of $\Gamma_3(x)$ has value 1 or 2 and that the subgraph induced by $\Gamma_3(x)$ must be bipartite. But in Theorem 6.68, we have shown that there exists a path of length 7 in $\Gamma_3(x)$. Hence it is impossible that $|O_f| = 1$. □

Lemma 6.73. *The case* $|O_f| = 3$ *cannot occur.*

Proof. If $|O_f| = 3$, then O_f is an ovoid in a quad Q of $\mathbb{E}_1$. Let O_1 be an ovoid of Q which intersects O_f in a unique point, let x_1 be a point of $\mathbb{E}_1$ at distance 2 from Q such that $\Gamma_2(x_1) \cap Q = O_1$, let $L = \{x_1, x_2, x_3\}$ be a line through x_1 completely contained in $\Gamma_2(Q)$ and let O_i, $i \in \{2, 3\}$, denote the ovoid $\Gamma_2(x_i) \cap Q$ of Q. Then $\{O_1, O_2, O_3\}$ is a partition of Q in ovoids. Hence, $|O_i \cap O_f| = 1$ for every $i \in \{1, 2, 3\}$. It follows that the point x_i has distance 2 from O_f. By Theorem 5.12, it now follows that every point of L has value 2, contradicting property (V_2). □

By Lemmas 6.70, 6.71, 6.72 and 6.73, it follows that $|O_f| = 9$. Let T denote the set of all quads of the form $\mathcal{C}(x_1, x_2)$ where x_1 and x_2 are two points of O_f. Since $\theta(O_f)$ is a subplane of $\mathrm{AG}(6, 3)$, the incidence structure G_f defined by O_f and T is an affine plane of order 3. So, $|T| = 12$.

Lemma 6.74. $\Gamma_3(O_f) = \emptyset$.

Proof. Suppose that u is a point of $\Gamma_3(O_f)$, then u has distance 3 to every point of O_f and distance 2 to every quad of T. There are precisely $|T| \cdot 6 = 72$ pairs (Q, L) with Q a quad of T and L a line through u contained in $\Gamma_2(Q)$. For every line L through u, let $T(L)$ denote the set of all quads $Q \in T$ for which $L \subset \Gamma_2(Q)$. Since u is contained in precisely 12 lines, there exists a line $L = \{u, u_1, u_2\}$ through u for which $|T(L)| \geq 6$. For every $x \in O_f$, L contains a unique point nearest to x of O_f and hence $\{\mathrm{d}(x, u_1), \mathrm{d}(x, u_2)\} = \{2, 3\}$.

Let Q denote a quad of T and put $Q \cap O_f = \{x_1, x_2, x_3\}$. If $Q \in T(L)$, then the points u, u_1 and u_2 determine three mutually disjoint ovoids O_u, O_{u_1} and O_{u_2} of Q. Since O_u is disjoint from the ovoid $\{x_1, x_2, x_3\}$, either O_{u_1} or O_{u_2} coincides with $\{x_1, x_2, x_3\}$. In any case, we have $\mathrm{d}(u_1, x_1) = \mathrm{d}(u_1, x_2) = \mathrm{d}(u_1, x_3)$

and $d(u_2, x_1) = d(u_2, x_2) = d(u_2, x_3)$. If $Q \notin T(L)$, then $u_i \in \Gamma_1(Q)$ for a certain $i \in \{1, 2\}$. Since (u_i, Q) is classical, $|\{d(u_i, x_1), d(u_i, x_2), d(u_i, x_3)\}| = 2$ and hence also $|\{d(u_{3-i}, x_1), d(u_{3-i}, x_2), d(u_{3-i}, x_3)\}| = 2$.

Now, consider the partial linear space with point set O_f and line set $T(L)$ (natural incidence). Since $|T(L)| \geq 6$, we have two possibilities.

- The partial linear space is connected. Then $d(x, u_i)$, $i \in \{1, 2\}$, is independent of the chosen point $x \in O_f$. So, $T = T(L)$. Suppose that $d(u_1, x) = 2$ for every point of $x \in O_f$. If x_1 and x_2 are two different points of O_f, then $\mathcal{C}(u_1, x_1) \cap \mathcal{C}(u_1, x_2) = \{u_1\}$ and so the quads $\mathcal{C}(u_1, x_1)$ and $\mathcal{C}(u_1, x_2)$ determine four different lines through u_1. Now, the nine quads $\mathcal{C}(u_1, x)$, $x \in O_f$, will determine 18 different lines of $\mathbb{E}_1$ through u_1, a contradiction, since there are only twelve such lines.

- The partial linear space is disconnected: there is one connected component O'_f of size 8 and one single point x^*. Then $d(x, u_i)$, $i \in \{1, 2\}$, is independent of the chosen point $x \in O'_f$ and different from $d(x^*, u_i)$. So, $|T(L)| = 8$ and $T \setminus T(L) = \{Q_1, Q_2, Q_3, Q_4\}$, where Q_1, Q_2, Q_3 and Q_4 are the four quads of T through x^*. By reasons of symmetry we may suppose that $d(u_1, x) = 2$ for every point of $x \in O'_f$. For every $i \in \{1, 2, 3, 4\}$, $d(u_1, Q_i) = 1$ and hence there exists a unique point v_i in Q_i collinear with u_1. Let w_{i1} and w_{i2} denote the two points of $Q_i \cap O_f$ collinear with v_i. For every $i \in \{1, 2, 3, 4\}$ and every $j \in \{1, 2\}$, let A_{ij} denote the set of two lines through u_1 contained in the quad $\mathcal{C}(u_1, w_{ij})$. Since $A_{i1} \cap A_{i2} = u_1 v_i$, $|A_{i1} \cup A_{i2}| = 3$ for every $i \in \{1, 2, 3, 4\}$. Now, for all $i, i' \in \{1, 2, 3, 4\}$ and all $j, j' \in \{1, 2\}$ with $i \neq i'$, $\mathcal{C}(w_{ij}, w_{i'j'}) \in T(L)$ and hence $A_{ij} \cap A_{i'j'} = \emptyset$. It now easily follows that $|\bigcup A_{ij}| = 12$. So, $L \in A_{i^*j^*}$ for a certain $i^* \in \{1, 2, 3, 4\}$ and a certain $j^* \in \{1, 2\}$. Now, $u \in L$ and $d(w_{i^*j^*}, L) = 1$ and so $d(u, O_f) \leq 2$, a contradiction. □

Put $Y := \Gamma_1(O_f)$ and $Z := \Gamma_2(O_f)$. Let Y_i, $i \in \mathbb{N} \setminus \{0\}$, denote the set of points of Y which are collinear with precisely i points of O_f. We have $\sum |Y_i| \cdot i = |O_f| \cdot 24 = 216$.

Lemma 6.75. *$|Y_1| = |Y_2| = 72$ and $|Y_i| = 0$ for all $i \geq 3$. So, $|Y| = 144$ and $|Z| = 576$.*

Proof. Suppose that the point u of $\Gamma_1(O_f)$ is collinear with three different points x_1, x_2 and x_3 of O_f. Then u is contained in the quad $\mathcal{C}(x_1, x_2)$. Since x_1 and x_2 are the only points of $\mathcal{C}(x_1, x_2) \cap O_f$ collinear with u, we have $x_3 \notin \mathcal{C}(x_1, x_2)$. Now, u is also contained in $\mathcal{C}(x_1, x_3)$ and so the quads $\mathcal{C}(x_1, x_3)$ and $\mathcal{C}(x_1, x_2)$ intersect in a line ux_1. If x_4 denotes the unique point of $\mathcal{C}(x_1, x_3) \cap O_f$ not collinear with u, then $d(x_2, x_4) = d(x_2, u) + d(u, x_4) = 1 + 2 = 3$, a contradiction. So, $|Y_i| = 0$ if $i \geq 3$. Since every two points of O_f have precisely two common neighbours, we have $|Y_2| = 2 \cdot \binom{|O_f|}{2} = 72$, $|Y_1| = 216 - 2|Y_2| = 72$ and $|Y| = |Y_1| + |Y_2| = 144$. Since $\mathbb{E}_1$ has precisely 729 points, $|Z| = 729 - |O_f| - |Y| = 576$. □

Using the notation of Theorem 5.13, we have that $m_0 = 9$, $m_1 = 144$, $m_2 = 576$ and $m_i = 0$ if $i \geq 3$. Since $\sum_{i=0}^{\infty}(-\frac{1}{s})^i m_i = 81$, we have found a contradiction.

6.6 The near hexagon $\mathbb{E}_2$

This near hexagon is due to Shult and Yanushka [91].

6.6.1 Definition and properties of $\mathbb{E}_2$

By [102], there is a unique Steiner system $S(5,8,24)$. (There are 24 points in such a Steiner system, each block contains eight points and every five different points are contained in a unique block.) We will now provide a model of $S(5,8,24)$. The projective plane $PG(2,4)$ has the following properties.

- The set of 168 hyperovals (i.e. sets of six points no three of which are collinear) can be divided into three classes $\mathcal{O}_1$, $\mathcal{O}_2$ and $\mathcal{O}_3$ of equal size. Two hyperovals are in the same class if and only if they intersect in an even number of points.
- The set of 360 Baer subplanes (i.e. subplanes of order 2) can be divided into three classes $\mathcal{B}_1$, $\mathcal{B}_2$ and $\mathcal{B}_3$ of equal size. Two Baer subplanes are in the same class if and only if they intersect in an odd number of points
- The indices i and j can be chosen in such a way that for $O \in \mathcal{O}_i$ and $S \in \mathcal{B}_j$, $|O \cap S|$ is even if and only if $i = j$.

Define now the following design $\mathcal{D}$. The points of $\mathcal{D}$ are the points of $PG(2,4)$ together with three new symbols ∞_1, ∞_2, and ∞_3. There are four types of blocks:

- $L \cup \{\infty_1, \infty_2, \infty_3\}$ where L is a line of $PG(2,4)$;
- $(O \cup \{\infty_1, \infty_2, \infty_3\}) \setminus \{\infty_i\}$ for each $O \in \mathcal{O}_i$;
- $S \cup \{\infty_i\}$ for each $S \in \mathcal{B}_i$;
- the symmetric difference $L \triangle L'$ of two distinct lines L and L' of $PG(2,4)$.

Taking the natural incidence, we find the unique Steiner system $S(5,8,24)$. Following [26], we will call the blocks of $S(5,8,24)$ *octads*. The Steiner system $S(5,8,24)$ has the following properties:

- Every $i \in \{1,2,3,4,5\}$ different points of $S(5,8,24)$ are contained in n_i octads. Here, $n_1 = 253$, $n_2 = 77$, $n_3 = 21$, $n_4 = 5$ and $n_5 = 1$.
- Every two different octads of $S(5,8,24)$ meet in either 0, 2 or 4 points.
- If B_1 and B_2 are two disjoint octads, then the complement of $B_1 \cup B_2$ is also an octad.
- If B_1 and B_2 are two octads which meet in four points, then the symmetric difference $B_1 \Delta B_2$ is again an octad of $S(5,8,24)$.

From $S(5,8,24)$, we can construct the following incidence structure $\mathbb{E}_2$:

- the points of $\mathbb{E}_2$ are the octads of $S(5,8,24)$;
- the lines of $\mathbb{E}_2$ are the triples of mutually disjoint octads;
- incidence is containment.

Theorem 6.76. • *If B_1 and B_2 are two octads of $S(5,8,24)$, then $d(B_1,B_2)$ is equal to* 0, 1, 2, *respectively* 3, *if and only if $|B_1 \cap B_2|$ is equal to* 8, 0, 4, *respectively* 2.

- *$\mathbb{E}_2$ is a dense near hexagon and every quad of $\mathbb{E}_2$ is isomorphic to $W(2)$.*
- *$\mathbb{E}_2$ is a regular near hexagon with parameters $(s,t_2,t) = (2,2,14)$. Every local space is isomorphic to* $\mathrm{PG}(3,2)$.

Proof. • Let B_1 and B_2 be two octads of $S(5,8,24)$. Obviously, $\mathrm{d}(B_1,B_2) = 0$ if and only if $|B_1 \cap B_2| = 8$.

Suppose $|B_1 \cap B_2| = 4$ and let B_3, B_4, B_5 denote the remaining octads through $B_1 \cap B_2$. If B is a common neighbour of B_1 and B_2, then B has at least three points in common with one of the octads B_3, B_4 or B_5. Suppose $|B \cap B_3| \geq 3$. Then $|B \cap B_3| = 4$ and $B \cap B_3 = B_3 \setminus (B_1 \cap B_2)$. The five octads through $B_3 \setminus (B_1 \cap B_2)$ are the octads B_3, $B_3 \Delta B_i$, $i \in \{1,2,4,5\}$. As a consequence, $B_3 \Delta B_4$, $B_3 \Delta B_5$ and $B_4 \Delta B_5$ are three common neighbours of B_1 and B_2.

Conversely, suppose that $\mathrm{d}(B_1,B_2) = 2$. Let B denote a common neighbour of B_1 and B_2 and let B' denote the complement of $B \cup B_1$. The octad B_2 has at most four points in common with B_1 and B'. Hence, $|B_2 \cap B_1| = 4$ and $|B' \cap B_2| = 4$.

Finally, suppose that $|B_1 \cap B_2| = 2$. Let B and B' denote two octads of $S(4,5,24)$ such that $\{B_1, B, B'\}$ is a line of $\mathbb{E}_2$. Then B_2 has at least three points in common with one of the octads B, B'. If $|B_2 \cap B| \geq 3$, then $|B_2 \cap B| = 4$. From $\mathrm{d}(B_2,B) = 2$ and $\mathrm{d}(B,B_1) = 1$, $\mathrm{d}(B_1,B_2) = 3$.

- Let $L = \{B_1, B_2, B_3\}$ be a line of $\mathbb{E}_2$ and let B denote a point of $\mathbb{E}_2$. In view of the possible intersections of two octads of $S(5,8,24)$, the multiset $\{|B \cap B_1|, |B \cap B_2|, |B \cap B_2|\}$ is equal to either $\{8,0^2\}$, $\{0,4^2\}$ or $\{4,2^2\}$. In any case we have that L contains a unique point nearest B. So, $\mathbb{E}_2$ is a near hexagon. Since every two points at distance 2 have precisely three common neighbours, $\mathbb{E}_2$ is a dense near hexagon and every quad is isomorphic to $W(2)$.
- By an easy counting, one finds that for every octad B, there are 30 other octads which are disjoint from B. It follows that $\mathbb{E}_2$ is a regular near hexagon with parameters $(s,t_2,t) = (2,2,14)$. By the Corollary on page 55 of [8], every local space of $\mathbb{E}_2$ is isomorphic to $\mathrm{PG}(3,2)$. □

Remark. Let B_1 and B_2 be two points at distance 2 from each other. Then $|B_1 \cap B_2| = 4$. Let B_3, B_4 and B_5 denote the other octads through $B_1 \cap B_2$. Put $X_0 := B_1 \cap B_2$ and $X_i := B_i \setminus X_0$ for every $i \in \{1,2,3,4,5\}$. Then the 15 octads of the $W(2)$-quad through B_1 and B_2 consists of the octads $X_i \cup X_j$, where i and j are two different elements of $\{0,1,2,3,4,5\}$. In this way, we rediscover Sylvester's model for $W(2)$.

Theorem 6.77 ([8]). *Every regular near hexagon with parameters* $(s,t_2,t) = (2,2,14)$ *is isomorphic to* $\mathbb{E}_2$.

Theorem 6.78. $\mathbb{E}_2$ *has no spread of symmetry.*

Proof. Similarly as in the proof of Theorem 6.3, this follows from the fact that every line of $\mathbb{E}_2$ is contained in a $W(2)$-quad. □

Theorem 6.79. *Let x be a point of $S(5,8,24)$. Then the 253 octads through x define an ovoid O_x of $\mathbb{E}_2$.*

Proof. If $\{B_1, B_2, B_3\}$ is a line of $\mathbb{E}_2$, then precisely one of the octads B_1, B_2, B_3 contains x, proving the theorem. □

In Section 6.6.2 we will show that every ovoid of $\mathbb{E}_2$ is of the form O_x for a certain point x of $S(5,8,24)$. This proves that $\mathbb{E}_2$ has 24 ovoids and hence also 24 ovoidal valuations.

Theorem 6.80. *For every point x of $\mathbb{E}_2$, there exists a path of length 5 in $\Gamma_3(x)$. As a consequence, $\mathbb{E}_2$ has no semiclassical ovoids.*

Proof. Take a quad Q at distance 2 from x. Then $\Gamma_2(x) \cap Q$ is an ovoid of Q and by Lemma 1.39 there exists a path of length 5 in Q not containing points of O. This path of length 5 is also a path of length 5 in $\Gamma_3(x)$. □

Theorem 6.81. *Every valuation f of $\mathbb{E}_2$ is either classical or ovoidal.*

Proof. Let f be a valuation of $\mathbb{E}_2$ which is neither classical nor ovoidal. By Theorem 6.80, f is not semiclassical. So, $O_f \geq 2$ and G_f is a linear space by Theorems 5.33 and 5.34.

We will now show that $\mathrm{d}(x,O_f) \leq 2$ and $f(x) = \mathrm{d}(x,O_f)$ for every point x of $\mathbb{E}_2$. Let Q be a special quad. If $x \in Q \cup \Gamma_1(Q)$, then $\mathrm{d}(x,O_f) \leq 2$. If $x \in \Gamma_2(Q)$, then the ovoids $\Gamma_2(x) \cap Q$ and $O_f \cap Q$ of Q meet in at least one point, proving that $\mathrm{d}(x,O_f) \leq 2$. By Theorem 5.12, it follows that $f(x) = \mathrm{d}(x,O_f)$.

Suppose $|O_f| = 5$. Let Q denote the unique $W(2)$-quad containing all points of O_f, let x be a point in $\Gamma_1(Q)$ not collinear with a point of O_f and let $L = \{x, y_1, y_2\}$ be a line through x such that $y_1, y_2 \in \Gamma_2(Q)$. Since $\mathrm{d}(x,O_f) = \mathrm{d}(y_1,O_f) = \mathrm{d}(y_2,O_f) = 2$, every point of L has value 2 by Theorem 5.12. This contradicts Property (V_2) in the definition of valuation.

Suppose $|O_f| > 5$. Considering an antiflag in G_f, we see that there exists a point in G_f which is incident with at least five lines. Hence, $|O_f| \geq 21$. On the other hand, since no two special quads meet in a line, $|O_f| \leq 21$. Hence, $|O_f| = 21$ and $G_f \cong \mathrm{PG}(2,4)$. Using the notation of Theorem 5.13, $m_0 = 21$, $m_1 = |\Gamma_1(O_f)|$ and $m_2 = |\Gamma_2(O_f)|$. Since every point of $\Gamma_1(O_f)$ is collinear with three points of O_f, $m_1 = \frac{m_0 \cdot 30}{3} = 210$. Since $\mathbb{E}_2$ has 759 points, $m_2 = 528$. So, $m_0 - \frac{m_1}{2} + \frac{m_2}{4} = 48$, contradicting Theorem 5.13. □

6.6.2 The ovoids of $\mathbb{E}_2$

The ovoids of $\mathbb{E}_2$ were classified in [15] with the aid of the Erdös–Ko–Rado Theorem ([70]). We will give an alternative proof. Let X denote the set of 24 points of the Steiner system $S(5,8,24)$. Recall that $\mathbb{E}_2$ is a regular near hexagon on $v = 759$ points with parameters $(s, t_2, t) = (2, 2, 14)$. Let O be an ovoid of $\mathbb{E}_2$.

Lemma 6.82. *The ovoid O has 253 octads. If α is a octad of O, then $m_2 := |\Gamma_2(\alpha) \cap O| = 140$ and $m_3 := |\Gamma_3(\alpha) \cap O| = 112$.*

Proof. The number of lines of $\mathbb{E}_2$ is equal to $\frac{v(t+1)}{s+1} = (t+1) \cdot |O|$. Hence $|O| = 253$. Counting the number of connections between points of $\Gamma_1(\alpha)$ and $\Gamma_2(\alpha) \cap O$ yields $|\Gamma_1(\alpha)| \cdot t = (t_2 + 1) \cdot |\Gamma_2(\alpha) \cap O|$, from which it follows that $m_2 = 140$ and $m_3 = |O| - 1 - m_2 = 112$. □

Lemma 6.83. *Four different elements of X are contained in one or five elements of O.*

Proof. The five octads through four elements of X form an ovoid O' in a quad Q. Let O'' be the ovoid of Q, induced by O, then O' and O'' have one or five points in common. □

Lemma 6.84. *Three different elements of X are contained in at least five elements and at most 21 elements of O.*

Proof. Take three elements a, b, c of X. Let d_1 be a fourth element of X and let α be an octad of O through $\{a, b, c, d_1\}$. Let d_2 be an element of X not in α and let β be an octad of O through $\{a, b, c, d_2\}$. By Lemma 6.83, there are now five octads of O through $\alpha \cap \beta$. The upper bound is immediate since $n_3 = 21$. □

Lemma 6.85. *Three different elements of X are contained in five or 21 elements of O.*

Proof. Let $\alpha \in \binom{X}{3}$ be contained in t_α octads of O. Double counting yields

$$\begin{aligned}
\sum_{\alpha \in \binom{X}{3}} 1 &= \binom{24}{3}, \\
\sum_{\alpha \in \binom{X}{3}} t_\alpha &= |O| \cdot \binom{8}{3}, \\
\sum_{\alpha \in \binom{X}{3}} t_\alpha(t_\alpha - 1) &= |O| \cdot m_2 \cdot \binom{4}{3}.
\end{aligned}$$

From Lemma 6.84 and $\sum_{\alpha \in \binom{X}{3}} (t_\alpha - 5)(21 - t_\alpha) = 0$, $t_\alpha = 5$ or $t_\alpha = 21$. □

Remarks. (1) Let T_1, respectively T_2, be the number of $\alpha \in \binom{X}{3}$ for which $t_\alpha = 5$, respectively $t_\alpha = 21$. From

$$\begin{aligned}
T_1 + T_2 &= \binom{24}{3}, \\
5T_1 + 21T_2 &= |O| \cdot \binom{8}{3},
\end{aligned}$$

$T_1 = \binom{23}{3}$ and $T_2 = \binom{23}{2}$.

(2) Let R_1, respectively R_2, denote the number of $\alpha \in \binom{X}{4}$, which are contained in exactly one, respectively exactly five, elements of O. From

$$\begin{aligned}
R_1 + R_2 &= \binom{24}{4}, \\
R_1 + 5R_2 &= |O| \cdot \binom{8}{4},
\end{aligned}$$

$R_1 = \binom{23}{4}$ and $R_2 = \binom{23}{3}$.

Lemma 6.86. *For every $A = \{p_1, p_2, p_3, p_4\} \subseteq X$, there exists an $i \in \{1, 2, 3, 4\}$ such that $A \setminus \{p_i\}$ is contained in exactly five octads of O.*

Proof. Suppose the contrary. Let α and β be two different elements of O through A. Take two different points x and y in $\alpha \setminus \beta$ and let z be a point of $\alpha \cap \beta$. Let $\gamma \neq \alpha$ be an octad of O through $\{x, y, z\}$, then γ intersects α in a fourth point u. Through $\{x, y, z, u\}$, there are five octads of O. If $u \in \alpha \setminus \beta$, then $\alpha \setminus \{x, y, z, u\}$ is contained in only one octad of O, a contradiction. Hence, $u \in \alpha \cap \beta$. Take a point $v \in \beta \setminus \alpha$. Let D_1 be the octad through $[(\alpha \cap \beta) - \{z\}] \cup \{v, y\}$ and D_2 be the octad through $[(\alpha \cap \beta) - \{u\}] \cup \{v, x\}$. D_1 and D_2 are two octads of the ovoid meeting in four points. The octad $D_1 \triangle D_2$ is not an element of O, a contradiction, since the set $\{x, y, z, u\}$ is contained in it. □

Lemma 6.87. *If $\{p_1, p_2, p_3, p_4\} \subseteq X$ is contained in exactly five elements of O, then there are exactly three subsets of size 3 in $\{p_1, p_2, p_3, p_4\}$ which are contained in 21 elements of O.*

Proof. Counting pairs (A, B) with (i) $A \in \binom{X}{3}$, (ii) $B \in \binom{X}{4}$, (iii) $A \subseteq B$ and (iv) A is contained in 21 elements of O, yields $21 \cdot T_2 = \sum k_B$, where the summation ranges over all R_2 elements B of $\binom{X}{4}$ through which there are five octads of O. The number k_B denotes the number of subsets of size 3 of B through which there are precisely 21 octads of O. By Lemma 6.86, $k_B \leq 3$. Since $21T_2 = 3R_2$, all values of k_B are equal to 3. □

Lemma 6.88. *Every two elements x and y of X are contained in at least 21 elements and in at most 77 elements of O.*

Proof. Let $z \in X$ be a third element of X. From the proof of Lemma 6.84, it follows that there exists a fourth element $u \in X$ such that $\{x, y, z, u\}$ is contained in five elements of O. By Lemma 6.87, at least one of $\{x, y, z\}$, $\{x, y, u\}$ is contained in 21 octads of O. The upper bound is immediate since $n_2 = 77$. □

Lemma 6.89. *Two different elements of X are contained in 21 or 77 octads of O.*

Proof. Let $\alpha \in \binom{X}{2}$ be contained in s_α octads of O. Double counting yields

$$\sum_{\alpha \in \binom{X}{2}} 1 = \binom{24}{2},$$

$$\sum_{\alpha \in \binom{X}{2}} s_\alpha = |O| \cdot \binom{8}{2},$$

$$\sum_{\alpha \in \binom{X}{2}} s_\alpha(s_\alpha - 1) = |O| \cdot m_2 \cdot \binom{4}{2} + |O| \cdot m_3.$$

From Lemma 6.88 and $\sum_\alpha (s_\alpha - 21)(77 - s_\alpha) = 0$, it follows that $s_\alpha = 21$ or $s_\alpha = 77$. □

Remark. Let S_1, respectively S_2, be the number of $\alpha \in \binom{X}{2}$ for which $s_\alpha = 21$, respectively $s_\alpha = 77$. From

$$S_1 + S_2 = \binom{24}{2},$$

$$21S_1 + 77S_2 = |O| \cdot \binom{8}{2},$$

it follows that $S_1 = \binom{23}{2}$ and $S_2 = \binom{23}{1}$.

Lemma 6.90. *For every $A = \{p_1, p_2, p_3\} \subseteq X$, there exists an $i \in \{1, 2, 3\}$ such that $A \setminus \{p_i\}$ is contained in exactly 21 octads of O.*

Proof. Suppose that all octads through $\{p_1, p_2\}$ or through $\{p_2, p_3\}$ belong to O. Choose $q_1, q_2 \notin A$ with $q_1 \neq q_2$ and q_3 not in the unique octad through $\{p_1, p_2, p_3, q_1, q_2\}$. Let B_1, respectively B_2, be the octad through $\{p_1, p_2, q_1, q_2, q_3\}$, respectively $\{p_2, p_3, q_1, q_2, q_3\}$. Then $B_1 \triangle B_2 \notin O$, since $B_1, B_2 \in O$. Since $\{p_1, p_3\} \subseteq B_1 \triangle B_2$, the lemma follows. □

Lemma 6.91. *If $\{p_1, p_2, p_3\} \subseteq X$ is contained in 21 octads of O, then there are exactly two subsets of size 2 in $\{p_1, p_2, p_3\}$ which are contained in 77 octads of O.*

Proof. Counting pairs (A, B) with (i) $A \in \binom{X}{2}$, (ii) $B \in \binom{X}{3}$, (iii) $A \subseteq B$ and (iv) A is contained in 77 elements of O, yields $22 \cdot S_2 = \sum l_B$, where the summation ranges over all T_2 elements B of $\binom{X}{3}$, through which there are 21 octads of O. The number l_B denotes the number of subsets of size 2 of B through which there are 77 octads of O. By Lemma 6.90, $l_B \leq 2$. Since $22 \cdot S_2 = T_2 \cdot 2$, all values of l_B must be equal to 2. □

Lemma 6.92. *There are at least 77 octads of O through every point.*

Proof. Let x denote the point. The element x is contained in a set $P \subseteq X$ of size 4, through which there are five elements of O. By Lemma 6.87, there exists a subset P' of size 3 in P containing x and contained in 21 octads of O. By Lemma 6.91, there exists a subset P'' of size 2 in P' containing x and contained in 77 octads of O. This proves the lemma. □

Lemma 6.93. *Every point of X is contained in 77 or 253 octads of O.*

Proof. Let $\alpha \in X$ be contained in μ_α octads of O. Double counting yields

$$\begin{aligned}
\sum_{\alpha \in X} 1 &= \binom{24}{1}, \\
\sum_{\alpha \in X} \mu_\alpha &= |O| \cdot \binom{8}{1}, \\
\sum_{\alpha \in X} \mu_\alpha(\mu_\alpha - 1) &= |O| \cdot m_2 \cdot \binom{4}{1} + |O| \cdot m_3 \cdot \binom{2}{1}.
\end{aligned}$$

By Lemma 6.92 and $\sum_\alpha (\mu_\alpha - 77)(253 - \mu_\alpha) = 0$, $\mu_\alpha = 77$ or $\mu_\alpha = 253$. □

Theorem 6.94. *There exists an $x \in X$, such that $O = O_x$.*

Proof. It suffices to prove that there exists a point $x \in X$ with $\mu_x = 253$. Let U_1, respectively U_2, be the number of $x \in X$ for which $\mu_x = 77$, respectively $\mu_x = 253$. From

$$\begin{aligned}
U_1 + U_2 &= 24, \\
5U_1 + 21U_2 &= 8 \cdot |O|,
\end{aligned}$$

it follows that $U_1 = 23$ and $U_2 = 1$. This proves the theorem. □

6.7 The near hexagon $\mathbb{E}_3$

A nonempty set X of points of a partial linear space $\mathcal{S}$ is called a *hyperoval* if every line of $\mathcal{S}$ has either 0 or two points in common with X. Pasechnik [78] showed that the point-line system of the hermitian variety $H(5,4)$ has two isomorphism classes of hyperovals. The hyperovals of one class contain 126 points and the hyperovals of the other class contain 162 points. (Regarding the hyperoval on 126 points, see also the discussion in Section 2.2 of [19]).

Let X be a hyperoval of size 126 in $H(5,4)$. One easily verifies that the following holds:

(1) each generator of $H(5,4)$ has 0 or six points in common with X;

(2) if π_1 and π_2 are two generators such that $\dim(\pi_1 \cap \pi_2) = 1$, $|\pi_1 \cap X| = |\pi_2 \cap X| = 6$, then $|\pi_1 \cap \pi_2 \cap X| = 2$.

Let $\mathbb{E}_3$ be the following incidence structure:

- the points of $\mathbb{E}_3$ are the generators of $H(5,4)$ intersecting X in six points;
- the lines of $\mathbb{E}_3$ are the lines of $H(5,4)$ intersecting X in two points;
- incidence is reverse containment.

The incidence structure $\mathbb{E}_3$ has 567 points. Each line of $\mathbb{E}_3$ is incident with three points and each point is incident with 15 lines. The incidence structure $\mathbb{E}_3$ is embedded in $DH(5,4)$. Let $\mathrm{d}'(\cdot,\cdot)$ be the distance function in $\mathbb{E}_3$ and let $\mathrm{d}(\cdot,\cdot)$ denote the distance function in $DH(5,4)$.

Theorem 6.95. *If π_1 and π_2 are two points of $\mathbb{E}_3$, then $\mathrm{d}'(\pi_1,\pi_2) = \mathrm{d}(\pi_1,\pi_2)$. As a consequence, $\mathbb{E}_3$ is a near hexagon.*

Proof. Notice that $\mathrm{d}'(\pi_1,\pi_2) \geq \mathrm{d}(\pi_1,\pi_2)$; so, the theorem holds if $\mathrm{d}(\pi_1,\pi_2) \leq 1$. Suppose $\mathrm{d}(\pi_1,\pi_2) = 2$. Then π_1 and π_2 intersect in a point p. Let q be a point of $\pi_1 \cap X$ different from p and let π_3 denote the unique generator of $DH(5,4)$ through q intersecting π_2 in a line. Obviously, π_3 is a common neighbour of π_1 and π_2 in the geometry $\mathbb{E}_3$, proving that $\mathrm{d}'(\pi_1,\pi_2) = 2$. Finally, suppose $\mathrm{d}(\pi_1,\pi_2) = 3$. Then π_1 and π_2 are disjoint. Let q be a point of $\pi_1 \cap X$ and let π_3 be the unique generator through q intersecting π_2 in a line. Since $\mathrm{d}'(\pi_1,\pi_3) = 2$ and $\mathrm{d}'(\pi_3,\pi_2) = 1$, $\mathrm{d}'(\pi_1,\pi_2) = 3$. □

Theorem 6.96. *$\mathbb{E}_3$ is a dense near hexagon. Every quad is isomorphic to either $W(2)$ or $Q(5,2)$.*

Proof. Let π_1 and π_2 be two points of $\mathbb{E}_3$ at distance 2 from each other. Then $\pi_1 \cap \pi_2$ is a point p. We have the following possibilities.

- $p \in X$. Then π_1 and π_2 have five common neighbours.

- $p \notin X$. Let L_1, L_3 and L_3 denote the three lines of π_1 through p intersecting X in two points. Let α_i, $i \in \{1,2,3\}$, denote the unique plane through L_i intersecting π_2 in a line. Then α_1, α_2 and α_3 are the three common neighbours of π_1 and π_2.

The theorem now readily follows. □

If Q is a quad of $\mathbb{E}_3$, then there exists a point x_Q on $H(5,4)$ such that Q consists of all generators of $H(5,4)$ through x_Q containing six points of X. If $x_Q \in X$, then Q is a $Q(5,2)$-quad.

Theorem 6.97. *If x is a point of $H(5,4) \setminus X$, then the generators of $H(5,4)$ through x containing six points of X define a $W(2)$-quad of $\mathbb{E}_3$. As a consequence, the map $Q \mapsto x_Q$ defines a bijection between the set of quads of $\mathbb{E}_3$ and the set of points of $H(5,4)$.*

Proof. We first prove that there is a generator through x containing six points of X. Suppose the contrary. Count in two different ways the triples (π_1, π_2, y), where y is a point of X and where π_1 and π_2 are two generators of $H(5,4)$ satisfying (i) $x \in \pi_1$, (ii) $y \in \pi_2$ and (iii) $\pi_1 \cap \pi_2$ is a line. There are 126 possibilities for y and for given y, there are 27 possibilities for π_2. The generator π_1 is completely determined by π_2. So, the number of triples is equal to $126 \cdot 27$. On the other hand, there are 27 possibilities for π_1. For given π_1, there are 16 possibilities for $L := \pi_1 \cap \pi_2$, and for given L, there is at most one possibility for π_2, see the property (2) mentioned above. For given π_2, there are at most six possibilities for y. Hence, we have $126 \cdot 27 \leq 27 \cdot 16 \cdot 1 \cdot 6$, a contradiction. Hence, there exists a generator π through x containing six points of X. Now, let L_1 and L_2 denote two lines of π through x containing two points of X. Let α_i, $i \in \{1,2\}$, denote a generator through L_i such that $\alpha_1 \cap \alpha_2 = \{x\}$. Then α_1 and α_2 are two points of $\mathbb{E}_3$ at distance 2 from each other. If Q denotes the quad through α_1 and α_2, then $x_Q = x$. This proves the theorem. □

Theorem 6.98. *Every local space of $\mathbb{E}_3$ is isomorphic to $\overline{W(2)}$.*

Proof. Since every line of $\mathbb{E}_3$ contains two points of X, every line of $\mathbb{E}_3$ is contained in two $Q(5,2)$-quads and three $W(2)$-quads by Theorem 6.97. The theorem now follows from Lemma 6.44. □

Theorem 6.99. *$\mathbb{E}_3$ has no spread of symmetry.*

Proof. Similarly as in the proof of Theorem 6.3, this follows from the fact that every line is contained in a $W(2)$-quad. □

Theorem 6.100. *If f is a nonclassical valuation of $\mathbb{E}_3$, then $G_f \cong \mathrm{PG}(2,4)$ and f is induced by a (classical) valuation of $DH(5,4)$.*

Proof. Since f is not classical, $f(x) \in \{0,1,2\}$ for every point x of $\mathbb{E}_3$. Since $Q(5,2)$ has no ovoids, every $Q(5,2)$-quad contains a unique point of O_f. Since there are 126 $Q(5,2)$-quads and since every point is contained in six $Q(5,2)$-quads, $|O_f| = \frac{126}{6} = 21$. Since every special quad is isomorphic to $W(2)$, G_f is a Steiner system $S(2,5,21)$ or a projective plane of order 4. Now, let x be a point of O_f and let Q be a special quad such that $x \not\in Q$. Let Q' denote the unique $Q(5,2)$-quad of $DH(5,4)$ containing Q, let x' denote the unique point in Q' collinear with x, let g' denote the classical valuation of $DH(5,4)$ determined by $g'(x') = 0$ and let g denote the nonclassical valuation of $\mathbb{E}_3$ induced by g'. Since $|O_f| = |O_g| = 21$, $G_f \cong G_g \cong \mathrm{PG}(2,4)$ and $\{x\} \cup (\Gamma_2(x) \cap Q) \subseteq O_f \cap O_g$, we must have $O_f = O_g$. Since every point of $\mathbb{E}_3$ is contained in a $Q(5,2)$-quad, $\mathrm{d}(x, O_f) \leq 2$ and $f(x) = \mathrm{d}(x, O_f) = \mathrm{d}(x, O_g) = g(x)$ by Theorem 5.12. This proves the theorem. □

Remark. Consider in $\mathrm{PG}(6,3)$ a nonsingular quadric $Q(6,3)$ and a nontangent hyperplane π intersecting $Q(6,3)$ in a nonsingular elliptic quadric $Q^-(5,3)$. There is a polarity associated with $Q(6,3)$ and we call two points orthogonal when one of them is contained in the polar hyperplane of the other. Let N denote the set of 126 internal points of $Q(6,3)$ which are contained in π, i.e. the set of all 126 points in π for which the polar hyperplane intersects $Q(6,3)$ in a nonsingular elliptic quadric. Let $\mathcal{S}$ be the following incidence structure:

- the points of $\mathcal{S}$ are the 6-tuples of mutually orthogonal points of N;
- the lines of $\mathcal{S}$ are the pairs of mutually orthogonal points of N;
- incidence is reverse containment.

Then $\mathcal{S}$ is isomorphic to $\mathbb{E}_3$, see [16]. A group-theoretical construction for $\mathbb{E}_3$ can be found in [2].

6.8 The known slim dense near polygons

Above we described five infinite classes of near polygons and three sporadic examples. In the case of generalized quadrangles we have the following isomorphisms: $DH(3,4) \cong \mathbb{G}_2 \cong Q(5,2)$, $\mathbb{H}_2 \cong DQ(4,2) \cong W(2)$ and $\mathbb{I}_2 \cong \mathbb{L}_3 \times \mathbb{L}_3$. Looking to the structure of the local spaces, we see that the near hexagons $DQ(6,2)$, $DH(5,4)$, $\mathbb{G}_3$, $\mathbb{H}_3$, $\mathbb{E}_1$, $\mathbb{E}_2$ and $\mathbb{E}_3$ are mutually nonisomorphic. The local spaces of $\mathbb{I}_3$ are isomorphic to the local spaces of $\mathbb{H}_3$, i.e. to $\mathrm{PG}(2,2)'$. We will see in Chapter 7 that $\mathbb{I}_3$ is isomorphic to $\mathbb{H}_3$. Looking at the structure of the local spaces, we see that the near $2n$-gons $DQ(2n,2)$, $DH(2n-1,4)$, $\mathbb{G}_n$, $\mathbb{H}_n$ and $\mathbb{I}_n$ are mutually nonisomorphic if $n \geq 4$.

Every known slim dense near polygon which is not a product near polygon nor a glued near polygon of type 1 belongs to one of the above-mentioned classes. Among the known slim dense near $2n$-gons, $n \geq 2$, only $DH(2n-1,4)$, $\mathbb{G}_n$ and $\mathbb{E}_1$ ($n = 3$) have spreads of symmetry. This observation allows us to construct the following class of near polygons (see also Theorem 4.59).

If Z_1 and Z_2 are two sets of near polygons, then $Z_1 \otimes Z_2$ denotes the (possibly empty) set of all near polygons of type 1 obtained by glueing an element of Z_1 with one of Z_2. We introduce the following sets.

$$\begin{aligned}
C_2 &= \{Q(5,2)\}; \\
C_3 &= \{\mathbb{G}_3, DH(5,4), \mathbb{E}_1\} \cup \Big(C_2 \otimes C_2\Big); \\
C_n &= \{\mathbb{G}_n, DH(2n-1,4)\} \cup \left(\bigcup_{2 \le i \le n-1} C_i \otimes C_{n+1-i}\right) \text{ for every } n \ge 4; \\
\mathcal{C} &= C_2 \cup C_3 \cup \cdots; \\
\mathcal{M} &= \{\mathbb{O}, \mathbb{L}_3, \mathbb{E}_2, \mathbb{E}_3\} \cup \mathcal{C} \cup \{DQ(2n,2) | n \ge 2\} \cup \{\mathbb{H}_n | n \ge 3\} \cup \{\mathbb{I}_n | n \ge 4\}.
\end{aligned}$$

Let $\mathcal{M}^\times$ denote the set of all near polygons obtained by taking the direct product of $k \ge 1$ elements of $\mathcal{M}$. Every known slim dense near polygon is isomorphic to an element of the class $\mathcal{M}^\times$. In [58], the following conjecture was introduced.

Conjecture. *Every slim dense near polygon is isomorphic to an element of the class* $\mathcal{M}^\times$.

6.9 The elements of C_3 and C_4

6.9.1 Spreads of symmetry of $Q(5,2)$

Consider a nonsingular hermitian variety $H(3,4)$ in $\mathrm{PG}(3,4)$. By Theorem 6.10, every spread of symmetry of $DH(3,4)$ arises by intersecting $H(3,4)$ with a non-tangent plane. Since every line of $H(3,4)$ intersects $H(3,4)$ in a point or a Baer subline, we have:

Lemma 6.101. *If S_1 and S_2 are two different spreads of symmetry of $Q(5,2)$, then $|S_1 \cap S_2| \in \{1,3\}$.*

Let S be a spread of symmetry of $Q(5,2)$. If K_1 and K_2 are two different lines of S, then K_1 and K_2 are contained in a unique subgrid G and there exists a unique line $K_3 \in S$ in G disjoint from K_1 and K_2. We call $R = \{K_1, K_2, K_3\}$ a *regulus* of S. The lines and reguli of S define an affine plane $\mathcal{A}_S$ of order 3. If R is a regulus of S, then $R^\perp$ denotes the set of three lines meeting each line of R. If R_1, R_2 and R_3 are three mutually disjoint reguli of S, then $R_1 \cup R_2^\perp \cup R_3^\perp$ is a spread of symmetry of $Q(5,2)$ intersecting S in three lines. Conversely, if S' is a spread of symmetry of $Q(5,2)$ such that $|S \cap S'| = 3$, then there exist three disjoint reguli R_1, R_2, R_3 of S such that $S' = R_1 \cup R_2^\perp \cup R_3^\perp$.

6.9.2 Another model for $Q(5,2)$

Let Γ be the following graph on the vertex set $\mathbb{F}_3 \times \mathbb{F}_3 \times \mathbb{F}_3$: two vertices (x_1, y_1, z_1) and (x_2, y_2, z_2) are adjacent if

- $x_1 \neq x_2$, $(y_1, z_1) = (y_2, z_2)$, or
- $x_2 = x_1 + y_1 z_2 - y_2 z_1$, $(y_1, z_1) \neq (y_2, z_2)$.

Lemma 6.102. *For every two vertices (x_1, y_1, z_1) and (x_2, y_2, z_2) of Γ, there exists a unique vertex (x_3, y_3, z_3) collinear with (x_1, y_1, z_1) and (x_2, y_2, z_2). As a consequence, every maximal clique of Γ has size 3.*

Proof. We need to distinguish two cases.

- $x_1 \neq x_2$ and $(y_1, z_1) = (y_2, z_2)$.
  If $(y_3, z_3) \neq (y_1, z_1)$, then we have $x_3 = x_1 + y_1 z_3 - z_1 y_3$ on the one hand and $x_3 = x_2 + y_1 z_3 - z_1 y_3$ on the other hand. This is impossible. Hence, $(y_3, z_3) = (y_1, z_1) = (y_2, z_2)$. It follows that x_3 is the unique element of $\mathbb{F}_3 \setminus \{x_1, x_2\}$.
- $x_2 = x_1 + y_1 z_2 - y_2 z_1$ and $(y_1, z_1) \neq (y_2, z_2)$.
  By the first case, we know that (y_1, z_1), (y_2, z_2) and (y_3, z_3) are mutually different. From $x_3 = x_2 + y_2 z_3 - z_2 y_3$ and $x_3 = x_1 + y_1 z_3 - z_1 y_3$, it follows that
  $$\begin{vmatrix} 1 & y_1 & z_1 \\ 1 & y_2 & z_2 \\ 1 & y_3 & z_3 \end{vmatrix} = 0.$$
  So, $(y_3, z_3) = (-y_1 - y_2, -z_1 - z_2)$. It also follows that $x_3 = -x_1 - x_2$. □

Let $\mathcal{S}$ be the incidence structure with points the vertices of Γ and with lines the maximal cliques (of size 3) of Γ (natural incidence).

Lemma 6.103. *The incidence structure $\mathcal{S}$ is isomorphic to $Q(5, 2)$.*

Proof. Every line of $\mathcal{S}$ contains three points. Since every vertex of Γ is incident with 10 other vertices, every point of $\mathcal{S}$ is incident with five lines. Now, let L be an arbitrary point. Since $|\Gamma_1(L)| = 24$, every point of $\mathcal{S}$ not contained on L is collinear with a unique point of L. This proves that $\mathcal{S}$ is a generalized quadrangle of order $(2, 4)$. So, $\mathcal{S}$ is isomorphic to $Q(5, 2)$. □

For all $a_{11}, a_{12}, a_{21}, a_{22}, b, c, d \in \mathbb{F}_3$ such that $a_{11}a_{22} - a_{12}a_{21} \neq 0$, consider the following permutation of the point set of $\mathcal{S}$:

$$\begin{bmatrix} x^\theta \\ y^\theta \\ z^\theta \end{bmatrix} = \begin{bmatrix} a_{11}a_{22} - a_{12}a_{21} & ba_{21} - ca_{11} & ba_{22} - ca_{12} \\ 0 & a_{11} & a_{12} \\ 0 & a_{21} & a_{22} \end{bmatrix} \begin{bmatrix} x \\ y \\ z \end{bmatrix} + \begin{bmatrix} d \\ b \\ c \end{bmatrix}.$$

We denote this permutation by $\theta(a_{11}, a_{22}, a_{12}, a_{21}, b, c, d)$. One easily verifies that $\theta(a_{11}, a_{22}, a_{12}, a_{21}, b, c, d)$ defines an automorphism of $\mathcal{S}$. Let S be the set of lines of the form $\{(\sigma, a, b) \,|\, \sigma \in \mathbb{F}_3\}$, $a, b \in \mathbb{F}_3$. The group $G := \{\theta(1, 1, 0, 0, 0, 0, f) \,|\, f \in \mathbb{F}_3\}$ of automorphisms of $\mathcal{S}$ fix each line of S and acts regularly on each line of S. It follows that S is a spread of symmetry and that G is the full group of automorphisms of $\mathcal{S}$ fixing each line of S. Since there is up to isomorphism only one spread of symmetry in $Q(5, 2)$, we can now easily show the following.

Lemma 6.104. *Let S be a spread of symmetry of $Q(5,2)$ and L a line of S. Then the group of automorphisms of $Q(5,2)$ stabilizing S and fixing the line $L \in S$ induces the full group of permutations of this line L.*

Lemma 6.105. *Let S_1 and S_2 be two different spreads of symmetry of $Q(5,2)$ and let G_{S_i}, $i \in \{1,2\}$, denote the group of automorphisms of $Q(5,2)$ fixing each line of S_i. Then G_{S_1} commutes with G_{S_2} if and only if $|S_1 \cap S_2| = 3$.*

Proof. By Theorem 4.56, G_{S_1} commutes with G_{S_2} if and only if $S_1 \cup S_2$ is a disjoint union of lines and grids. The lemma now easily follows. □

We can also show the following.

Lemma 6.106. *Let S be a spread of symmetry of $Q(5,2)$ and let G_S denote the group of automorphisms of $Q(5,2)$ fixing each line of S. For every two reguli R_1 and R_2 of S, there exists an automorphism θ of $Q(5,2)$ which satisfies the following properties:*

(a) $S^\theta = S$;

(b) $R_1^\theta = R_2$;

(c) θ *commutes with every element of* G_S.

Proof. Since all spreads of symmetry of $Q(5,2)$ are equivalent, we may consider the above-mentioned spread of symmetry S in $\mathcal{S} \cong Q(5,2)$. The elements of G_S are precisely the elements $\theta(1,1,0,0,0,0,f)$ with $f \in \mathbb{F}_3$. An automorphism $\theta(a_{11},a_{22},a_{12},a_{21},b,c,d)$ commutes with every element of G_S if and only if $a_{11}a_{22} - a_{12}a_{21} = 1$. The automorphism $\theta(a_{11},a_{22},a_{12},a_{21},b,c,d)$ induces the following automorphism of the affine plane $\mathcal{A}_S$:

$$\begin{bmatrix} y' \\ z' \end{bmatrix} = \begin{bmatrix} a_{11} & a_{12} \\ a_{21} & a_{22} \end{bmatrix} \begin{bmatrix} y \\ z \end{bmatrix} + \begin{bmatrix} b \\ c \end{bmatrix}.$$

It is now easily seen that we can take elements a_{11}, a_{22}, a_{12}, a_{21}, b, c and d in $\mathbb{F}_3$ such that $\theta(a_{11},a_{22},a_{12},a_{21},b,c,d)$ satisfies all required conditions. □

6.9.3 The near polygons $DH(2n-1,4) \otimes Q(5,2)$, $\mathbb{G}_n \otimes Q(5,2)$ and $\mathbb{E}_1 \otimes Q(5,2)$

By Theorem 6.10, $DH(2n-1,4)$ has up to isomorphism a unique spread of symmetry. By Corollary 6.66, $\mathbb{E}_1$ has up to isomorphism a unique spread of symmetry. By Theorem 6.38, Lemma 6.39 and Corollary 6.42, $\mathbb{G}_n$ has up to isomorphism a unique spread of symmetry.

Theorem 6.107. *Let $\mathcal{S}$ be a near polygon which is isomorphic to either $DH(2n-1,4)$, $\mathbb{G}_n$ or $\mathbb{E}_1$. Then there exists a unique glued near polygon of type $\mathcal{S} \otimes Q(5,2)$.*

Proof. We will use the notation of Chapter 4. Put $\mathcal{A}_1 = \mathcal{S}$ and $\mathcal{A}_2 = Q(5,2)$. Up to equivalence, $\mathcal{A}_1$ and $\mathcal{A}_2$ have unique spreads of symmetry. So, we may choose arbitrary spreads of symmetry S_1 and S_2 in $\mathcal{A}_1$ and $\mathcal{A}_2$, respectively. By Theorem 4.16, we may also choose arbitrary base lines $L_1^{(1)} \in S_1$ and $L_1^{(2)} \in S_2$. Let $\mathcal{B}$ be the line of size 3. By Theorem 4.48, every permutation $\theta_1 : \mathcal{B} \to L_1^{(1)}$ and every permutation $\theta_2 : \mathcal{B} \to L_1^{(2)}$ gives rise to a glued near hexagon of type $\mathcal{S} \otimes Q(5,2)$. All these glued near hexagons are isomorphic since the group of automorphisms of $\mathcal{A}_2$ fixing S_2 and the line $L_1^{(2)}$ of S_2 induces the full group of six permutations of the line $L_1^{(2)}$, see Lemma 6.104. □

In the sequel, we will use the notations $DH(2n-1,4) \otimes Q(5,2)$, $\mathbb{G}_n \otimes Q(5,2)$ and $\mathbb{E}_1 \otimes Q(5,2)$ to denote the unique glued near polygons of these types.

6.9.4 Spreads of symmetry of $Q(5,2) \otimes Q(5,2)$

Let T_1 and T_2 denote the two partitions of $Q(5,2) \otimes Q(5,2)$ in $Q(5,2)$-quads. Every element of $T_1 \cup T_2$ is big in $Q(5,2) \otimes Q(5,2)$. Every element of T_1 intersects every element of T_2 in a line and $S^* := \{F_1 \cap F_2 | F_1 \in T_1, F_2 \in T_2\}$ is a spread of $Q(5,2) \otimes Q(5,2)$. If $Q \in T_i$, $i \in \{1,2\}$, then $S_Q = \{Q \cap F \,|\, F \in T_{3-i}\}$ is a spread of symmetry of Q.

Theorem 6.108. *Up to equivalence, $Q(5,2) \otimes Q(5,2)$ has two spreads of symmetry.*

Proof. Let B be a spread of symmetry of $Q(5,2) \otimes Q(5,2)$. By Theorems 4.54 (b), 4.55 and Lemma 6.105, there exists a quad Q of $T_1 \cup T_2$ and a spread of symmetry S in Q such that $|S \cap S_Q| \in \{3,9\}$ and $B = \bar{S}$. Conversely, let Q be a quad of T_1 and let S be a spread of symmetry of Q such that $|S \cap S_Q| \in \{3,9\}$. By Theorem 4.53 (b) and Lemma 6.105, $\bar{S}$ is a spread of symmetry of $Q(5,2) \otimes Q(5,2)$. This proves that S^* is a spread of symmetry of $Q(5,2) \otimes Q(5,2)$. If Q is a quad of T_1 and if S_1 and S_2 are two spreads of symmetry of Q intersecting S_Q in three lines, then by Theorem 4.53 (b) and Lemma 6.106, there exists an automorphism of $Q(5,2) \otimes Q(5,2)$ mapping S_1 to S_2 (and fixing each element of T_1). Since there exists an automorphism of $Q(5,2) \otimes Q(5,2)$ interchanging T_1 and T_2, there exists, up to equivalence, only one spread of symmetry in $Q(5,2) \otimes Q(5,2)$ different from S^*. □

Definition. We call S^* the spread of symmetry of type (a). The 24 remaining spreads of symmetry of $Q(5,2) \otimes Q(5,2)$ are called spreads of symmetry of type (b).

6.9.5 Near polygons of type $(Q(5,2) \otimes Q(5,2)) \otimes Q(5,2)$

Let $\mathcal{S}$ be a glued near octagon of type $(Q(5,2) \otimes Q(5,2)) \otimes Q(5,2)$. We say that $\mathcal{S}$ is of type $(Q(5,2) \otimes Q(5,2)) \otimes_a Q(5,2)$, respectively of type $(Q(5,2) \otimes Q(5,2)) \otimes_b$

$Q(5,2)$, if it arises from a spread of symmetry of type (a), respectively of type (b), of $Q(5,2) \otimes Q(5,2)$.

Theorem 6.109. *There exist unique glued near octagons of type* $(Q(5,2) \otimes (5,2)) \otimes_a Q(5,2)$ *and* $(Q(5,2) \otimes Q(5,2)) \otimes_b Q(5,2)$.

Proof. We show that there exists a unique glued near octagon of type $(Q(5,2) \otimes (5,2)) \otimes_b Q(5,2)$. The proof for the uniqueness of the glued near octagon of type $(Q(5,2) \otimes (5,2)) \otimes_a Q(5,2)$ is completely similar. All spreads of symmetry of type (b) in $Q(5,2) \otimes Q(5,2)$ are equivalent and all spreads of symmetry of $Q(5,2)$ are equivalent. As a consequence, we may choose an arbitrary spread of symmetry of type (b) in $Q(5,2) \otimes Q(5,2)$, which we call S_1, and an arbitrary spread of symmetry S_2 in $Q(5,2)$. Every glued near polygon which can be obtained for a certain choice of the base lines can always be obtained for any other choice of the base lines. Hence, we may also choose arbitrary base lines $L_1^{(1)} \in S_1$ and $L_1^{(2)} \in S_2$. Let $\mathcal{B}$ be the line of size 3. By Theorem 4.48, every permutation $\theta_1 : \mathcal{B} \rightarrow L_1^{(1)}$ and every permutation $\theta_2 : \mathcal{B} \rightarrow L_1^{(2)}$ gives rise to a glued near hexagon of type $(Q(5,2) \otimes Q(5,2)) \otimes_b Q(5,2)$. All these glued near hexagons are isomorphic since the group of automorphisms of $Q(5,2)$ fixing S_2 and the line $L_1^{(2)} \in S_2$ induces the full group of six permutations of the line $L_1^{(2)}$, see Lemma 6.104. □

Corollary 6.110. *We have*

$$\begin{aligned} C_3 &= \{DH(5,4), \mathbb{G}_3, Q(5,2) \otimes Q(5,2)\}, \\ C_4 &= \{DH(7,4), \mathbb{G}_4, DH(5,4) \otimes Q(5,2), \mathbb{G}_3 \otimes Q(5,2), \mathbb{E}_1 \otimes Q(5,2), \\ &\qquad (Q(5,2) \otimes Q(5,2)) \otimes_a Q(5,2), (Q(5,2) \otimes Q(5,2)) \otimes_b Q(5,2)\}. \end{aligned}$$

Chapter 7

Slim dense near hexagons

7.1 Introduction

In this chapter, we classify all slim dense near hexagons. This classification is mainly based on the paper [12], which itself was a compilation of the reports [9], [14] and [17]. The classification in [12] relies on Fisher's theory on groups generated by 3-transpositions ([72]) and Buekenhout's geometric interpretation of that theory ([18]). The proof which we will present here avoids the use of group theory and is purely combinatorial. The results of our classification are summarized in the following theorem.

Theorem 7.1. *If $\mathcal{S}$ is a dense near hexagon of order $(2,t)$, then $\mathcal{S}$ is isomorphic to one of the 11 near hexagons mentioned in the following table.*

near hexagon	v	$t+1$	*big quads*	*other quads*	*local spaces*
$\mathbb{L}_3 \times \mathbb{L}_3 \times \mathbb{L}_3$	27	3	$\mathbb{L}_3 \times \mathbb{L}_3$	–	$C_{2,2}$
$W(2) \times \mathbb{L}_3$	45	4	$\mathbb{L}_3 \times \mathbb{L}_3$, $W(2)$	–	$C_{3,2}$
$Q(5,2) \times \mathbb{L}_3$	81	6	$\mathbb{L}_3 \times \mathbb{L}_3$, $Q(5,2)$	–	$C_{5,2}$
$\mathbb{H}_3 \cong \mathbb{I}_3$	105	6	$W(2)$	$\mathbb{L}_3 \times \mathbb{L}_3$	$\mathrm{PG}(2,2)'$
$DQ(6,2)$	135	7	$W(2)$	–	$\mathrm{PG}(2,2)$
$Q(5,2) \otimes Q(5,2)$	243	9	$Q(5,2)$	$\mathbb{L}_3 \times \mathbb{L}_3$	$C_{5,5}$
$\mathbb{G}_3$	405	12	$Q(5,2)$	$\mathbb{L}_3 \times \mathbb{L}_3$, $W(2)$	$\mathcal{L}_{\mathbb{G}_3}$
$\mathbb{E}_1$	729	12	–	$\mathbb{L}_3 \times \mathbb{L}_3$	K_{12}
$\mathbb{E}_2$	759	15	–	$W(2)$	$\mathrm{PG}(3,2)$
$\mathbb{E}_3$	567	15	$Q(5,2)$	$W(2)$	$\overline{W(2)}$
$DH(5,4)$	891	21	$Q(5,2)$	–	$\mathrm{PG}(2,4)$

In the table $Q(5,2) \otimes Q(5,2)$ denotes the unique glued near hexagon of type $Q(5,2) \otimes Q(5,2)$. The number v denotes the total number of points of the near hexagon and $t+1$ denotes the constant number of lines through a point. $\mathcal{L}_{\mathbb{G}_3}$ denotes a linear space such that all local spaces of $\mathbb{G}_3$ are isomorphic to $\mathcal{L}_{\mathbb{G}_3}$. K_{12} denotes the complete graph on 12 vertices.

For a given slim dense near hexagon, all local spaces are isomorphic. We can use this property to show the following result regarding the structure of slim dense near polygons. Recall that every slim dense near polygon is finite by Theorem 2.35.

Theorem 7.2. *For every slim dense near polygon $\mathcal{S}$ there exist constants $a_{\mathcal{S}}$, $b_{\mathcal{S}}$ and $c_{\mathcal{S}}$ such that every point x of $\mathcal{S}$ is contained in precisely $a_{\mathcal{S}}$ grid-quads, $b_{\mathcal{S}}$ $W(2)$-quads and $c_{\mathcal{S}}$ $Q(5,2)$-quads. Moreover, we have $t_{\mathcal{S}}(t_{\mathcal{S}}+1) = 2a_{\mathcal{S}}+6b_{\mathcal{S}}+20c_{\mathcal{S}}$.*

Proof. For each of the 11 slim dense near hexagons H, let a_H, b_H, respectively c_H, denote the constant number of grid-quads, $W(2)$-quads, respectively $Q(5,2)$-quads through a given point of H. Consider two collinear points x and y of $\mathcal{S}$ and let μ denote the number of grid-quads through xy. For every hex H through xy, let λ_H denote the number of grid-quads through xy. The total number of grid-quads through x is equal to $\mu + \sum(a_H - \lambda_H)$ where the summation ranges over all hexes H through the line xy. Similarly, the number of grid-quads through y is also equal to $\mu + \sum(a_H - \lambda_H)$. Hence every two collinear points are contained in the same number of grid-quads. By connectedness of $\mathcal{S}$, it follows that every point of $\mathcal{S}$ is contained in the same number of grid-quads, say $a_{\mathcal{S}}$. In a similar way, one shows that there exist constants $b_{\mathcal{S}}$ and $c_{\mathcal{S}}$ such that every point of $\mathcal{S}$ is contained in $b_{\mathcal{S}}$ $W(2)$-quads and $c_{\mathcal{S}}$ $Q(5,2)$-quads. Since every two intersecting lines of $\mathcal{S}$ are contained in a unique quad, $t_{\mathcal{S}}(t_{\mathcal{S}}+1) = 2a_{\mathcal{S}} + 6b_{\mathcal{S}} + 20c_{\mathcal{S}}$. □

7.2 Elementary properties of slim dense near hexagons

In this section we will collect some elementary properties of slim dense near hexagons. Let $\mathcal{S}$ be a slim dense near hexagon, let v denote the total number of points of $\mathcal{S}$ and let $t+1$ denote the constant number of lines through a point. By Theorem 2.6, there exist constants n_i, $i \in \{0,1,2,3\}$, such that $|\Gamma_i(x)| = n_i$ for every point x of $\mathcal{S}$. We have

$$\begin{aligned} n_0 &= 1, \\ n_1 &= 2(t+1), \\ n_3 &= 2(n_2 - 4t). \end{aligned}$$

Every two points at distance 2 are contained in a unique quad, which is either a grid-quad, a $W(2)$-quad or a $Q(5,2)$-quad.

Lemma 7.3. *If Q_1 and Q_2 are two $W(2)$-quads of a slim dense near hexagon, then precisely one of the following holds:*

(1) $Q_1 = Q_2$;

(2) *$Q_1 \cap Q_2$ is a line;*

(3) *$Q_1 \cap Q_2$ is a point;*

(4) *$Q_2 \subseteq \Gamma_1(Q_1)$;*

(5) *$Q_1 \cap Q_2 = \emptyset$ and $Q_2 \cap \Gamma_1(Q_1)$ induces a (3×3)-grid;*

(6) *$Q_1 \cap Q_2 = \emptyset$ and $Q_2 \cap \Gamma_1(Q)$ consists of three intersecting lines.*

As a corollary, $Q_2 \cap \Gamma_2(Q_1) \in \{0, 6, 8\}$.

Proof. If $Q_1 \cap Q_2 \neq \emptyset$, then we have one of the cases (1), (2) or (3). So, suppose that $Q_1 \cap Q_2 = \emptyset$ and $Q_2 \cap \Gamma_2(Q_1) \neq \emptyset$. Put $X := Q_2 \cap \Gamma_1(Q_1)$. Since $W(2)$ has no partitions in ovoids, $\Gamma_2(Q_1)$ does not contain lines by Theorem 1.23 (4). As a consequence, every line of Q_2 meets X. If x_1 and x_2 are two collinear points of X, then x_1x_2 is completely contained in X by Theorem 1.23 (1). Hence, X is either a set of noncollinear points or induces a possibly degenerate subquadrangle of Q_2. For the latter possibility, either case (5) or (6) occurs. We prove that the first possibility cannot occur. Suppose $Q_1 \cap Q_2 = \emptyset$ and $Q_2 \cap \Gamma_1(Q_1)$ is an ovoid O_2 of Q_2. Put $\Gamma_1(O_2) \cap Q_1 = O_1$. Let x be a point of $Q_2 \setminus O_2$. Then $\mathrm{d}(x, Q_1) = 2$. The point x has distance 1 from three points of O_2 and hence distance 2 from at least three points of O_1. It follows that $\Gamma_2(x) \cap Q_1 = O_1$. So, x must be collinear with all points of O_2, a contradiction. □

For every point x of $\mathcal{S}$, let a_x, b_x, respectively c_x, denote the total number of grid-quads, $W(2)$-quads, respectively $Q(5,2)$-quads, through x. Since every two lines through x are contained in a unique grid, we have

$$2a_x + 6b_x + 20c_x = t(t+1). \tag{7.1}$$

Since every two points at distance 2 are contained in a unique quad, we have

$$4a_x + 8b_x + 16c_x = n_2. \tag{7.2}$$

From equations (7.1) and (7.2), it follows that if $\mathcal{S}$ contains only two types of quads, then the number of quads of each type through a point is a constant.

If x is a point of $\mathcal{S}$ and if Q is a quad of $\mathcal{S}$, then $\mathrm{d}(x, Q) \leq 2$. If $\mathrm{d}(x, Q) \leq 1$, then by Theorem 1.5, x is classical with respect to Q. If $\mathrm{d}(x, Q) = 2$, then x is ovoidal with respect to Q since $\mathrm{d}(x, y) \in \{2, 3\}$ for every point y of Q.

Suppose Q is a big quad of $\mathcal{S}$ and let x denote a point of Q. By Theorem 1.7, every quad of $\mathcal{S}$ different from Q either is disjoint from Q or intersects Q in a line. The quad Q corresponds to a line L_Q in the local space $\mathcal{L}_x := \mathcal{L}(\mathcal{S}, x)$ which meets every other line of the local space. If Q has order $(2, t_2)$, then $v = |Q| + |\Gamma_1(Q)|$ or

$$v = 3(1 + 2t_2)[1 + 2(t - t_2)].$$

So, if a quad of order $(2, t_2)$ is big, then every quad of order $(2, t_2)$ is big.

Suppose that $\mathcal{S}$ contains a $Q(5,2)$-quad Q. Since $Q(5,2)$ has no ovoids, every point of $\mathcal{S}$ is classical with respect to Q and it follows that Q is a big quad. The total number of points is then equal to $v = 27(2t-7)$.

Suppose that $\mathcal{S}$ contains a $W(2)$-quad Q'. If Q' is big in $\mathcal{S}$, then the total number of points is equal to $v = 15(2t-3)$. Suppose now that Q' is not big in $\mathcal{S}$, then there exists a point $x \in \Gamma_2(Q)$. The points in Q' at distance 2 from x form an ovoid $\{x_1, \ldots, x_5\}$. Since $d(x, Q) = 2$, the quad $\mathcal{C}(x, x_i)$, $i \in \{1, \ldots, 5\}$, intersects Q in $\{x_i\}$ and cannot be isomorphic to $Q(5,2)$. So, $\mathcal{C}(x, x_i)$ is isomorphic to $\mathbb{L}_3 \times \mathbb{L}_3$ or $W(2)$ for every $i \in \{1, \ldots, 5\}$. If L is a line through x, then L has at most one point in common with $\Gamma_1(Q)$ by Theorem 1.23 (1). It follows that the quads $\mathcal{C}(x, x_i)$, $i \in \{1, \ldots, 5\}$, two by two intersect in the point x. Since $W(2)$ has no partition in ovoids, $\Gamma_2(Q)$ does not contain lines by Theorem 1.23 (4). Hence every line through x contains a unique point of $\Gamma_1(Q)$. It follows that the quads $\mathcal{C}(x, x_1), \ldots, \mathcal{C}(x, x_5)$ determine a partition of the lines through x. So, $10 \leq t+1 \leq 15$.

We will perform the classification for each of the following cases.

(I) $\mathcal{S}$ is a regular near hexagon.

(II) $\mathcal{S}$ contains grid-quads and $W(2)$-quads, but no $Q(5,2)$-quads.

(III) $\mathcal{S}$ contains grid-quads and $Q(5,2)$-quads, but no $W(2)$-quads.

(IV) $\mathcal{S}$ contains $W(2)$-quads and $Q(5,2)$-quads, but no grid-quads.

(V) $\mathcal{S}$ contains grid-quads, $W(2)$-quads and $Q(5,2)$-quads.

7.3 Case I: $\mathcal{S}$ is a regular near hexagon

Suppose that $\mathcal{S}$ is a regular near hexagon with parameters $s = 2$, t_2 and t. We distinguish the following cases.

- If $t_2 = 1$, then from eigenvalue techniques, see Chapter 3, it follows that $t = 2$ or $t = 11$. If $t = 2$, then $\mathcal{S}$ is isomorphic to $\mathbb{L}_3 \times \mathbb{L}_3 \times \mathbb{L}_3$ by Theorem 4.4. If $t = 11$, then $\mathcal{S}$ is isomorphic to $\mathbb{E}_1$ by Theorem 6.60.

- If $t_2 = 2$, then from eigenvalue techniques, see Chapter 3, it follows that $t = 6$ or $t = 14$. If $t = 6$, then $\mathcal{S}$ is the dual polar space $DQ(6,2)$ by Corollary 3.7. If $t = 14$, then $\mathcal{S}$ is isomorphic to $\mathbb{E}_2$ by Theorem 6.77.

- If $t_2 = 4$, then every quad of $\mathcal{S}$ is isomorphic to $Q(5,2)$ and hence big. So, $\mathcal{S}$ is a classical near hexagon. It follows that $\mathcal{S} \cong DH(5,4)$.

7.4 Case II: $\mathcal{S}$ contains grid-quads and $W(2)$-quads but no $Q(5,2)$-quads

In this case there exist constants a and b such that every point is contained in a grid-quads and b $W(2)$-quads. We have $2a + 6b = t(t+1)$, $n_0 = 1$, $n_1 = 2(t+1)$, $n_2 = 4a + 8b = 4(a+2b)$, $n_3 = 2(4a + 8b - 4t) = 8(a + 2b - t)$ and $v = 3 - 6t + 12(a + 2b)$. We distinguish two cases.

7.4.1 There exists a big $W(2)$-quad

Then $v = 15(2t - 3)$ and every $W(2)$-quad is big. We can express a and b in terms of t. We find that

$$a = \frac{-2t^2 + 16t - 24}{2},$$
$$b = \frac{t^2 - 5t + 8}{2}.$$

Lemma 7.4. *There exists a constant α such that for every $W(2)$-quad Q and every line L such that $|L \cap Q| = 1$, there are α grid-quads and $3 - \alpha$ $W(2)$-quads through L.*

Proof. Since Q is big, every quad through L intersects Q in a line by Theorem 1.7. Suppose that there are α_L grid-quads and $3 - \alpha_L$ $W(2)$-quads through L. Since the quads through L partition the set of lines through $L \cap Q$ different from L, we find that $t = \alpha_L + 2(3 - \alpha_L) = 6 - \alpha_L$. Hence α_L is equal to $\alpha := 6 - t$. □

Lemma 7.5. *If $\alpha = 3$, then $\mathcal{S}$ is isomorphic to $W(2) \times \mathbb{L}_3$.*

Proof. We find that $t + 1 = 4$. The lemma now follows from Theorem 4.1. □

Lemma 7.6. *The case $\alpha = 2$ cannot occur.*

Proof. One calculates that $t + 1 = 5$, $a = 4$ and $b = 2$. Hence, every local space is isomorphic to the $(3,3)$-cross. By Theorem 4.24, it follows that $\mathcal{S}$ is a glued near hexagon of type $W(2) \otimes W(2)$. But this is impossible, because by Theorem 4.42, $W(2)$ has no spreads of symmetry. □

Lemma 7.7. *If $\alpha = 1$, then $\mathcal{S}$ is isomorphic to $\mathbb{H}_3$.*

Proof. One calculates that $t + 1 = 6$, $a = 3$, $b = 4$ and $v = 105$. It easily follows that every local space is isomorphic to $\mathrm{PG}(2,2)'$. Let Γ be the graph with vertices the $W(2)$-quads of $\mathcal{S}$ with two $W(2)$-quads adjacent whenever they are disjoint. The graph Γ has $w = \frac{v \cdot 4}{15} = 28$ vertices. There are 15 $W(2)$-quads intersecting a given $W(2)$-quad in a line. It follows that the graph Γ is regular with valency $k = w - 1 - 15 = 12$.

Let R and S be two disjoint quads of order 2 and let $T = \mathcal{R}_R(S)$. There are 15 lines meeting R, S and T, and every such line is contained in two quads of order 2. Hence there are exactly 10 quads of order 2 meeting R, S and T. From the three quads ($\neq T$) of order 2 through a point of T, there is only one disjoint from R and S. Hence there are five quads of order 2, disjoint from R and S, meeting T in a line. Similarly, there are five quads of order 2, disjoint from T, meeting R (respectively S) in a line. Together with R, S and T, we counted all 28 quads of order 2. It easily follows that every maximal clique of Γ contains either three or seven vertices. Moreover, every two adjacent vertices are contained in a unique maximal clique of size 3 and a unique maximal clique of size 7.

Let X be a set of size 8. Since there are $\frac{w\times k}{7\times 6} = 8$ maximal cliques of size 7, every element of X can be associated to a maximal clique. Since every quad of order 2 is contained in $\frac{k}{6} = 2$ maximal cliques of size 7, every quad Q of order 2 corresponds to a set $A_Q \subseteq X$ of order 2. Since there are 28 quads of order 2 and 28 subsets of order 2, the correspondence is bijective. Every line of $\mathcal{S}$ is contained in two quads Q_1 and Q_2 of order 2, and since $A_{Q_1} \cap A_{Q_2} = \emptyset$, every line corresponds to a partition $\{A_{Q_1}, A_{Q_2}, X \setminus (A_{Q_1} \cup A_{Q_2})\}$ of X. Since there are 210 lines and 210 partitions of X in two subsets of size 2 and one subset of size 4, the correspondence is bijective. Every point of $\mathcal{S}$ is contained in four quads Q_i, $i \in \{1,2,3,4\}$, of order 2, and hence corresponds to a partition $\{A_{Q_1}, A_{Q_2}, A_{Q_3}, A_{Q_4}\}$ of X. Since there are 105 points and 105 partitions of X in four subsets of order 2, the correspondence is bijective. It is now clear that $\mathcal{S} \cong \mathbb{H}_3$. □

7.4.2 No $W(2)$-quad is big

In this case, we have

$$10 \leq t+1 \leq 15. \tag{7.3}$$

From $2a + 6b = t(t+1)$, it follows that

$$1 \leq a \leq \frac{t(t+1)}{2} - 3. \tag{7.4}$$

Each pair (t, a) which is possible by equations (7.3) and (7.4) will be ruled out by at least one of the Lemmas 7.8, 7.10, 7.11, 7.12, 7.13, 7.14. Notice that b and v can be written as functions of t and a since $2a+6b = t(t+1)$ and $v = 3-6t+12(a+2b)$. The following lemma is straightforward.

Lemma 7.8.

- *The number of grid-quads is equal to $N_1 := \frac{av}{9}$ and the number of $W(2)$-quads is equal to $N_2 := \frac{bv}{15}$. As a consequence, these numbers are integral.*
- *If Q is a $W(2)$-quad, then $|\Gamma_2(Q)| = N$, where $N := v - 15 - 30(t-2)$.*

Lemma 7.9. *Let x be a point of $\mathcal{S}$ which is ovoidal with respect to a $W(2)$-quad Q. Then x is contained in $t-9$ $W(2)$-quads and $14-t$ grid-quads which meet Q.*

Proof. Suppose that x is contained in k_1 grid-quads and k_2 $W(2)$-quads meeting Q. Since there are five quads meeting Q, we have $k_1 + k_2 = 5$. We have seen earlier that the five quads through x partition the set of lines through x. So, we also have $2k_1 + 3k_2 = t + 1$. The lemma now immediately follows. □

Lemma 7.10. *Let Q be a given $W(2)$-quad. The total number of grid-quads, respectively $W(2)$-quads, which meet Q in a unique point is given by $\frac{1}{4}(14-t)N$, respectively $\frac{1}{8}(t-9)N$. If $t \neq 10$, then the total number of $W(2)$-quads which intersect Q in a given point is equal to $\frac{1}{120}(t-9)N$. As a consequence,*

- *$\frac{1}{4}(14-t)N$ is integral and at most equal to $15a$,*
- *if $t \neq 10$, then $\frac{1}{120}(t-9)N$ is integral and at most equal to $b-1$.*

Proof. Let A_x denote the total number of $W(2)$-quads which meet Q in the point $x \in Q$. Then $A := \sum_{x \in Q} A_x$ denotes the the total number of $W(2)$-quads which meet Q in a unique point. Counting in two ways the number of pairs (u, Q'), where $|Q \cap Q'| = 1$ and $u \in Q' \cap \Gamma_2(Q)$ gives $A \cdot 8 = N \cdot (t-9)$. Hence, $A = \frac{1}{8}(t-9)N$. In a similar way one shows that there are precisely $\frac{1}{4}(14-t)N$ grid-quads which intersect Q in a point. Suppose now that $t \neq 10$. We will show that $A_x = A_y$ for all points x and y of Q. Since the noncollinearity graph of Q is connected it suffices to prove this if x and y are not collinear. Counting the number of pairs (Q_x, Q_y), where Q_x and Q_y are $W(2)$-quads satisfying $Q_x \cap Q = \{x\}$, $Q_y \cap Q = \{y\}$ and $|Q_x \cap Q_y| = 1$, we find that $A_x \cdot (t-10) = A_y \cdot (t-10)$ or $A_x = A_y$. Hence if $t \neq 10$, then $A_x = \frac{1}{120}(t-9)N$ for every point x of Q. □

Lemma 7.11. *The case $(t, a, b) = (11, 15, 17)$ cannot occur.*

Proof. Let x be a point of $\mathcal{S}$ and consider the local space $\mathcal{L}_x$. Let L be a line of size 3 in $\mathcal{L}_x$. Since $t = 11$, every point of $\mathcal{L}_x$ is contained in at most five lines of size 3. Hence, there are at most 13 lines of size 3 which have nonempty intersection with L. Let Q denote the $W(2)$-quad corresponding with L. By Lemma 7.10, there are precisely four $W(2)$-quads R such that $R \cap Q = \{x\}$. Hence, there are precisely four lines of size 3 in $\mathcal{L}_x$ which are disjoint with L. Since there are precisely 17 lines of size 3, we find that every point of L is contained in precisely five lines of size 3. Hence, every point of $\mathcal{L}_x$ is incident with either 0 or five lines of size 3. So, the number of points of $\mathcal{L}_x$ which are contained in five lines of size 3 is given by $\frac{3b}{5}$, a contradiction, since this number is not an integer. □

Remark. If Q is a $W(2)$-quad, then the number of $W(2)$-quads intersecting Q in a line is equal to $\frac{1}{3}[15(b-1) - \frac{1}{8}(t-9)N]$.

Lemma 7.12. $\frac{1}{3}(b - 1 - \frac{1}{120}(t-9)N) + 1 \geq \frac{3b}{t+1}$.

Proof. For every line L of $\mathcal{S}$, let i_L denote the number of $W(2)$-quads through L.

We have

$$\begin{aligned}\sum_L 1 &= \frac{1}{3}(t+1)v,\\ \sum_L i_L &= 15N_2 = vb,\\ \sum_L i_L(i_L-1) &= \frac{1}{3}N_2[15(b-1)-\frac{1}{8}(t-9)N].\end{aligned}$$

The inequality $\sum_L (i_L - \frac{3b}{t+1})^2$ now gives the desired inequality. □

From Lemma 7.3 we get the following result.

Lemma 7.13. *Let Q be a $W(2)$-quad and let m_i, $i \in \{0,6,8\}$, denote the number of $W(2)$-quads Q' for which $|\Gamma_2(Q) \cap Q'| = i$. Then the following (in)equalities hold:*

$$\begin{aligned}m_0+m_6+m_8 &= N_2,\\ 6m_6+8m_8 &= bN,\\ m_0 &\geq 1+\frac{1}{3}[15(b-1)-\frac{1}{8}(t-9)N].\end{aligned}$$

As a consequence, $\frac{1}{8}bN \leq N_2 - 1 - 5(b-1) + \frac{1}{24}(t-9)N$.

Lemma 7.14. *If $t = 9$, then $b \leq 6$.*

Proof. Let x be a point of $\mathcal{S}$ and consider the local space $\mathcal{L}_x$. By Lemma 7.10, every two lines of size 3 in $\mathcal{L}_x$ meet each other.

If $\mathcal{L}_x$ contains a point u which is collinear with four lines of size 3, then every other line of size 3 in $\mathcal{L}_x$ necessarily goes through u. It follows that $b \leq 4$.

Suppose $b \geq 5$ and let L_1 and L_2 denote two lines of size 3. The number of lines of size 3 intersecting L_1 and L_2 in a point different from $L_1 \cap L_2$ is at most 4. The number of lines of size 3 through $L_1 \cap L_2$ is at most 3. Hence, $b \leq 7$.

Suppose that $b = 7$, then the lines of size 3 determine a Fano-plane in each local space. Hence, there are two types of lines in $\mathcal{S}$:

- lines of type I are contained in three grid-quads and three $W(2)$-quads;
- lines of type II are contained in 0 $W(2)$-quads and nine grid-quads.

Let Q denote a $W(2)$-quad of $\mathcal{S}$ and define the numbers m_0, m_6 and m_8 as in Lemma 7.13. From $a = 24$, $v = 405$ and $N = 180$, we find that

$$\begin{aligned}m_0+m_6+m_8 &= 189,\\ 6m_6+8m_8 &= 1260,\\ m_0 &= 1+2\cdot 15+\epsilon,\end{aligned}$$

where ϵ denotes the number of $W(2)$-quads contained in $\Gamma_1(Q)$. The only possible solution is $\epsilon = 0$, $m_0 = 31$, $m_6 = 2$ and $m_8 = 156$. Each point of Q is contained in four lines of type I which are not contained in Q. Hence, there are 120 points x in $\Gamma_1(Q)$ for which $x\,\pi_Q(x)$ is a line of type I. Let x_1 be one of these points. Every $W(2)$-quad R through $x_1\,\pi_Q(x_1)$ intersects Q in a line and the two lines of R through x_1 different from $x_1\,\pi_Q(x_1)$ are also of type I. In this way, we find six lines of type I through x_1 which are contained in $\Gamma_1(Q)$. As a consequence, there are four $W(2)$-quads through x_1 which contain three of these six lines. Since $\epsilon = 0$, each of these quads contributes to m_8. Hence, $m_8 \geq 4 \cdot 120$, a contradiction. □

7.5 Case III: $\mathcal{S}$ contains grid-quads and $Q(5,2)$-quads but no $W(2)$-quads

In this case, there exist constants a and c such that every point is contained in a grid-quads and c $Q(5,2)$-quads. We have $t(t+1) = 2a + 20c$, $n_0 = 1$, $n_1 = 2(t+1)$, $n_2 = 4a + 16c = 4(a+4c)$, $n_3 = 2(4a + 16c - 4t) = 8(a + 4c - t)$ and $v = 3 - 6t + 12(a + 4c)$. Since $\mathcal{S}$ has a big $Q(5,2)$-quad, we have $v = 27(2t-7)$. Equating both expressions for v, we find that $a + 4c = 5t - 16$. We can now express a and c in terms of t. We find

$$\begin{aligned} a &= \frac{-t^2 + 24t - 80}{3}, \\ c &= \frac{t^2 - 9t + 32}{12}. \end{aligned}$$

Lemma 7.15. *There exists a constant α such that for every $Q(5,2)$-quad Q and every line L such that $|L \cap Q| = 1$, there are α grid-quads and $5 - \alpha$ $Q(5,2)$-quads through L.*

Proof. Since Q is big, every quad through L intersects Q in a line by Theorem 1.7. Suppose that there are α_L grid-quads and $5 - \alpha_L$ $Q(5,2)$-quads through L. Since the quads through L partition the set of lines through $L \cap Q$ different from L, we find that $t = \alpha_L + 4(5 - \alpha_L) = 20 - 3\alpha_L$. Hence α_L is equal to $\alpha := \frac{20-t}{3}$. □

Lemma 7.16. *If $\alpha = 5$, then $\mathcal{S} \cong Q(5,2) \times \mathbb{L}_3$. If $\alpha = 4$, then $\mathcal{S}$ is isomorphic to $Q(5,2) \otimes Q(5,2)$. The case $\alpha \in \{0,1,2,3\}$ cannot occur.*

Proof.
- If $\alpha = 5$, then $t = 5$, $c = 1$ and $a = 5$. By Theorem 4.1, $\mathcal{S} \cong Q(5,2) \times \mathbb{L}_3$.
- If $\alpha = 4$, then $t = 8$, $c = 2$ and $a = 16$. Every local space is a $(5,5)$-cross. By Theorem 4.24, $\mathcal{S}$ is a glued near hexagon of type $Q(5,2) \otimes Q(5,2)$. By Theorem 6.107, there exists a unique glued near hexagon of type $Q(5,2) \otimes Q(5,2)$.
- If $\alpha = 3$, then $t = 11$ and $c = \frac{t^2-9t+32}{12} \notin \mathbb{N}$. So, this case cannot occur.
- If $\alpha = 2$, then $t = 14$ and $c = \frac{t^2-9t+32}{12} \notin \mathbb{N}$. So, this case cannot occur.

- If $\alpha = 1$, then $t = 17$, $c = 14$ and $a = 13$. We will show that this case cannot occur. Let x be an arbitrary point of $\mathcal{S}$. Since $a = 13$, there exists a line through x which is contained in at least two grid-quads R_1 and R_2. Since $c = 14$, there exists a $Q(5, 2)$-quad Q through x not containing the line L. Now, R_1 and R_2 are two grid-quads through L intersecting Q in a line, contradicting $\alpha = 1$.
- Since there exist grid-quads through every point, α cannot be equal to 0. □

7.6 Case IV: $\mathcal{S}$ contains $W(2)$-quads and $Q(5, 2)$-quads but no grid-quads

There exist constants b and c such that every point is contained in b $W(2)$-quads and c $Q(5, 2)$-quads. We have $6b+20c = t(t+1)$, $n_0 = 1$, $n_1 = 2(t+1)$, $n_2 = 8b+16c$, $n_3 = 2(8b+16c-4t) = 8(2b+4c-t)$ and $v = 3-6t+12(2b+4c)$. Since $\mathcal{S}$ contains a big $Q(5, 2)$-quad, we have $v = 27(2t-7)$. Equating both expressions for v, we find that $2b + 4c = 5t - 16$.

Lemma 7.17. *No $W(2)$-quad is big.*

Proof. Suppose the contrary. Then $v = 15(2t-3) = 30t-45$. Together with $v = 27(2t-7)$ this implies that $t = 6$. From $6b+20c = t(t+1)$ and $2b+4c = 5t-16$, it now follows that $c = 0$, a contradiction. □

Lemma 7.18. $t + 1 = 15$, $b = 15$ *and* $c = 6$.

Proof. Let Q be a $W(2)$-quad and let x be a point at distance 2 from Q. Then there are five quads through x which intersect Q in a point. All these quads are isomorphic to $W(2)$ and partition the set of lines through x. Hence $t + 1 = 15$. From $6b + 20c = t(t + 1)$ and $2b + 4c = 5t - 16$, it now follows that $b = 15$ and $c = 6$. □

Lemma 7.19. *Every local space is isomorphic to $\overline{W(2)}$.*

Proof. Every local space $\mathcal{L}$ has 15 points, 15 lines of size 3 and six lines of size 5. Every line of size 5 meets every other line of the local space. For every point x of $\mathcal{L}$ there exists a line L of size 5 not containing x and every line through x meets L. As a consequence, every point of $\mathcal{L}$ is contained in precisely five lines. Since all these lines cover all the points of $\mathcal{L}$, it easily follows that every point is contained in three lines of size 3 and two lines of size 5. The lemma now follows from Lemma 6.44. □

Definition. Let Ω denote the incidence structure whose points are the pairs $(x, \{K_1, K_2\})$ where (x, K_1) and (x, K_2) are two different flags of $Q(5, 2)$, and whose lines are the pairs (N, R) with N a line of $Q(5, 2)$ and R a proper subquadrangle of $Q(5, 2)$ through N. A point $(x, \{K_1, K_2\})$ is incident with a line (N, R) if and only if the following conditions are satisfied:

(i) $x \in N$,

(ii) K_1 and K_2 are lines of R,

(iii) $K_1 \neq N \neq K_2$ if $R \cong W(2)$.

Recall that with every big convex subpolygon F of a slim dense near polygon $\mathcal{S}$, there is associated a partial linear space $\Omega(\mathcal{S}, F)$, see Section 2.6.

Lemma 7.20. *If Q is a $Q(5,2)$-quad of $\mathcal{S}$, then $\Omega(\mathcal{S}, Q) \cong \Omega$.*

Proof. Every POINT K of $\Omega(\mathcal{S}, Q)$ is contained in two big quads Q_1 and Q_2 intersecting Q in the respective lines K_1 and K_2. Clearly Q_i, $i \in \{1,2\}$, is the only big quad through K_i different from Q, and $K = Q_1 \cap Q_2$. Hence, the map $\theta : K \mapsto (K \cap Q, \{K_1, K_2\})$ is a bijection between the point sets of $\Omega(\mathcal{S}, Q)$ and Ω. Now, let G be a subgrid of $\mathcal{S}$ intersecting Q in a line N, and let K, L and M be the three POINTS of $\Omega(\mathcal{S}, Q)$ incident with G. If G is contained in a $Q(5,2)$-quad, then we may suppose that $K_1 = L_1 = M_1 = N$. If Q_A, $A \in \{K, L, M\}$, denotes the big quad through A and A_2, then the reflection of Q_K about Q_L is equal to Q_M. Hence K_2, L_2 and M_2 are contained in a subgrid R of $\mathcal{S}$. Suppose now that G is contained in a $W(2)$-quad. As before, the reflection of K_i, $i \in \{1,2\}$, around L_j, $j \in \{1,2\}$, belongs to $\{M_1, M_2\}$. This proves that K_1, K_2, L_1, L_2, M_1 and M_2 are contained in a subquadrangle R isomorphic to $W(2)$. We have just shown that the incidence is preserved by θ. The lemma now follows since $\Omega(\mathcal{S}, Q)$ and Ω have the same number of lines. □

Corollary 7.21. *$\mathcal{S}$ is isomorphic to $\mathbb{E}_3$.*

Proof. Let Q' denote an arbitrary $Q(5,2)$-quad of $\mathbb{E}_3$. Since $\mathbb{E}_3$ has $W(2)$-quads and $Q(5,2)$-quads, but no grid-quads, $\Omega(\mathbb{E}_3, Q') \cong \Omega(\mathcal{S}, Q)$. By Theorem 2.40, $\mathcal{S} \cong \mathbb{E}_3$. □

7.7 Case V: $\mathcal{S}$ contains grid-quads, $W(2)$-quads and $Q(5,2)$-quads

In this case, we have $v = 27(2t-7)$.

Lemma 7.22. *No quad isomorphic to $W(2)$ is big. As a consequence, $10 \leq t+1 \leq 15$.*

Proof. Suppose the contrary. Then $v = 30t - 45$. Hence $t = 6$. Let x be a point which is contained in a $Q(5,2)$-quad Q. Let L_1 and L_2 be the two lines through x not contained in Q. Then $\mathcal{C}(L_1, L_2)$ is a $W(2)$-quad which intersects Q in a line. Hence, the local space at x is isomorphic to the $(3,5)$-cross. By Theorem 4.24 it follows that $\mathcal{S}$ is a glued near hexagon of type $W(2) \otimes Q(5,2)$. This is impossible since $W(2)$ has no spreads of symmetry by Theorem 4.42. □

Lemma 7.23. *If $t+1 \geq 12$, then every line of $\mathcal{S}$ is contained in a $Q(5,2)$-quad.*

Proof. Let L be a line which meets a $Q(5,2)$-quad Q. If L were not contained in a $Q(5,2)$-quad, then since every quad through L meets Q, we would have that $t \leq 5 \cdot 2$, a contradiction. Hence, every line meeting a $Q(5,2)$-quad is contained in a $Q(5,2)$-quad. Since there exists at least one $Q(5,2)$-quad, every point of $\mathcal{S}$ is contained in a $Q(5,2)$-quad. So, every line of $\mathcal{S}$ meets a $Q(5,2)$-quad and is contained in a $Q(5,2)$-quad. □

Lemma 7.24. *The case $t+1 = 13$ cannot occur.*

Proof. The total number of points is equal to $v = 27(2t-7) = 27 \cdot 17$. Consider a local space $\mathcal{L}_x$. By Lemma 7.23, every point of $\mathcal{L}_x$ is contained in a line of size 5. Hence, $\mathcal{L}_x$ contains at least three lines of size 5, say L_1, L_2 and L_3. We distinguish two cases.

- Suppose $L_1 \cap L_2 \cap L_3 = \emptyset$.
  Then there exists a unique point in $\mathcal{L}_x$ not contained in $L_1 \cup L_2 \cup L_3$. This point is contained in a line L_4 of size 5. The line L_4 has at most one point in common with L_1, L_2 and L_3. Hence, $|L_4| \leq 4$, a contradiction.
- Suppose $L_1 \cap L_2 \cap L_3 = \{u\}$.
  Then every point of $\mathcal{L}_x$ is contained in $L_1 \cup L_2 \cup L_3$. Since every line of $\mathcal{L}_x$ meets L_1, L_2 and L_3, there are precisely 16 lines of size 3 and no line of size 2.

Hence, the total number of $W(2)$-quads is equal to $\frac{16v}{15}$, which is not an integer. So, we have a contradiction. □

For every point x of $\mathcal{S}$, let $I(x)$ denote the intersection of all $Q(5,2)$-quads through x.

Lemma 7.25. *If $t+1 \in \{12, 14, 15\}$, then $I(x) = \{x\}$ for every point x of $\mathcal{S}$.*

Proof. Suppose the contrary. Let L be a line in $I(x)$. By Lemma 7.23, the $Q(5,2)$-quads through L partition the set of lines through x. So, t is a multiple of 4, a contradiction. □

Lemma 7.26. *If $t+1 \in \{12, 14, 15\}$, then there exist constants a, b and c, such that every point of $\mathcal{S}$ is contained in a grid-quads, b $W(2)$-quads and c $Q(5,2)$-quads.*

Proof. By the connectedness of $\mathcal{S}$, it suffices to prove that $a_x = a_y$, $b_x = b_y$ and $c_x = c_y$ for every two collinear points x and y. Let z denote the third point on the line xy. By Lemma 7.25, $I(z) = \{z\}$. So, there exists a $Q(5,2)$-quad through z not containing xy. The reflection about Q is an automorphism of $\mathcal{S}$ mapping x to y. The lemma now follows immediately. □

Lemma 7.27. *If $t+1 \in \{14, 15\}$, then the following holds for every point x of $\mathcal{S}$.*

(i) *The number b is a multiple of* 5.

(ii) *Every point of $\mathcal{L}_x$ is contained in at most three lines of size* 5.

(iii) *If there exists a point in $\mathcal{L}_x$ which is contained in precisely three lines of size 5, then $t+1=15$ and $c=4$.*

(iv) *If every point of $\mathcal{L}_x$ is contained in at most two lines of size 5, then $c=4$ if $t+1=14$ and $c\in\{5,6\}$ if $t+1=15$.*

Proof. (i) The total number of $W(2)$-quads is equal to $\frac{vb}{15}=\frac{27(2t-7)b}{15}$. The statement now follows from the fact that this number must be integral.

(ii) If there were more than three lines of size 5 through a point of $\mathcal{L}_x$, then we would have $t+1\geq 17$.

(iii) Let L_1, L_2 and L_3 denote three lines of size 5 through the same point of $\mathcal{L}_x$. If $t+1=14$, then there exists a unique point u in $\mathcal{L}_x$ not contained in $L_1\cup L_2\cup L_3$. By Lemma 7.23, the point u is contained in a line of size 5 which has at most one point in common with each of the lines L_1, L_2 and L_3. This is impossible. If $t+1=15$, then there exist two points u_1 and u_2 in $\mathcal{L}_x$ not contained in $L_1\cup L_2\cup L_3$. The point u_1 is contained in a line of size 5 which has at most one point in common with each of the lines L_1, L_2 and L_3. Hence, this line coincides with u_1u_2. It now immediately follows that $c=4$.

(iv) Let L_1, L_2 and L_3 denote three lines of size 5 in $\mathcal{L}_x$. If $t+1=14$, let u_1 and u_2 denote the two points of $\mathcal{L}_x$ not contained in $L_1\cup L_2\cup L_3$. The point u_1 is contained in a line of size 5 which intersects each of the lines L_1, L_2 and L_3 in at most one point and hence coincides with u_1u_2. So, $c=4$. If $t+1=15$, let u_1, u_2 and u_3 denote the three points of $\mathcal{L}_x$ not contained in $L_1\cup L_2\cup L_3$. Every line of size 5 different from L_1, L_2 and L_3 has two points outside $L_1\cup L_2\cup L_3$ and hence is equal to either u_1u_2, u_1u_3 or u_2u_3. Hence, $c\leq 6$. Since, each of the points u_1, u_2 and u_3 is contained in a line of size 5, we must also have that $c\geq 5$. □

Lemma 7.28. *The cases $t+1=14$ and $t+1=15$ cannot occur.*

Proof. Suppose $t+1=14$. From $(t-4)(t-5)=2b+12(c-1)$ and Lemma 7.27 (i), it follows that c is a 5-fold+2, contradicting properties (iii) and (iv) of the same lemma. If $t+1=15$, then from $(t-4)(t-5)=2b+12(c-1)$, it follows that c is a 5-fold+1. Hence, $c=6$ and $b=15$. From $2a+6b+20c=(t+1)t$ it now follows that $a=0$, a contradiction, since grid-quads do occur. □

Lemma 7.29. *We have $n_2=20t-64$.*

Proof. Let x be a point of $\mathcal{S}$ which is contained in a $Q(5,2)$-quad. Then $\Gamma_2(x)=(\Gamma_2(x)\cap Q)\cup(\Gamma_2(x)\cap\Gamma_1(Q))$. A point y of $\Gamma_1(Q)$ lies at distance 2 from x if and only if the unique point of Q collinear with x is collinear with x. It follows that $n_2=16+10\cdot 2\cdot(t-4)=20t-64$. □

Lemma 7.30. *If $t+1=10$, then one of the following cases occurs for a point x of $\mathcal{S}$:*

(i) $(a_x, b_x, c_x) = (5, 10, 1)$;

(ii) $(a_x, b_x, c_x) = (13, 4, 2)$.

If $t + 1 = 11$, *then one of the following cases occurs:*

(i) $(a_x, b_x, c_x) = (0, 15, 1)$;

(ii) $(a_x, b_x, c_x) = (8, 9, 2)$.

Proof. Obviously, $c_x \leq 2$. The lemma now follows from the fact that $2a_x + 6b_x + 20c_x = (t+1)t$ and $4a_x + 8b_x + 16c_x = n_2 = 20t - 64$. □

Lemma 7.31. *The case* $t + 1 = 10$ *cannot occur.*

Proof. By the proof of Lemma 7.22, no two $W(2)$-quads intersect in a unique point. Let x be a point of type (i). Let L denote the unique line of size 5 in $\mathcal{L}_x$ and let a, b, c, d and e denote the five points of $\mathcal{L}_x$ not contained in L. Let f be the intersection points of the lines ab and L. The line cd intersects ab. Hence f belongs to cd. In a similar way one shows that f belongs to ce. But this is impossible. Hence, every point is of type (ii). But then the total number of $W(2)$-quads is equal to $\frac{v \cdot 4}{15} = \frac{297 \cdot 4}{15}$, which is not an integer. So, the case $t + 1 = 10$ cannot occur. □

Lemma 7.32. *The case* $t + 1 = 11$ *cannot occur.*

Proof. In this case $\mathcal{S}$ contains $v = 351$ points. If all points were of type (ii), then $\mathcal{S}$ would contain $\frac{351 \cdot 9}{15}$ $W(2)$-quads, a contradiction. Hence, there exists a point x of type (i). Since there exists no grid-quad through x, every $W(2)$-quad either contains x or contains a unique point collinear with x (see the proof of Lemma 7.22).

Step 1: Every point of $\Gamma_3(x)$ *has type (ii).*

Proof. Suppose the contrary. Let y be a point of type (i) in $\Gamma_3(x)$. The unique $Q(5,2)$-quad through y contains a unique point of $\Gamma_1(x)$. Hence, at most 10 $W(2)$-quads through y contain a point at distance 1 from x, a contradiction, since each of the 15 $W(2)$-quads through y must contain such a point. □

Step 2: There are precisely 81 *points of type (i) and* 270 *points of type (ii). There are* 23 $Q(5,2)$*-quads.*

Proof. Suppose that there are α points of type (ii) and $351 - \alpha$ points of type (i). The total number of $Q(5,2)$-quads is equal to $\frac{2\alpha + (351-\alpha)}{27}$. Hence α is a multiple of 27. The total number of $W(2)$-quads is equal to $\frac{9\alpha + 15(351-\alpha)}{15}$. Hence α is a multiple of 5. By Step 1, we know that $\alpha \geq n_3 = 192$. The statement now readily follows. □

Step 3: 12 *points in* $\Gamma_1(x)$ *have type (i) and* 10 *have type (ii).*

Proof. Suppose A' points in $\Gamma_1(x)$ are of type (i) and B' points are of type (ii). Counting pairs (w, Q), where w is a point collinear with x and where Q is a $Q(5,2)$-quad through w gives $A' + 2B' = 1 \cdot 10 + 22 \cdot 1 = 32$. Together with $A' + B' = 22$, this implies that $A' = 12$ and $B' = 10$. □

We will now derive a contradiction. Let x' be one of the 68 points of type (i) at distance 2 from x. By Step 3, there exists a point x'' of type (i) collinear with x not contained in the quad through x and x'. But then $\mathrm{d}(x', x'') = 3$, contradicting Step 1. □

Lemma 7.33. *If $t + 1 = 12$, then $\mathcal{S}$ is isomorphic to $\mathbb{G}_3$.*

Proof. Every local space contains precisely three lines of size 5, nine lines of size 3 and nine lines of size 2 and hence is isomorphic to $\mathcal{L}_{\mathbb{G}_3}$. The lemma now follows from Theorem 7.34 which we will prove in the appendix of this chapter. □

7.8 Appendix

The aim of this appendix is to prove the following theorem.

Theorem 7.34. *Let $\mathcal{S} = (\mathcal{P}, \mathcal{L}, \mathrm{I})$ be a slim dense near $2n$-gon, $n \geq 3$, with the property that every local space is isomorphic to $\mathcal{L}_{\mathbb{G}_n}$. Call a line of $\mathcal{S}$ special if it is not contained in a $W(2)$-quad and ordinary otherwise. Suppose that $\mathcal{C}(L_1, \ldots, L_k) \cong \mathbb{G}_k$ for every $k \in \{1, \ldots, n-1\}$, for every point x and for every k different special lines $L_1, \ldots, L_k$ through x. Then $\mathcal{S} \cong \mathbb{G}_n$.*

In Theorem 7.34, $\mathcal{L}_{\mathbb{G}_n}$ denotes a linear space such that $\mathcal{L}(\mathbb{G}_n, x) \cong \mathcal{L}_{\mathbb{G}_n}$ for every point x of $\mathbb{G}_n$. We prove the theorem in a sequence of lemmas. Let V_k, $k \in \{1, \ldots, n-1\}$, denote the set of all convex sub-$2k$-gons generated by k special lines through a given point of $\mathcal{S}$. Every element of V_k, $k \in \{1, \ldots, n-1\}$, is isomorphic to $\mathbb{G}_k$ and hence contains $m_k := \frac{3^k (2k)!}{2^k k!}$ points. By Theorem 2.31 and the fact that every local space is isomorphic to $\mathcal{L}_{\mathbb{G}_n}$, it follows that every element of V_{n-1} is big.

Lemma 7.35. *Let M_1, M_2 and M_3 be three mutually disjoint lines in a subgrid G of $\mathcal{S}$. If M_1 and M_2 are special, then also M_3 is special.*

Proof. There exists an element $F \in V_{n-1}$ through M_2 not containing G. Since $\mathcal{R}_F \in \mathrm{Aut}(\mathcal{S})$, $M_3 = \mathcal{R}_F(M_1)$ is special. □

Lemma 7.36. *Every $Q(5,2)$-quad Q of $\mathcal{S}$ can be partitioned into three grids, such that a line of Q is special if and only if it is contained in one of these grids.*

Proof. If $x \in Q$, then $\mathcal{L}(\mathcal{S}, x) \cong \mathcal{L}_{\mathbb{G}_n}$ and hence exactly two from the five lines of $Q \cong Q(5,2)$ through x are special. Since Q contains 27 points, it has exactly $\frac{27 \cdot 2}{3} = 18$ special lines. Consider a special line $L \subseteq Q$ and let M_1, M_2 and M_3 denote the three special lines of Q intersecting L in a point. By Lemma 7.35,

M_1, M_2 and M_3 are contained in a grid G_1. Let G_2 and G_3 denote the subgrids of Q as in Lemma 1.38. At most 10 from the 18 special lines meet G_1; hence $G_2 \cup G_3$ contains two intersecting special lines N_1 and N_2. We may suppose that $N_1, N_2 \subseteq G_3$. For every line P of G_2, there exists a unique $i \in \{1,2,3\}$ and a unique $j \in \{1,2\}$ such that P, M_i and N_j are contained in a grid. Hence by Lemma 7.35, every line of G_2 is special. Since Q contains exactly 12 special lines disjoint from G_2, all lines of G_1 and G_3 are special. This proves the lemma. □

Define the following relation R on the set $V := V_{n-1}$. For two elements $v_1, v_2 \in V$, we say that $(v_1, v_2) \in R$ if exactly one of the following holds:

(i) $v_1 = v_2$,

(ii) $v_1 \cap v_2 = \emptyset$ and every line meeting v_1 and v_2 is special.

Lemma 7.37. *The relation R is an equivalence relation and every equivalence class contains exactly three elements.*

Proof. Let $v \in V$ be arbitrary. Every point $a \in v$ is contained in a unique special line $L_a = \{a, a_1, a_2\}$ not contained in v, and we define $\Omega_a := \{v_{a_1}, v_{a_2}\}$ where v_{a_i} denotes the unique element of V through a_i not containing L_a. It suffices to prove that $\Omega_a = \Omega_b$ for all $a, b \in v$.

Suppose first that $\mathrm{d}(a,b) = 1$. Let c denote the unique third point on the line ab and let v' denote an element of V through c not containing ab. Since $\mathcal{R}_{v'} \in \mathrm{Aut}(\mathcal{S})$, $\mathcal{R}_{v'}(L_a)$ is a special line through b and hence equal to L_b. As a consequence L_b is contained in the quad $Q := \mathcal{C}(b, L_a)$. Since L_a is special, Q is not isomorphic to $W(2)$. Suppose that Q is a grid. Since v_{a_i} is big, $Q \cap v_{a_i}$ is a line that meets L_b. Since $L_b \cap v_{a_i} \neq \emptyset$, $i \in \{1,2\}$, $\Omega_b = \{v_{a_1}, v_{a_2}\} = \Omega_a$. Suppose that Q is a $Q(5,2)$-quad. Since $Q \in V_2$ and $v, v_{a_1}, v_{a_2} \in V$, $Q \cap v$, $Q \cap v_{a_1}$ and $Q \cap v_{a_2}$ are special lines. By Lemma 7.36, the unique line through b intersecting $Q \cap v_{a_i}$ is special and hence equal to L_b. Since $L_b \cap v_{a_i} \neq \emptyset$, $i \in \{1,2\}$, $\Omega_b = \{v_{a_1}, v_{a_2}\} = \Omega_a$.

If a and b are not collinear, consider then a path $a = c_0, \ldots, c_k = b$ of length $k = d(a,b)$ between a and b. Then $\Omega_a = \Omega_{c_0} = \cdots = \Omega_{c_k} = \Omega_b$. □

Lemma 7.38. *Let v_1, v_2 and v_3 be three different elements of V for which $(v_1, v_2) \in R$. Then $v_1 \cap v_3 \neq \emptyset$ if and only if $v_2 \cap v_3 \neq \emptyset$.*

Proof. If $a \in v_1 \cap v_3$, then v_3 necessarily contains the unique special line L_a through a not contained in v_1. Since $L_a \cap v_2 \neq \emptyset$, the lemma follows. □

Lemma 7.39. *Let $v_1, v_2, v_3, v_4 \in V$ such that $(v_i, v_j) \notin R$ for all $i, j \in \{1,2,3,4\}$ with $i \neq j$. If $v_1 \cap v_2 = \emptyset$ and $v_3 = \mathcal{R}_{v_2}(v_1)$, then v_4 intersects at least one of v_1, v_2 and v_3.*

Proof. Since every point of $\mathcal{S}$ is contained in n elements of V, we have $|V| = \frac{m_n \cdot n}{m_{n-1}} = 3n(2n-1)$.

(i) Let N_1 denote the number of elements of V intersecting v_1, v_2 and v_3. Every line intersecting v_1 and v_2 is ordinary and hence is contained in $n-2$ elements of V. Each of these $n-2$ elements intersects v_1 in an element of V_{n-2}. Hence $N_1 = \frac{m_{n-1}\cdot(n-2)}{m_{n-2}} = 3(n-2)(2n-3)$.

(ii) Let N_2 denote the number of elements of $V \setminus \{v_1\}$ meeting v_1 and disjoint from v_2 and v_3. By (i), every point of v_1 is contained in $n-2$ elements of V which intersect v_2 and v_3. Hence every point of v_1 is contained in a unique element of $V \setminus \{v_1\}$ disjoint from v_2 and v_3. This element intersects v_1 in an element of V_{n-2}. Hence $N_2 = \frac{m_{n-1}}{m_{n-2}} = 3(2n-3)$.

(iii) There are $N_3 = 9$ elements of V belonging to one of the equivalence classes determined by v_1, v_2 and v_3.

The lemma now follows since $N_1 + 3N_2 + N_3 = |V|$. □

Let Γ be the graph whose vertices are the equivalence classes determined by R with two classes γ_1 and γ_2 adjacent if and only if $v_1 \cap v_2 = \emptyset$ for every $v_1 \in \gamma_1$ and every $v_2 \in \gamma_2$. The graph Γ has $\frac{|V|}{3} = \binom{2n}{2}$ vertices.

Lemma 7.40. *The graph Γ is regular with valency $k(\Gamma) = 4(n-1)$.*

Proof. Let v be a fixed element of V. From the $3n(2n-1)$ elements in V, three are contained in the equivalence class of v, and $\frac{m_{n-1}\cdot(n-1)}{m_{n-2}} = 3(n-1)(2n-3)$ intersect v in an element of V_{n-2}. By Lemma 7.38 it then follows that $k(\Gamma) = \frac{3n(2n-1)-3(n-1)(2n-3)-3}{3} = 4(n-1)$. □

Lemma 7.41. *Every two adjacent vertices γ_1 and γ_2 of Γ are contained in two maximal cliques, one of size 3 and one of size $2n-1$.*

Proof. Let $v_1 \in \gamma_1$, $v_2 \in \gamma_2$, let v_3 denote the reflection of v_2 about v_1 and let γ_3 denote the equivalence class of v_3. By Lemma 7.39, $\{\gamma_1, \gamma_2, \gamma_3\}$ is a maximal clique. Let $C \neq \{\gamma_1, \gamma_2, \gamma_3\}$ denote another maximal clique through γ_1 and γ_2. If $\gamma_4 \in C \setminus \{\gamma_1, \gamma_2\}$, then every $v_4 \in \gamma_4$ intersects v_3. By the proof of Lemma 7.39, there are $N_2 = 3(2n-3)$ mutually disjoint elements in $V \setminus \{v_3\}$ which intersect v_3 and are disjoint from $v_1 \cup v_2$. By Lemma 7.38, these elements of V correspond to $\frac{N_2}{3} = 2n-3$ vertices of Γ. The maximal clique C necessarily consists of γ_1, γ_2 and these $2n-3$ vertices of Γ. This proves our lemma. □

Lemma 7.42. *There is a bijective correspondence between the maximal cliques of size $2n-1$ in Γ and the elements of $B = \{\bar{e}_0, \ldots, \bar{e}_{2n-1}\}$. There is a bijective correspondence between the vertices of Γ and the pairs of the set B.*

Proof. The graph Γ has $\frac{|\Gamma|\cdot k(\Gamma)}{(2n-1)\cdot(2n-2)} = 2n$ maximal cliques of size $2n-1$, proving the first part of the lemma. Since every vertex of Γ is contained in $\frac{k(\Gamma)}{2n-2} = 2$ maximal cliques, it corresponds with a subset of size 2 of B. By Lemma 7.41, every pair of B corresponds to at most one vertex of Γ. The second part of the lemma now follows since there are as many vertices in Γ as there are pairs in B. □

Lemma 7.43. *Let v_1, v_2 denote two nonequivalent disjoint elements of V, let v_3 denote the reflection of v_2 around v_1, and let γ_k, $k \in \{1,2,3\}$, denote the equivalence class determined by v_k. Then there exist $\bar{f}_1, \bar{f}_2, \bar{f}_3 \in B$ such that γ_j, $j \in \{1,2,3\}$, corresponds to $\{\bar{f}_j, \bar{f}_{j+1}\}$, where indices are taken modulo 3.*

Proof. Let γ_1 correspond to $\{\bar{f}_1, \bar{f}_2\} \subseteq B$, γ_2 to $\{\bar{g}_1, \bar{g}_2\} \subseteq B$ and γ_3 to $\{\bar{h}_1, \bar{h}_2\} \subseteq B$. Since γ_1, γ_2 and γ_3 are not contained in a maximal clique of size $2n-1$, $\{\bar{f}_1, \bar{f}_2\} \cap \{\bar{g}_1, \bar{g}_2\} \cap \{\bar{h}_1, \bar{h}_2\} = \emptyset$. Since there is a unique maximal clique of size $2n-1$ through γ_1 and γ_2, $|\{\bar{f}_1, \bar{f}_2\} \cap \{\bar{g}_1, \bar{g}_2\}| = 1$. Similarly, $|\{\bar{f}_1, \bar{f}_2\} \cap \{\bar{h}_1, \bar{h}_2\}| = 1$ and $|\{\bar{g}_1, \bar{g}_2\} \cap \{\bar{h}_1, \bar{h}_2\}| = 1$. The lemma now immediately follows. □

We define X as the set of all points of weight 2 in $\mathrm{PG}(2n-1,4)$ with respect to a given reference system.

Lemma 7.44. *The point-line geometry Δ with point set V and line set*

$$\{\{v_1, v_2, \mathcal{R}_{v_2}(v_1)\} | v_1, v_2 \in V,\ v_1 \cap v_2 = \emptyset\}$$

is isomorphic to the point-line geometry Δ' whose points are the elements of X and whose lines are those lines L of $\mathrm{PG}(2n-1,4)$ for which $|L \cap X| = 3$ (natural incidence).

Proof. We first construct a bijection between V and X. For every $i \in \{1,\ldots,2n-1\}$, the equivalence class corresponding to $\{\bar{e}_0, \bar{e}_i\}$ contains three elements of V which can be labeled with the three elements of the set $\{\langle \bar{e}_0 + \alpha\bar{e}_i\rangle | \alpha \in \mathbb{F}_4^*\} \subseteq X$. For all $i,j \in \{1,2,\ldots,2n-1\}$ with $i<j$ and every $\alpha \in \mathbb{F}_4^*$, the reflection of $\langle \bar{e}_0 + \alpha\bar{e}_j\rangle$ (regarded as an element of V) around $\langle \bar{e}_0 + \bar{e}_i\rangle$ is labeled with the element $\langle \bar{e}_i + \alpha\bar{e}_j\rangle$ of X. In this way, we have a bijection between V and X.

For all $i,j \in \{1,2,\ldots,2n-1\}$ with $i<j$, we now define a binary operation $\otimes_{ij}$ on $\mathbb{F}_4^*$ in the following way: $\langle \bar{e}_i + (\alpha \otimes_{ij} \beta)\bar{e}_j\rangle$ is the reflection of $\langle \bar{e}_0 + \beta\bar{e}_j\rangle$ about $\langle \bar{e}_0 + \alpha\bar{e}_i\rangle$. Clearly $\otimes_{ij}$ determines a latin square of order 3 on the set $\mathbb{F}_4^*$. Since $1 \otimes_{ij} \alpha = \alpha$ for every $\alpha \in GF(4)^*$, we necessarily have $\alpha \otimes_{ij} \beta = \alpha^{\epsilon_{ij}} \cdot \beta$ for some $\epsilon_{ij} \in \{+1,-1\}$.

Let $i,j,k \in \{1,\ldots,2n-1\}$ such that $i<j<k$ and let $\alpha,\beta,\gamma \in \mathbb{F}_4^*$. Put $v = \langle \bar{e}_0 + \gamma\bar{e}_i\rangle$, $v_1 = \langle \bar{e}_0 + \alpha\bar{e}_j\rangle$, $v_2 = \langle \bar{e}_0 + \beta\bar{e}_k\rangle$ and $v_3 = \langle \bar{e}_j + (\alpha^{\epsilon_{jk}} \cdot \beta)\bar{e}_k\rangle$. Since $v_3 = \mathcal{R}_{v_1}(v_2)$ and $\mathcal{R}_v \in \mathrm{Aut}(\mathcal{S})$, the reflection of $\mathcal{R}_v(v_2)$ around $\mathcal{R}_v(v_1)$ equals $\mathcal{R}_v(v_3)$. Hence, the reflection of $\langle \bar{e}_i + (\gamma^{\epsilon_{ij}} \cdot \alpha)\bar{e}_j\rangle$ around $\langle \bar{e}_i + (\gamma^{\epsilon_{ik}} \cdot \beta)\bar{e}_k\rangle$ equals $\langle \bar{e}_j + (\alpha^{\epsilon_{jk}} \cdot \beta)\bar{e}_k\rangle$. In particular, the reflection of $\langle \bar{e}_i + \alpha\bar{e}_j\rangle$ around $\langle \bar{e}_i + \beta\bar{e}_k\rangle$ equals $\langle \bar{e}_j + (\alpha^{\epsilon_{jk}} \cdot \beta)\bar{e}_k\rangle$. Hence $(\gamma^{\epsilon_{ij}} \cdot \alpha)^{\epsilon_{jk}} \cdot (\gamma^{\epsilon_{ik}} \cdot \beta) = (\alpha^{\epsilon_{jk}} \cdot \beta)$ or $\epsilon_{ij}\epsilon_{jk} = -\epsilon_{ik}$. Putting $\epsilon_{11} = -1$, we have that $\epsilon_{1j}\epsilon_{jk} = -\epsilon_{1k}$ for all $j,k \in \{1,\ldots,2n-1\}$ with $j<k$.

For a point $v \in V$ with label $\langle \bar{e}_i + \alpha\bar{e}_j\rangle$, $i<j$, we put $\theta(v) := \langle \bar{e}_i + \alpha^{\epsilon_{1j}}\bar{e}_j\rangle$. Clearly θ is a bijection between V and X. Now, choose i, j and k such that $0 \le i < j < k \le 2n-1$, and let $\alpha,\beta \in \mathbb{F}_4^*$. Since $v_1 := \theta^{-1}(\langle \bar{e}_i + \alpha\bar{e}_j\rangle)$

and $v_2 := \theta^{-1}(\langle \bar{e}_i + \beta\bar{e}_k\rangle)$ have respective labels $\langle \bar{e}_i + \alpha^{\epsilon_{1j}}\bar{e}_j\rangle$ and $\langle \bar{e}_i + \beta^{\epsilon_{1k}}\bar{e}_k\rangle$, the reflection v_3 of v_2 around v_1 has label $\langle \bar{e}_j + (\alpha^{\epsilon_{1j}\epsilon_{jk}}\beta^{\epsilon_{1k}})\bar{e}_k\rangle$. Hence $\theta(v_3) = \langle \bar{e}_j + (\alpha^{\epsilon_{1j}\epsilon_{jk}\epsilon_{1k}}\beta^{\epsilon_{1k}\epsilon_{1k}})\bar{e}_k\rangle = \langle \bar{e}_j + (\alpha^{-1}\beta)\bar{e}_k\rangle$. It is now easily seen that θ is an isomorphism between Δ and Δ'. □

Recall that $\mathbb{G}_n = (Y, Y', \mathrm{I})$, where Y, respectively Y', is the set of all good subspaces of $H(2n-1,4)$ with dimension $n-1$, respectively $n-2$. If $\pi \in Y$, then $\mathcal{G}_\pi$ consists of n elements of X; if $\pi \in Y'$ is a special line of $\mathbb{G}_n$, then $\mathcal{G}_\pi$ consists of $n-1$ elements of X; if $\pi \in Y'$ is an ordinary line of $\mathbb{G}_n$, then $\mathcal{G}_\pi$ consists of $n-2$ elements of X and one point of weight 4. We refer to Section 6.3.3 for more details.

Every point x of $\mathcal{S}$ is contained in n elements $v_1, \ldots, v_n$ of V. Since $v_i \cap v_j \neq \emptyset$, the supports of $\theta(v_i)$ and $\theta(v_j)$ are disjoint. We define $\phi(x) := \langle \theta(v_1), \ldots, \theta(v_n)\rangle$. Clearly $\phi(x) \in Y$.

Lemma 7.45. *The map $\phi : \mathcal{P} \mapsto Y$ is bijective.*

Proof. Let $\pi \in Y$, then $\{v_1, \ldots, v_n\} := \theta^{-1}(X \cap \pi)$ is a set of n elements of V and $v_1 \cap \cdots \cap v_n$ is a convex subpolygon. Since a line of $\mathcal{S}$ is contained in at most $n-1$ elements of V, $|v_1 \cap \cdots \cap v_n| \leq 1$. If $\pi = \phi(x)$, then $\{x\} = v_1 \cap \cdots \cap v_n$, proving that ϕ is injective. Since $|Y| = |\mathcal{P}| = \frac{3^n \cdot (2n)!}{2^n \cdot n!}$, ϕ necessarily is bijective. □

For a line $L = \{x_1, x_2, x_3\}$ of $\mathcal{S}$, we put $\phi'(L) = \phi(x_1) \cap \phi(x_2) \cap \phi(x_3)$.

Lemma 7.46. *For every line L of $\mathcal{S}$, $\phi'(L) \in Y'$.*

Proof. (A) Suppose that L is special. Let $v_1, \ldots, v_{n-1}$ denote the $n-1$ elements of V through L, and let w_i, $i \in \{1,2,3\}$, denote the unique element of V through x_i not containing L. Clearly $\phi'(L) = \langle \theta(v_1), \ldots, \theta(v_{n-1})\rangle \in Y'$.
(B) Suppose that L is an ordinary line. Let $v_1, \ldots, v_{n-2}$ denote those elements of V through L, and let u_i and w_i denote the two elements of V through x_i not containing L. We may suppose that $u_3 = \mathcal{R}_{u_1}(u_2)$. Then $w_2 = \mathcal{R}_{w_1}(u_3)$ and $w_3 = \mathcal{R}_{u_1}(w_2)$. Putting $\theta(u_1) = \langle \bar{e}_0 + \alpha\bar{e}_1\rangle$, $\theta(w_1) = \langle \bar{e}_2 + \beta\bar{e}_3\rangle$ and $\theta(u_2) = \langle \bar{e}_1 + \gamma\bar{e}_2\rangle$, we find $\theta(u_3) = \langle \bar{e}_0 + \alpha\gamma\bar{e}_2\rangle$, $\theta(w_2) = \langle \bar{e}_0 + \alpha\beta\gamma\bar{e}_3\rangle$ and $\theta(w_3) = \langle \bar{e}_1 + \beta\gamma\bar{e}_3\rangle$. One easily calculates that $\phi'(L) = \langle \theta(v_1), \ldots, \theta(v_{n-2}), \langle \bar{e}_0 + \alpha\bar{e}_1 + \alpha\gamma\bar{e}_2 + \alpha\beta\gamma\bar{e}_3\rangle\rangle \in Y'$. □

Lemma 7.47. *The map $\phi' : \mathcal{L} \mapsto Y'$ is bijective.*

Proof. Let $\pi' \in Y'$. If $\pi' = \phi'(L)$, then necessarily $L = \{\phi^{-1}(\pi) \,|\, \pi \in Y \text{ and } \pi' \subset \pi\}$. Hence ϕ is injective. Since $|\mathcal{L}| = |Y'| = \frac{3^{n-1}(2n)!(3n-1)}{2^{n+1}(n-1)!}$, ϕ' is bijective. □

Now, a point x and a line L of $\mathcal{S}$ are incident if and only if $\phi(x)$ and $\phi'(L)$ are incident in $\mathbb{G}_n$. This proves that $\mathcal{S} \cong \mathbb{G}_n$.

Chapter 8

Slim dense near polygons with a nice chain of convex subpolygons

8.1 Overview

We give an overview of the theorems which we will prove in this chapter. These results are based on the papers [36], [40], [58], [59] and [62].

Theorem 8.1. *Let $\mathcal{S}$ be a slim dense near $2n$-gon, $n \geq 3$, containing a big convex subpolygon F isomorphic to $DQ(2n-2,2)$. Then $\mathcal{S}$ is isomorphic to one of the following near polygons:*

- $DQ(2n,2)$;
- $DQ(2n-2,2) \times \mathbb{L}_3$;
- $\mathbb{I}_n$.

Theorem 8.2. *Let $\mathcal{S}$ be a slim dense near $2n$-gon, $n \geq 4$, containing a big convex subpolygon F isomorphic to $DH(2n-3,4)$. Then $\mathcal{S}$ is isomorphic to one of the following near polygons:*

- $DH(2n-1,4)$;
- $DH(2n-3,4) \otimes Q(5,2)$;
- $DH(2n-3,4) \times \mathbb{L}_3$.

In the previous theorem, $DH(2n-3,4) \otimes Q(5,2)$ denotes the unique glued near polygon of type $DH(2n-3,4) \otimes Q(5,2)$, see Section 6.9.3.

Theorem 8.3. *Let $\mathcal{S}$ be a slim dense near $2n$-gon, $n \geq 4$, containing a big convex subpolygon F isomorphic to $\mathbb{H}_{n-1}$. Then $\mathcal{S}$ is isomorphic to either $\mathbb{H}_n$ or $\mathbb{H}_{n-1} \times \mathbb{L}_3$.*

Theorem 8.4. *Let $\mathcal{S}$ be a slim dense near $2n$-gon, $n \geq 5$, containing a big convex subpolygon F isomorphic to $\mathbb{I}_{n-1}$. Then $\mathcal{S}$ is isomorphic to $\mathbb{I}_{n-1} \times \mathbb{L}_3$.*

Remark. Since $\mathbb{I}_3 \cong \mathbb{H}_3$, the case $n = 4$ of Theorem 8.4 has already been treated in Theorem 8.3.

Theorem 8.5. *Let $\mathcal{S}$ be a slim dense near $2n$-gon, $n \geq 4$, containing a big convex subpolygon F isomorphic to $\mathbb{G}_{n-1}$. Then $\mathcal{S}$ is isomorphic to either $\mathbb{G}_n$, $\mathbb{G}_{n-1} \times \mathbb{L}_3$, or $\mathbb{G}_{n-1} \otimes Q(5,2)$.*

In the previous theorem, $\mathbb{G}_{n-1} \otimes Q(5,2)$ denotes the unique glued near polygon of type $\mathbb{G}_{n-1} \otimes Q(5,2)$, see Section 6.9.3.

Theorem 8.6. *Let $\mathcal{S}$ be a slim dense near octagon containing a hex H isomorphic to $\mathbb{E}_3$, then $\mathcal{S} \cong \mathbb{E}_3 \times \mathbb{L}_3$.*

Theorem 8.7. *Let $\mathcal{S}$ be a slim dense near polygon containing a big convex subpolygon F isomorphic to the direct product $\mathcal{S}_1 \times \mathcal{S}_2$ of two near polygons $\mathcal{S}_1$ and $\mathcal{S}_2$ of diameter at least 1. Then there exists an $i \in \{1,2\}$ and a dense near polygon $\mathcal{S}'_i$ such that the following holds:*

- *$\mathcal{S}'_i$ has a big convex subpolygon isomorphic to $\mathcal{S}_i$;*
- *$\mathcal{S} \cong \mathcal{S}'_i \times \mathcal{S}_{3-i}$.*

Define now the following classes of near polygons:

$$\begin{aligned} D_2 &= \{Q(5,2)\}; \\ D_n &= \{\mathbb{G}_n, DH(2n-1,4)\} \cup \left(\bigcup_{2 \leq i \leq n-1} D_i \otimes D_{n+1-i} \right) \quad \text{for every } n \geq 3; \\ \mathcal{D} &= D_2 \cup D_3 \cup \cdots . \end{aligned}$$

Remark that $\mathcal{D}$ consists of those near polygons of the class $\mathcal{C}$ (see Section 6.8) which do not contain $\mathbb{E}_1$ as a hex.

Theorem 8.8. *Let $\mathcal{S}$ be a slim dense near $2n$-gon, $n \geq 4$, containing a big convex subpolygon F which is isomorphic to an element of $\mathcal{D}$. Then one of the following possibilities occurs:*

- *$\mathcal{S} \cong F \times \mathbb{L}_3$;*
- *$\mathcal{S}$ is isomorphic to an element of $\mathcal{D}$.*

Consider a chain $F_0 \subset F_1 \subset \cdots \subset F_n$ of convex subpolygons of a dense near $2n$-gon $\mathcal{S}$. Such a chain is called *nice* if it satisfies the following properties:

- $\mathrm{diam}(F_i) = i$ for every $i \in \{0, \ldots, n\}$;
- F_i, $i \in \{0, \ldots, n-1\}$, is big in F_{i+1}.

Clearly $F_n = \mathcal{S}$.

Define now the following set:

$$\mathcal{N} = \{\mathbb{O}, \mathbb{L}_3, \mathbb{E}_3\} \cup \mathcal{D} \cup \{DQ(2n,2) | n \geq 2\} \cup \{\mathbb{H}_n | n \geq 3\} \cup \{\mathbb{I}_n | n \geq 4\}.$$

Let $\mathcal{N}^\times$ denote the set of all near polygons obtained by taking the direct product of at least one element of $\mathcal{N}$.

Theorem 8.9. *A slim dense near $2n$-gon $\mathcal{S}$ has a nice chain of subpolygons if and only if it is isomorphic to an element of $\mathcal{N}^\times$.*

8.2 Proof of Theorem 8.1

We will use induction on n. By Theorem 7.1 the theorem holds if n is equal to 3. Suppose therefore that $n \geq 4$ and that the theorem holds for any near $2n'$-gon with $n' \in \{3, \ldots, n-1\}$. Every convex sub-$2\delta$-gon, $\delta \in \{3, \ldots, n-1\}$, intersecting F, intersects F in a $DQ(2\delta - 2, 2)$ and hence is isomorphic to either $DQ(2\delta, 2)$, $DQ(2\delta - 2, 2) \times \mathbb{L}_3$ or $\mathbb{I}_\delta$ by Theorem 1.7 and the induction hypothesis. As a consequence no $Q(5,2)$-quad meets F. If a line of F is contained in two grid-quads Q_1 and Q_2, then the hex $\mathcal{C}(Q_1, Q_2)$ intersects F in a $W(2)$-quad and hence has a line which is incident with at least two grid-quads and at least one big $W(2)$-quad, contradicting Theorem 7.1. As a consequence there are at most $t_F + 1$ grid-quads through every point of F. For every line K intersecting F in a point x, let α_K denote the number of grid-quads through K. Since every quad through K meets F in a line, the number of $W(2)$-quads through K equals $t_F + 1 - \alpha_K$. Counting the number of lines through x, we have $t_{\mathcal{S}} = \alpha_K + 2(t_F + 1 - \alpha_K)$. Hence α_K is independent from the line K and equal to $\alpha := 2t_F + 2 - t_{\mathcal{S}}$. Since there are $t_{\mathcal{S}} - t_F = t_F + 2 - \alpha$ lines through x not contained in F, there are precisely $\alpha(t_F + 2 - \alpha)$ grid-quads through x. Since this number is at most $t_F + 1$, we necessarily have $\alpha \in \{0, 1, t_F + 1\}$.

If $\alpha = t_F + 1$, then $t_{\mathcal{S}} = t_F + 1$ and $\mathcal{S} \cong DQ(2n-2, 2) \times \mathbb{L}_3$ by Theorem 4.1.

If $\alpha = 0$, then every quad meeting F is isomorphic to $W(2)$. Hence, every convex sub-2δ-gon, $\delta \in \{2, \ldots, n-1\}$, intersecting F is isomorphic to $DQ(2\delta, 2)$. Now, consider an arbitrary point-quad pair (y, Q) and let F' denote an arbitrary convex sub-$2(n-1)$-gon through Q intersecting F. Since $F' \cong DQ(2n-2, 2)$, F' is big in $\mathcal{S}$ and hence $\mathrm{d}(y, z) = \mathrm{d}(y, \pi_{F'}(y)) + \mathrm{d}(\pi_{F'}(y), z)$ for every point $z \in Q$. Since $(\pi_{F'}(y), Q)$ is classical, also (y, Q) is classical. Hence, every point-quad relation is classical, and so $\mathcal{S}$ itself is classical. Since $\mathcal{S}$ has only $W(2)$-quads, it is necessarily isomorphic to $DQ(2n, 2)$.

We still need to consider the case $\alpha = 1$. Let $\Omega(\mathcal{S}, F)$ be the incidence structure whose points are the lines of $\mathcal{S}$ intersecting F in a point, whose lines are the (not necessarily convex) subgrids of $\mathcal{S}$ intersecting F in a line and whose incidence relation is the natural one. For every line L of $\mathcal{S}$ intersecting F in a point x, let L' denote the unique line of F through x for which $\mathcal{C}(L, L')$ is a grid. Since $t_F + 1 = \alpha(t_F + 2 - \alpha)$, every line of F is also contained in a unique grid-quad and hence there is a bijective correspondence between the points L of $\Omega(\mathcal{S}, F)$ and the flags (x, L') of F. The lines of $\Omega(\mathcal{S}, F)$ then correspond with certain triples of flags. We will prove that such a triple is either of the form $\{(x, K), (y, K), (z, K)\}$ with $K = \{x, y, z\}$ a line of F, or of the form $\{(L \cap K, L), (M \cap K, M), (N \cap K, N)\}$ with K, L, M, and N four distinct lines of F satisfying:

- $K = (L \cap K) \cup (M \cap K) \cup (N \cap K)$;
- $\mathcal{C}(L, M, N) \cong W(2)$;
- L, M and N are not contained in a subgrid of $\mathcal{C}(L, M, N)$.

Let G be a subgrid of $\mathcal{S}$ intersecting F in a line K, and let A, B and C be the three points of $\Omega(\mathcal{S}, F)$ incident with G. If G is a quad, then $A' = B' = C' = K$ and we obtain a triple of the first kind. If G is not a quad, then the hex $H := \mathcal{C}(K, A, A')$ contains a grid-quad $\mathcal{C}(A, A')$ and two big $W(2)$-quads $\mathcal{C}(A, K)$ and $\mathcal{C}(A', K)$ through the line K and hence is isomorphic to $\mathbb{I}_3$. As a consequence the unique grid-quads through B and C are contained in H. This implies that the three lines A', B' and C' are contained in the $W(2)$-quad $\mathcal{C}(A', K)$. Let B'' denote the unique line of $\mathcal{C}(A', K)$ through $B \cap B'$ different from K and B'. $\mathcal{C}(B, B'')$ is a $W(2)$-quad. The reflection (in H) of the grid-quad $\mathcal{C}(A, A')$ about $\mathcal{C}(B, B')$ is a grid-quad through C which necessarily coincides with $\mathcal{C}(C, C')$. So, A', B'' and C' are contained in a grid. Hence, A', B' and C' are not contained in a grid. Now, one counts that there are as many lines in $\Omega(\mathcal{S}, F)$ as there are triples of the first or second type. So, we have found a description of $\Omega(\mathcal{S}, F)$ only in terms of objects of F. By Theorem 2.40, the structure of a slim dense near polygon $\mathcal{S}'$ with a big convex subpolygon F' is completely determined by the structure of $\Omega(\mathcal{S}', F')$. Therefore $\mathcal{S}$ necessarily is isomorphic to $\mathbb{I}_n$.

8.3 Proof of Theorem 8.2

Lemma 8.10. *The slim near polygon $\mathcal{S}$ does not contain quads isomorphic to $W(2)$.*

Proof. Suppose the contrary, then by Theorem 7.2, there exists a $W(2)$-quad Q through a point x of F. By Theorem 1.7, this quad Q intersects F in a line K. Let L be a line of Q through x different from K and let $H \cong DH(5, 4)$ be a hex of F through K. Let X denote the set of points of $\mathcal{L}(H, x)$ (i.e. the set of lines of H through x) which are contained in a $W(2)$-quad together with L. We will now show that $|U \cap X| \in \{0, 3\}$ for every line U of $\mathcal{L}(H, x)$. Suppose that $|U \cap X| \geq 1$ and let Q_U denote the $Q(5, 2)$-quad corresponding with U. Since $|U \cap X| \geq 1$, there exists a $W(2)$-quad R through L which intersects Q_U in a line. By Theorem 7.1, $\mathcal{C}(Q_U, R)$ is isomorphic to either $\mathbb{G}_3$ or $\mathbb{E}_3$. In any case, exactly three lines of Q_U through x are contained in a $W(2)$-quad together with L, or equivalently, $|U \cap X| = 3$. Since H contains exactly five $Q(5,2)$-quads through K, we have $|X| = 1 + 5 \cdot 2 = 11$. Hence X is a set of 11 points in $\mathcal{L}(H,x) \cong \mathrm{PG}(2,4)$ such that every line meets the set in either 0 or three points. Such a set does not exist, otherwise, every point of $\mathcal{L}(H,x)$ outside X would be contained in $\frac{11}{3}$ lines meeting X, which is clearly impossible. $\square$

For every line L intersecting F in a point x, let A_L be the set of lines of F through x such that $\mathcal{C}(L,M) \cong Q(5,2)$. By Theorem 1.7 and Lemma 8.10, $\mathcal{C}(L,L') \cong Q(5,2)$ for every line $L' \neq L$ intersecting F in x. Hence $t_{\mathcal{S}} - t_F = 3|A_L| + 1$. As a consequence, $|A_L|$ is independent from the choice of L and equal to $\alpha := \frac{t_{\mathcal{S}}-t_F-1}{3}$.

Lemma 8.11. *A_L is a subspace of $\mathcal{L}(F,x)$.*

Proof. Let K_1 and K_2 be two different lines of A_L and let K_3 be an arbitrary line through x contained in $\mathcal{C}(K_1,K_2)$. Since the hex $\mathcal{C}(L,K_1,K_2)$ has three $Q(5,2)$-quads through the same point x and no $W(2)$-quads, it must be isomorphic to $DH(5,4)$. Hence, $\mathcal{C}(L,K_3) \cong Q(5,2)$ and $K_3 \in A_L$. $\square$

Lemma 8.12. $\alpha \in \{0, 1, t_F + 1\}$.

Proof. We suppose that $2 \leq \alpha \leq t_F$ and derive a contradiction.

(a) Suppose first that $n = 4$, so $F \cong DH(5,4)$. Let x be an arbitrary point of F and let L be an arbitrary line through x not contained in F. Since A_L is a subspace and $2 \leq |A_L| \leq t_F = 20$, there exists a $Q(5,2)$-quad Q_x through x such that A_L is the set of five lines of Q_x through x. Also, $\alpha = 5$, $t_{\mathcal{S}} = t_F + 3\alpha + 1 = 36$ and the hex $H_x := \mathcal{C}(L, Q_x)$ is isomorphic to $DH(5,4)$. If L' is a line through x different from L and not contained in F, then $\mathcal{C}(L,L')$ is isomorphic to $Q(5,2)$ and hence intersects F in a line of Q_x. It follows that every line through x not contained in F is contained in H_x. Since $H_x \cong DH(5,4)$, we then have that $H_x = \mathcal{C}(\Gamma_1(x) \setminus F)$. So, H_x and $Q_x = H_x \cap F$ only depend on x and not on the line L. If $y \in Q_x$, then $H_y = H_x$ and $Q_y = Q_x$. Hence, the quads Q_x, $x \in F$, partition the point set of F, and the hexes H_x, $x \in F$, partition the point set of $\mathcal{S}$. Since $F \cong DH(5,4)$ is big in $\mathcal{S}$, every hex isomorphic to $DH(5,4)$ is big. In particular, each of the hexes H_x is big. The total number of quads Q_x, $x \in F$, equals $\frac{|DH(5,4)|}{|Q(5,2)|} = 33$. Let X denote the set of 33 points of $H(5,4)$ corresponding to the quads Q_x, $x \in F$. If u is a point of X, then we denote the $Q(5,2)$-quad of F corresponding with it as u^ϕ and the unique hex of $\mathcal{S}$ intersecting F in u^ϕ by $H(u)$. We will now show that X is a subspace of the linear space $\mathcal{L}$ defined in Lemma 6.9. Consider two different points u_1 and u_2 in X. Since u_1^ϕ and u_2^ϕ are disjoint, $u_1u_2 \cap H(5,4)$ is a Baer subline $\{u_1, u_2, u_3\}$. Since every quad intersecting u_1^ϕ and u_2^ϕ also intersects u_3^ϕ, u_3^ϕ is the reflection of u_1^ϕ about u_2^ϕ (in F). The reflection of $H(u_1)$ about $H(u_2)$ (in $\mathcal{S}$) is a hex which meets F in the quad u_3^ϕ and hence coincides with $H(u_3)$. As a consequence $u_3 \in X$. This proves that X is a subspace of the linear space $\mathcal{L}$. Hence, there exists a subspace π such that $X = \pi \cap H(5,4)$. Since no two points of X are collinear on $H(5,4)$, we have $|X| \leq |H(2,4)| = 9$, contradicting $|X| = 33$.

(b) Suppose now that $n \geq 5$. Let x denote an arbitrary point of F and let L denote an arbitrary line through x not contained in F. Since $2 \leq \alpha \leq t_F$, there exist lines K_1, K_2 and K_3 through x such that $K_1, K_2 \in A_L$, $K_1 \neq K_2$ and $K_3 \notin A_L$. Now, the near octagon $\mathcal{C}(L, K_1, K_2, K_3)$ contradicts (a). □

Lemma 8.13.

- *If $\alpha = 0$, then $\mathcal{S} \cong DH(2n-3,4) \times \mathbb{L}_3$.*
- *If $\alpha = 1$, then $\mathcal{S} \cong DH(2n-3,4) \otimes Q(5,2)$.*
- *If $\alpha = t_F + 1$, then $\mathcal{S} \cong DH(2n-1,4)$.*

Proof. If $\alpha = 0$, then $t_{\mathcal{S}} - t_F = 1$ and hence $\mathcal{S} \cong DH(2n-3,4) \times \mathbb{L}_3$ by Theorem 4.1. If $\alpha = t_F + 1$, then no grid-quad intersects F and hence all quads are isomorphic to $Q(5,2)$ by Theorem 7.2 and Lemma 8.10. Since the generalized quadrangle $Q(5,2)$ has no ovoids, all point-quad relations must be classical and $\mathcal{S}$ is a classical near polygon. Now, $DH(2n-1,4)$ is the only classical near $2n$-gon in which all quads are isomorphic to $Q(5,2)$.

Suppose now that $\alpha = 1$, then $t_{\mathcal{S}} = t_F + 4$. If x is an arbitrary point of F and if L and M denote two lines through x not contained in F, then $\mathcal{C}(K, L)$ is a $Q(5,2)$-quad. Hence, every point $x \in F$ is contained in a unique $Q(5,2)$-quad Q_x which intersects F in a line L_x. The lines L_x, $x \in F$, determine a spread of F and the set $T_1 := \{Q_x | x \in F\}$ determines a partition of $\mathcal{S}$ in quads.

Now, consider a point x of $\mathcal{S}$ not contained in F, let y be the unique point of F collinear with x, and let $A := xy$, B, C, D and E be the lines of $Q := Q_y$ through x. Let L be a line intersecting Q in x and consider the hex $H := \mathcal{C}(L, Q)$. The hex H contains a grid-quad $\mathcal{C}(L, A)$, no $W(2)$-quads, and at least two $Q(5,2)$-quads through the line $L_y = Q \cap F$, namely Q and $H \cap F$. It follows from Theorem 7.1 that $H \cong Q(5,2) \otimes Q(5,2)$. Hence, exactly one line of Q through x, say B, is contained in a $Q(5,2)$-quad with L.

Now, consider a convex sub-$2(n-1)$-gon F' of $\mathcal{S}$ containing $Q' := \mathcal{C}(B, L)$ and intersecting Q in B. We will show that every quad of F' is isomorphic to $Q(5,2)$. Let L_1 and L_2 be two lines of F' through x. If $L_1 \neq B \neq L_2$, then the hex $H' := \mathcal{C}(A, L_1, L_2)$ contains grid-quads $\mathcal{C}(A, L_1)$ and $\mathcal{C}(A, L_2)$, a $Q(5,2)$-quad $\mathcal{C}(A, L_1, L_2) \cap F$, and no $W(2)$-quads. Hence H' must be isomorphic to either $Q(5,2) \times \mathbb{L}_3$ or $Q(5,2) \otimes Q(5,2)$, see Theorem 7.1. In any case $\mathcal{C}(L_1, L_2)$ is isomorphic to $Q(5,2)$. If $L_1 = B$ and if L_2 is not contained in Q', then $\mathcal{C}(L_2, M) \cong Q(5,2)$ for every line $M \neq B$ through x contained in Q'. This is only possible if

$\mathcal{C}(L_2, Q')$ is isomorphic to $DH(5,4)$, see Theorem 7.1. But then also $\mathcal{C}(L_1, L_2) = \mathcal{C}(B, L_2)$ is isomorphic to $Q(5,2)$. It now follows that every quad of F' through x is isomorphic to $Q(5,2)$. By Theorem 7.2 it then follows that every quad of F' is isomorphic to $Q(5,2)$. As before, we then know that F' is classical and isomorphic to $DH(2n-3,4)$. Obviously, F' is the only convex subpolygon through x isomorphic to $DH(2n-3,4)$.

Repeating the above construction for every point x outside F, we obtain a partition T_2 of $\mathcal{S}$ in convex subpolygons isomorphic to $DH(2n-3,4)$. We can now apply Theorem 4.28 and conclude that $\mathcal{S}$ is a glued near polygon of type $DH(2n-3,4) \otimes Q(5,2)$. In Theorem 6.107, we have shown that there exists a unique glued near polygon of such type. □

8.4 Proof of Theorem 8.3

Let $\mathcal{G}_n$ denote the rank $n-1$ geometry whose i-objects, $i \in \{1,\ldots,n-1\}$, are the partitions of $\{1,\ldots,n+1\}$ in $n+1-i$ subsets (natural incidence). Then every local geometry of $\mathbb{H}_n$ is isomorphic to $\mathcal{G}_n$ by Corollary 6.17. Let $\mathcal{M}_n$ denote the partial linear space with points, respectively lines, the subsets of size 2, respectively the subsets of size 3, of $\{1,\ldots,n+1\}$, with containment as incidence relation. Let $\mathcal{L}_n$ denote the linear space obtained from $\mathcal{M}_n$ by adding lines of size 2. Every 1-object of $\mathcal{G}_n$ is a partition of $\{1,\ldots,n+1\}$ in $n-1$ singletons and one subset of size 2. If we identify the 1-object with the subset of size 2 corresponding with it, we see that every local space of $\mathbb{H}_n$ is isomorphic to $\mathcal{L}_n$, see also Theorem 6.18.

Lemma 8.14. *Every proper subspace of $\mathcal{L}_n$, $n \geq 3$, corresponds to an object of $\mathcal{G}_n$.*

Proof. Suppose that A is a proper subspace of $\mathcal{L}_n$, then the following graph Λ_A can be defined. The vertices of Λ_A are the elements of $\{1,\ldots,n+1\}$, and two different vertices x_1 and x_2 are adjacent if and only if $\{x_1,x_2\}$ is a point of A. If x_1, x_2 and x_3 are three different vertices of Λ_A satisfying $x_1 \sim x_2$ and $x_1 \sim x_3$, then the line $\{x_1,x_2,x_3\}$ of $\mathcal{L}_n$ is incident with the points $\{x_1,x_2\}$ and $\{x_1,x_3\}$ of A. Hence also the third point $\{x_2,x_3\}$ of $\{x_1,x_2,x_3\}$ belongs to A, or equivalently, $x_2 \sim x_3$. As a consequence, Λ_A is a disjoint union of cliques. Hence the subspace A determines a partition of $\{1,\ldots,n+1\}$, or equivalently, an object O_A of $\mathcal{G}_n$. The 1-objects incident with O_A are in bijective correspondence with the points of A. □

Lemma 8.15. *The partial linear space $\mathcal{M}_n$ is connected.*

Proof. Take two arbitrary points $\{x_1,x_2\}$ and $\{x_3,x_4\}$ of $\mathcal{M}_n$. If $|\{x_1,x_2\} \cap \{x_3,x_4\}| = 1$, then the two points are contained in the line $\{x_1,x_2\} \cup \{x_3,x_4\}$ of $\mathcal{M}_n$. If $\{x_1,x_2\} \cap \{x_3,x_4\} = \emptyset$, then $\{x_1,x_2\}, \{x_2,x_3\}, \{x_3,x_4\}$ is a path in $\mathcal{M}_n$ connecting the points $\{x_1,x_2\}$ and $\{x_3,x_4\}$. □

We will now prove Theorem 8.3. Let $t+1$ denote the constant number of lines through a point of $\mathcal{S}$. If $t = t_F + 1$, then $\mathcal{S} \cong \mathbb{H}_{n-1} \times \mathbb{L}_3$ by Theorem 4.1.

In the sequel, we will suppose that $t \geq t_F + 2$. Then there exist two lines K and L not contained in F meeting each other in a point belonging to F. By Theorem 1.7, the quad $\mathcal{C}(K, L)$ meets F in a line.

Lemma 8.16. *The following properties hold:*

(a) $t + 1 = \frac{n(n+1)}{2}$;

(b) *the total number v of points in $\mathcal{S}$ is equal to* $\frac{(2n+2)!}{2^{n+1}(n+1)!}$;

(c) $\mathcal{G}(\mathcal{S}, x) \cong \mathcal{G}_n$ *for every point* $x \in F$.

Proof. Let Q denote an arbitrary quad intersecting F in a line. Since $F \cong \mathbb{H}_{n-1}$, there exists a $W(2)$-quad Q' through $Q \cap F$. The hex $\mathcal{C}(Q, Q')$ contains the big $W(2)$-quad Q' and can therefore not contain a $Q(5,2)$-quad, see Theorem 7.1. In particular, Q is not isomorphic to $Q(5,2)$. If Q is a $Q(5,2)$-quad disjoint from F, then $\pi_F(Q)$ would be a $Q(5,2)$-quad of F, which is impossible. Hence $\mathcal{S}$ has no quads isomorphic to $Q(5,2)$.

We will now show that from the six slim dense near hexagons without a $Q(5,2)$-quad, see Theorem 7.1, only $\mathbb{L}_3 \times \mathbb{L}_3 \times \mathbb{L}_3$, $\mathbb{L}_3 \times W(2)$ and $\mathbb{H}_3$ can occur as hex. Suppose the contrary and let H denote a hex of $\mathcal{S}$ isomorphic to $DQ(6,2)$, $\mathbb{E}_1$ or $\mathbb{E}_2$. We clearly have $|H| \geq 135$. If H is disjoint from F, then $\pi_F(H)$ is a subhexagon of F isomorphic to H and $\mathcal{C}(\pi_F(H))$ is a hex, which is, by Theorem 6.15, isomorphic to either $\mathbb{L}_3 \times \mathbb{L}_3 \times \mathbb{L}_3$, $\mathbb{L}_3 \times W(2)$ or $\mathbb{H}_3$. Hence $135 \leq |H| = |\pi_F(H)| \leq |\mathcal{C}(\pi_F(H))| \leq 105$, a contradiction. As a consequence H must meet F. Since F is big in $\mathcal{S}$, $Q := F \cap H$ is a big quad of H. Since $\mathbb{E}_1$ and $\mathbb{E}_2$ have no big quads, $H \cong DQ(6,2)$ and $Q \cong W(2)$. Since $F \cong \mathbb{H}_{n-1}$, F has a hex $\bar{H} \cong \mathbb{H}_3$ through Q. Since $\bar{H}$ is big in the suboctagon $G := \mathcal{C}(H, \bar{H})$, $|G| = |\bar{H}| \cdot (1 + 2(t_G - t_{\bar{H}})) = 210 \cdot t_G - 945$. On the other hand, the total number of points of G at distance at most 1 from H is equal to $|H| \cdot (1 + 2(t_G - t_H)) = 270 \cdot t_G - 1485$. Since this number is at most $|G|$, $t_G \leq 9$. Since $t_G \geq t_H + t_{\bar{H}} - t_Q = 9$, we must have equality. Let R and $\bar{R}$ denote $W(2)$-quads through x contained in H and $\bar{H}$, respectively, such that $R \cap Q = \bar{R} \cap Q$ is a line. Every line of $\mathcal{C}(R, \bar{R})$ through x is contained in $H \cup \bar{H}$. Since $\mathcal{C}(R, \bar{R}) \cap H = R$ and $\mathcal{C}(R, \bar{R}) \cap \bar{H} = \bar{R}$, the hex $\mathcal{C}(R, \bar{R})$ has order $(2, 4)$, a contradiction.

Now, let Q denote an arbitrary $W(2)$-quad intersecting F in a line K. If L denotes an arbitrary line through x not contained in $Q \cup F$, then the hex $\mathcal{C}(Q, L)$ necessarily is isomorphic to $\mathbb{H}_3$ and intersects F in a $W(2)$-quad through K. Conversely, if R is one of the $n - 2$ $W(2)$-quads of F through K, then the hex $\mathcal{C}(K, R)$ is isomorphic to $\mathbb{H}_3$ and contains one line through x not contained in $Q \cup F$. Hence, there are exactly $(t_F + 1) + 2 + (n - 2) = \frac{n(n+1)}{2}$ lines through x, n of which are not contained in F. Since F is big in $\mathcal{S}$, $\mathcal{S}$ has exactly $v = |F| + 2 \cdot |F| \cdot (t - t_F) = \frac{(2n+2)!}{2^{n+1}(n+1)!}$ points. If A and B are two lines through x not contained in F, then $\mathcal{C}(A, B)$ intersects F in a line and hence is isomorphic to $W(2)$. As a consequence the line A is contained in exactly $n - 1$ $W(2)$-quads.

Let V_A denote the set of all lines of F through x which are contained in one of these $W(2)$-quads. Since $\mathcal{L}(F,x) \cong \mathcal{L}_{n-1}$, every line C of F through x corresponds with a subset W_C of size 2 of $\{1,2,\ldots,n\}$. If $C_1, C_2 \in V_A$, then $\mathcal{C}(C_1,C_2)$ is a $W(2)$-quad, since $\mathcal{C}(A,C_1,C_2) \cong \mathbb{H}_3$; hence W_{C_1} and W_{C_2} have a unique point in common. As a consequence $\{W_C \,|\, C \in V_A\}$ consists of the $n-1$ pairs through a fixed element i_A of $\{1,\ldots,n\}$. We will now prove that for every $i \in \{1,\ldots,n\}$ there exists a unique line A through x not contained in F such that $i = i_A$. Suppose the contrary, then there are two different lines A and B satisfying $i_A = i_B$. If $C \in V_A \setminus \{\mathcal{C}(A,B) \cap F\}$, then the quads $\mathcal{C}(C,A)$, $\mathcal{C}(C,B)$ and $\mathcal{C}(C, \mathcal{C}(A,B) \cap F)$ are isomorphic to $W(2)$ (notice that $C, \mathcal{C}(A,B) \cap F \in V_A = V_B$). As a consequence the hex $H := \mathcal{C}(A,B,C)$ has at least three $W(2)$-quads through C, implying that $t_H \geq 6$, a contradiction. With every line D of $\mathcal{S}$ through x, there now corresponds a subset $\tilde{W}_D$ of size 2 of $\{1,\ldots,n+1\}$: if $D \subset F$, then we put $\tilde{W}_D := W_D$; if $D \not\subset F$, then we put $\tilde{W}_D := \{i_D, n+1\}$. If D_1, D_2 and D_3 are three lines through x which are contained in a $W(2)$-quad, then the set $\tilde{W}_{D_1} \cup \tilde{W}_{D_2} \cup \tilde{W}_{D_3}$ has size 3. Hence $\mathcal{L}(\mathcal{S},x) \cong \mathcal{L}_n$. An object of $\mathcal{G}(\mathcal{S},x)$ corresponds to a subspace of $\mathcal{L}(\mathcal{S},x)$ and hence corresponds to an object of $\mathcal{G}_n$, see Lemma 8.14. Since maximal flags of $\mathcal{G}(\mathcal{S},x)$ correspond to flags of $\mathcal{G}_n$, i-objects of $\mathcal{G}(\mathcal{S},x)$ correspond to i-objects of $\mathcal{G}_n$. Conversely, every i-object O of $\mathcal{G}_n$ corresponds to an i-object of $\mathcal{G}(\mathcal{S},x)$. [Since $\mathcal{L}(\mathcal{S},x) \cong \mathcal{L}_n$, this holds if $i \in \{1,2\}$; suppose therefore that $i \geq 3$. The i-object O of $\mathcal{G}_n$ is generated by an $(i-1)$-object O_1 and a 1-object O_2 of $\mathcal{G}_n$. By the induction hypothesis, O_1 and O_2 correspond to an $i-1$ object $\tilde{O}_1$ and a 1-object $\tilde{O}_2$, respectively, of $\mathcal{G}(\mathcal{S},x)$. Let $\tilde{O}$ denote the i-object of $\mathcal{G}(\mathcal{S},x)$ generated by $\tilde{O}_1$ and $\tilde{O}_2$. The i-object corresponding to $\tilde{O}$ contains O_1 and O_2 and hence coincides with O.] The bijective correspondence between the objects of $\mathcal{G}(\mathcal{S},x)$ and $\mathcal{G}_n$ clearly preserves the incidence relation. □

Lemma 8.17. *$\mathcal{G}(\mathcal{S},x) \cong \mathcal{G}_n$ for every point x in $\mathcal{S}$.*

Proof. By the previous lemma, we may suppose that $x \notin F$. Since $\mathcal{G}(\mathcal{S},\pi(x)) \cong \mathcal{G}_n$, there exists a chain $G^{(1)} \subset G^{(2)} \subset \cdots \subset G^{(n)}$ of convex subpolygons through $\pi(x)$ such that the following properties are satisfied:

(a) $G^{(i)}$, $i \in \{1,\ldots,n\}$, is a convex sub-$2i$-gon;

(b) $G^{(1)} = x\,\pi(x)$,

(c) $G^{(2)}$ is a $W(2)$-quad,

(d) $\mathcal{G}(G^{(i)},\pi(x)) \cong \mathcal{G}_i$ for every $i \in \{3,4,\ldots,n\}$,

(e) every quad of $G^{(i)}$ through $\pi(x)$ either is contained in $G^{(i-1)}$ or intersects $G^{(i-1)}$ in a line ($i \in \{2,3,\ldots,n\}$).

By the proof of Lemma 8.16, $G^{(3)}$ is isomorphic to either $\mathbb{L}_3 \times \mathbb{L}_3 \times \mathbb{L}_3$, $W(2) \times \mathbb{L}_3$ or $\mathbb{H}_3$. Since $\mathcal{G}(G^{(3)},\pi(x)) \cong \mathcal{G}_3$, we necessarily have $G^{(3)} \cong \mathbb{H}_3$. Suppose now that $G^{(i)} \cong \mathbb{H}_i$ for a certain $i \in \{3,\ldots,n-2\}$. By Theorem 2.31 and property (e), $G^{(j)}$ is big in $G^{(j+1)}$ for every $j \in \{1,\ldots,n-1\}$. We may suppose that

$G^{(i+1)} \cong \mathbb{H}_{i+1}$ or $\mathbb{G}^{(i+1)} \cong \mathbb{H}_i \times \mathbb{L}_3$ (otherwise prove Theorem 8.3 by induction). Since $\mathcal{G}(G^{(i+1)}, \pi(x)) \cong \mathcal{G}_{i+1}$, $G^{(i+1)} \cong \mathbb{H}_{i+1}$. As a consequence, $G^{(n-1)}$ is a big convex sub-$2(n-1)$-gon of $\mathcal{S}$ isomorphic to $\mathbb{H}_{n-1}$. Since $x \in G^{(n-1)}$, our lemma follows from Lemma 8.16. □

For every $i \geq 0$, let v_i denote the total number of points of the near polygon $\mathbb{H}_i$. Now, let W denote the set of all convex sub-$2(n-1)$-gons F such that $\mathcal{G}(F, x) \cong \mathcal{G}_{n-1}$ for a certain point $x \in F$. With a similar reasoning as in the proof of Lemma 8.17, we have that $F \cong \mathbb{H}_{n-1}$ for every $F \in W$. Since $\mathcal{G}(\mathcal{S}, x) \cong \mathcal{G}_n$ for every point x of $\mathcal{S}$, x is contained in $n+1$ elements of W. Hence $|W| = \frac{v \cdot (n+1)}{v_{n-1}} = (2n+1)(n+1)$. Consider now the graph Γ whose vertices are the elements of W, with two vertices adjacent whenever they are disjoint regarded as subpolygons.

Lemma 8.18. *The graph Γ is regular with degree $k(\Gamma) = 4n$.*

Proof. Let F be an element of W. Every point x outside F is contained in a unique line L_x meeting F. Since $\mathcal{G}(\mathcal{S}, x) \cong \mathcal{G}_n$, L_x is contained in $n-1$ elements of W. Hence x is contained in $[(n+1)-(n-1)] = 2$ elements of W disjoint from F. As a consequence Γ is regular with degree $k(\Gamma) = \frac{(v-v_{n-1}) \cdot 2}{v_{n-1}} = 4n$. □

Lemma 8.19. *Every maximal clique has size 3 or $2n+1$. Moreover, there are $2n+2$ maximal cliques of size $2n+1$.*

Proof. Let F_1 and F_2 be two adjacent elements of Γ, then $F_3 := \mathcal{R}_{F_1}(F_2) \in W$. Put $M_1 := \{F_1, F_2, F_3\}$. Every element of W which meets two elements of M_1 intersects each of the three elements of M_1 in an $\mathbb{H}_{n-2}$. Since every line is contained in $n-1$ elements of W, there are $\frac{v_{n-1} \cdot (n-1)}{v_{n-2}} = (2n-1)(n-1)$ elements of W meeting F_1, F_2 and F_3. Every point of F_3 is contained in a unique element of W different from F_3 and disjoint from F_1 and F_2. In total, we get $\frac{v_{n-1}}{v_{n-2}} = 2n-1$ such elements. If u is contained in two of these elements, then u is collinear with two different points of F_3, a contradiction, since F_3 is convex. The total number of elements of W meeting at least one of the subpolygons F_1, F_2 and F_3 is hence equal to $3+(2n-1)(n-1)+3 \cdot (2n-1)$, which is exactly the total number of elements of $|W|$. So, M_1 is a maximal clique. Let M_2 denote another maximal clique through F_1 and F_2. By the previous reasoning we know that M_2 consists of F_1, F_2 and the $2n-1$ elements of W disjoint from F_1 and F_2 and intersecting F_3 in an $\mathbb{H}_{n-2}$. Since every two adjacent vertices of Γ are contained in a unique maximal clique of size $2n+1$, there are $\frac{|W| \cdot k(\Gamma)}{(2n+1) \cdot (2n)} = 2n+2$ such maximal cliques. □

We are now ready to prove Theorem 8.3. If V denotes the set of all $(2n+1)$-cliques in Γ, then $|V| = 2n+2$. Every element of W is contained in $\frac{k(\Gamma)}{2n} = 2$ such cliques, so each element of W corresponds to a pair $\{v_1, v_2\} \in \binom{V}{2}$. This correspondence is injective, since there is at most one $(2n+1)$-clique through every two adjacent vertices of Γ. Since $|W| = (2n+1)(n+1) = \binom{|V|}{2}$, the correspondence is even bijective. Now choose a point x of $\mathcal{S}$. The point x is contained in $n+1$ elements of W. If $F_1 \in W$ and $F_2 \in W$ are two such elements, then the corresponding pairs are disjoint, otherwise there would be a maximal clique containing F_1 and F_2. Hence, points correspond to partitions of V in $n+1$ pairs. This correspondence is injective since the $n+1$ elements of W through a point have only that point in common. Since there are as many partitions of V in $n+1$ pairs as there are points in $\mathcal{S}$, namely $\frac{(2n+2)!}{2^{n+1}(n+1)!}$, the correspondence is bijective. Now, choose a line L of $\mathcal{S}$. The line L is contained in $n-1$ elements of W, and the corresponding pairs are mutually disjoint. Hence, every line of $\mathcal{S}$ corresponds to a partition of V in $n-1$ sets of size 2 and one set of size 4. This correspondence is injective since the $n-1$ elements of W through a line have only that line in common. Since there are as many lines as such partitions, the correspondence is bijective. Theorem 8.3 now immediately follows. □

8.5 Proof of Theorem 8.4

Suppose that Q is a $Q(5,2)$-quad intersecting F in a line and let Q' denote a $W(2)$-quad of F through $Q \cap F$. The hex $\mathcal{C}(Q,Q')$ then contains a $Q(5,2)$-quad Q and a big $W(2)$-quad Q', contradicting Theorem 7.1. Hence, no $Q(5,2)$-quad meets F. Suppose now that R is a $W(2)$-quad intersecting F in a line L, let x denote an arbitrary point of L and let L' denote the unique line of F through x for which $\mathcal{C}(L,L')$ is a grid. The number of $W(2)$-quads of F through L equals $2^{n-2}-2$. If R' is one of these quads, then by Theorem 7.1, the hex $\mathcal{C}(R,R')$ has a line through x not contained in $R \cup R'$. As a consequence $t_{\mathcal{S}} - t_F \geq 2^{n-2}$. Also by Theorem 7.1, $\mathcal{C}(R,L') \cong W(2) \times \mathbb{L}_3$ and hence $\mathcal{C}(R,L')$ has two grids through L' which intersect F in the line L'. As a consequence, the number of grid-quads intersecting F in L' is at least twice the number of $W(2)$-quads intersecting F in L. Suppose now that Q_1 and Q_2 are two grid-quads through L different from $\mathcal{C}(L,L')$. By Theorem 7.1, the hex $\mathcal{C}(Q_1,Q_2)$ cannot intersect F in a big $W(2)$-quad. Hence $\mathcal{C}(Q_1,Q_2) \cap F = \mathcal{C}(L,L')$ and $Q_2 \subset \mathcal{C}(Q_1,L')$. As a consequence, there are at most two grid-quads intersecting F in the line L. By symmetry, there are also at most two grid-quads intersecting F in the line L' and hence there is at most one $W(2)$-quad intersecting F in L. Since every line K through x not contained in F is contained in a quad together with L, we have $1+1+2 \geq t_{\mathcal{S}} - t_F \geq 2^{n-2}$ or $n \leq 4$, contradicting our assumption $n \geq 5$. As a consequence, every quad intersecting F in a line is a grid. The theorem now immediately follows from Theorem 4.1. □

8.6 Proof of Theorem 8.5

We will suppose that the theorem holds for any dense near polygon of diameter between 4 and $n-1$ (= induction hypotheses). Let $t+1$ denote the constant number of lines through a point of $\mathcal{S}$. If $t = t_F + 1$, then $\mathcal{S} \cong \mathbb{G}_{n-1} \times \mathbb{L}_3$ by

Theorem 4.1. In the sequel, we will suppose that $t > t_F + 1$.

Lemma 8.20. *If a $Q(5,2)$-quad Q intersects F in a line, then this line is a special line of $F \cong \mathbb{G}_{n-1}$.*

Proof. Suppose that $L := Q \cap F$ is an ordinary line of F. Then L is contained in a $W(2)$-quad $R \subset F$. By Theorem 1.7, the $W(2)$-quad R is big in the hex $H := \mathcal{C}(Q, R)$. By Theorem 7.1, none of the near hexagons with a big $W(2)$-quad contains a $Q(5,2)$-quad. This contradicts the fact that $Q \subset H$. Hence L is a special line of F. □

Lemma 8.21. *No hex H isomorphic to $\mathbb{E}_1$, $\mathbb{E}_2$, $\mathbb{E}_3$ or $DH(5,4)$ meets F.*

Proof. Suppose the contrary. By Theorem 1.7, $H \cap F$ is a big quad of H. By Theorem 7.1, we then have: (i) $H \cong \mathbb{E}_3$ or $H \cong DH(5,4)$, and (ii) $Q := H \cap F \cong Q(5,2)$. The $Q(5,2)$-quad Q contains an ordinary line K of F. By (i), H has a $Q(5,2)$-quad through K different from Q. This quad contradicts Lemma 8.20. □

Lemma 8.22. *Every point x of F is contained in a $Q(5,2)$-quad which intersects F in a line. Hence $t \geq t_F + 4$.*

Proof. Since $t > t_F + 1$, there exist two lines K and L through x not contained in F. Since F is big in $\mathcal{S}$, $\mathcal{C}(K, L)$ intersects F in a line M; hence $\mathcal{C}(K, L) \cong W(2)$ or $\mathcal{C}(K, L) \cong Q(5,2)$. Suppose that $\mathcal{C}(K, L) \cong W(2)$. There exists a $Q(5,2)$-quad $Q \subset F$ through M. The hex $H := \mathcal{C}(K, R)$ contains a $Q(5,2)$-quad and a $W(2)$-quad. By Theorem 7.1 and Lemma 8.21, H is isomorphic to $\mathbb{G}_3$ and hence contains a $Q(5,2)$-quad through x different from Q. This proves our lemma. □

First Case: $t = t_F + 4$

Let P_2 denote the set of all $Q(5,2)$-quads meeting F in a line. By Lemma 8.22 and the fact that $t = t_F + 4$, it follows that every point $x \in F$ is contained in a unique element of P_2. If y is an arbitrary point outside F, then $Q_y := Q_{\pi(y)}$ is the unique element of P_2 through y. Hence P_2 is a partition of the point set of $\mathcal{S}$ in $Q(5,2)$-quads. Clearly the set $S_1 := \{Q \cap F \mid Q \in P_2\}$ is a spread S_1 of F.

Lemma 8.23. *The spread S_1 is an admissible spread of F.*

Proof. Since $F \cong \mathbb{G}_{n-1}$, we need to verify conditions (a) and (b) of Lemma 6.41. Condition (a) is precisely Lemma 8.20. We now prove that also condition (b) is satisfied. Let K be an arbitrary line of S_1, let Q denote the unique quad of P_2 through K and let R be an arbitrary grid-quad of F through K. The hex $H := \mathcal{C}(Q, R)$ has a $Q(5,2)$-quad and a big grid-quad and hence is isomorphic to $Q(5,2) \times \mathbb{L}_3$ by Theorem 7.1. As a consequence H contains three quads of P_2 and the two lines of R disjoint from K also belong to S_1. □

Lemma 8.24. *Every convex sub-$2(n-1)$-gon isomorphic to $\mathbb{G}_2 \otimes \mathbb{G}_{n-2}$ meets F.*

Proof. Let F' be a convex sub-$2(n-1)$-gon isomorphic to $\mathbb{G}_2 \otimes \mathbb{G}_{n-2}$ and disjoint from F. The near hexagon $\mathcal{S}$ has $v_{\mathcal{S}} = (1+2\cdot(t-t_F))\cdot|F| = \frac{3^{n+1}\cdot(2n-2)!}{2^{n-1}\cdot(n-1)!}$ points. The total number of points at distance at most 1 from F' equals $(1+2(t-t_{F'}))\cdot|F'|$. Since this number is precisely $v_{\mathcal{S}}$, also F' is big in $\mathcal{S}$. By Theorem 1.10, $F \cong F'$. From $\frac{3(n-1)^2-(n-1)-2}{2} = t_F = t_{F'} = \frac{3(n-2)^2-(n-2)-2}{2} + 4$, it then follows that $n=3$, but this contradicts our assumption $n \geq 4$. □

Lemma 8.25. *Every point y of $\mathcal{S}$ is contained in a unique big convex subpolygon F_y satisfying:*

(i) $F_y \cong F$;

(ii) $F_y = F$ *or* $F_y \cap F = \emptyset$.

Proof. Suppose that y is contained in two such subpolygons F_1 and F_2. Since $F_3 := F_1 \cap F_2$ is big in F_1, $F_3 \cong \mathbb{G}_{n-2}$ by Theorem 6.35. Hence $t \geq t_{F_1} + t_{F_2} - t_{F_3}$ or $t_{F_2} - t_{F_3} \leq 4$. Since $t_{F_2} - t_{F_3} = 3n-5$, $n \leq 3$, a contradiction. So, it suffices to show that y is contained in at least one big convex subpolygon satisfying (i) and (ii). This trivially holds if $y \in F$, so we suppose that $y \notin F$. By Lemma 8.20, Q_y intersects F in a special line K. If $L_1, \ldots, L_{n-2}$ denote the other special lines of F through $\pi(y)$, then $F_4 := \mathcal{C}(L_1, \ldots, L_{n-2})$ is isomorphic to $\mathbb{G}_{n-2}$. Put $F_5 := \mathcal{C}(L_1, \ldots, L_{n-2}, y\,\pi(y))$. Since $t_{F_5} = t_{F_4} + 1$, $F_5 \cong F_4 \times \mathbb{L}_3$. Hence y is contained in a convex sub-$2(n-2)$-gon F'_y isomorphic to $\mathbb{G}_{n-2}$. By Theorem 2.32 every convex sub-$2(n-1)$-gon through F'_y intersect Q_y in a line. Hence there are exactly five convex sub-$2(n-1)$-gons through F'_y. One of them is F_5. Let F_6 denote one of the four others. The projection of F_6 on F is distance-preserving and since the projection $\mathcal{C}(L_1, \ldots, L_{n-2})$ of F'_y is big in F, also F'_y is big in F_6. If $n=4$, then $F'_y \cong Q(5,2)$ and hence $F_6 \cong \mathbb{G}_3$ or $F_6 \cong \mathbb{G}_2 \times \mathbb{L}_3$ by Theorem 7.1, Lemma 8.21 and Lemma 8.24. If $n \geq 5$, then $F'_y \cong \mathbb{G}_{n-2}$ and hence $F_6 \cong \mathbb{G}_{n-1}$ or $F_6 \cong \mathbb{G}_{n-2} \times \mathbb{L}_3$ by the induction hypothesis and Lemma 8.24. Suppose now that all the five convex sub-$2(n-1)$-gons through F'_y are isomorphic to $\mathbb{G}_{n-2} \times \mathbb{L}_3$. Then $t = t_{F'_y} + 5$ or $t_F = t_{F'_y} + 1$, a contradiction since $t_F - t_{F'_y} = 3n-5$ and $n \geq 4$. Hence there exists a convex sub-$2(n-1)$-gon through F'_y isomorphic to $\mathbb{G}_{n-1}$. Our lemma now follows since $y \in F'_y$. □

The convex sub-$2(n-1)$-gons F_y, $y \in \mathcal{P}$, determine a partition P_1 of $\mathcal{S}$ in subpolygons isomorphic to $\mathbb{G}_{n-1}$. Every quad of P_2 intersects each subpolygon of P_1 in a line. All conditions of Theorem 4.28 are now satisfied and it follows that $\mathcal{S}$ is a glued near polygon of type $\mathbb{G}_2 \otimes \mathbb{G}_n$. By Theorem 6.107 there exists a unique glued near polygon of that type.

Second Case: $t > t_F + 4$

Put $\delta := t - t_F$.

Lemma 8.26. *We have $\delta \leq 3n-2$. If equality holds, then no hex isomorphic to $\mathbb{G}_2 \otimes \mathbb{G}_2$ meets F.*

Proof. By Lemmas 8.20 and 8.22 there exists a $Q(5,2)$-quad Q which intersects F in a special line K. By Theorem 7.1 and Lemma 8.21, every hex H through Q is isomorphic to either $\mathbb{G}_2 \times \mathbb{L}_3$, $\mathbb{G}_2 \otimes \mathbb{G}_2$ or $\mathbb{G}_3$. In the first case $H \cap F$ is a grid. In the two other cases $H \cap F$ is a $Q(5,2)$-quad. Let λ_1, respectively λ_2, denote the number of hexes through Q which are isomorphic to $\mathbb{G}_2 \otimes \mathbb{G}_2$, respectively $\mathbb{G}_3$. By Theorems 6.33 and 6.34, F has $n-2$ $Q(5,2)$-quads through K and hence $\lambda_1 + \lambda_2 = n-2$. Counting over all hexes H through Q, we find that $\delta = t_Q + \sum(t_H - t_Q - t_{H \cap F}) = 4 + 3\lambda_2 \leq 4 + 3(n-2) = 3n-2$. The lemma now immediately follows. □

Lemma 8.27. *If a $W(2)$-quad Q intersects F in a line, then this line is an ordinary line of $F \cong \mathbb{G}_{n-1}$.*

Proof. Suppose that $Q \cap F$ is a special line and let $x \in Q \cap F$. If R is one of the $n-2$ $Q(5,2)$-quads of F through $Q \cap F$, then the hex $\mathcal{C}(Q,R)$ has $W(2)$-quads and $Q(5,2)$-quads. By Theorem 7.1 and Lemma 8.21, it then follows that $\mathcal{C}(Q,R) \cong \mathbb{G}_3$. Hence the hex $\mathcal{C}(Q,R)$ contains exactly five lines through x which are not contained in $Q \cup R$. Summing over all possible R, we find that $\delta \geq 2 + 5(n-2) = 5n-8$. Together with $\delta \leq 3n-2$, this implies that $n \leq 3$, a contradiction. Hence $Q \cap F$ is an ordinary line. □

Lemma 8.28. *Every point x of F is contained in a $W(2)$-quad which intersects F in a line.*

Proof. By Lemma 8.22, there exists a $Q(5,2)$-quad Q through x intersecting F in a line. Since $t > t_F + 4$, there exists a line K through x not contained in $Q \cup F$. By Theorem 7.1 and Lemma 8.21, the hex $H = \mathcal{C}(Q,K)$, which intersects F in a big quad, is isomorphic to $\mathbb{G}_3$. The required $W(2)$-quad can now be chosen in the hex H. □

Lemma 8.29. *We have $\delta \geq 3n-2$. If equality holds, then no hex isomorphic to $DQ(6,2)$ meets F.*

Proof. Let Q denote a $W(2)$-quad intersecting F in an ordinary line K. By Theorem 6.31, K is contained in a unique $Q(5,2)$-quad and $3(n-3)$ $W(2)$-quads of F. If T is the unique $Q(5,2)$-quad, then the hex $H := \mathcal{C}(Q,T)$ is isomorphic to $\mathbb{G}_3$. If T is one of the $3(n-3)$ $W(2)$-quads of F through K, then $H = \mathcal{C}(Q,T)$ is isomorphic to either $\mathbb{H}_3$ or $DQ(6,2)$. Hence $\delta = t_Q + \sum(t_H - t_Q - t_{H \cap F}) \geq 2 + 5 + 3(n-3) = 3n-2$. The lemma now immediately follows. □

From Lemmas 8.26 and 8.29, we obtain:

Corollary 8.30. *The following holds:*

- $\delta = 3n-2$, $t = \delta + t_F = \frac{3n^2-n-2}{2}$, $|\mathcal{P}| = (2\delta+1) \cdot |F| = \frac{3^n \cdot (2n)!}{2^n \cdot n!}$ *and* $|\mathcal{L}| = \frac{|\mathcal{P}| \cdot (t+1)}{3} = \frac{3^{n-1}(2n)!(3n-1)}{2^{n+1}(n-1)!}$;
- *no hex isomorphic to $\mathbb{G}_2 \otimes \mathbb{G}_2$ meets F;*
- *no hex isomorphic to $DQ(6,2)$ meets F.*

Lemma 8.31. (a) *Every special line L of $F \cong \mathbb{G}_{n-1}$ is contained in a unique $Q(5,2)$-quad which is not contained in F.*

(b) *Let $x \in F$. All the $Q(5,2)$-quads through x which are not contained in F have a line A_x in common.*

Proof. (a) Suppose that the line L is contained in two such $Q(5,2)$-quads Q and R. The hex $\mathcal{C}(Q,R)$ intersects F in a big quad, which is necessarily isomorphic to $Q(5,2)$. The line L of $\mathcal{C}(Q,R)$ is then contained in at least three $Q(5,2)$-quads and hence $\mathcal{C}(Q,R)$ must be isomorphic to $DH(5,4)$, contradicting Lemma 8.21. Hence L is contained in at most one $Q(5,2)$-quad which is not contained in F. We will now prove that L is contained in a unique such $Q(5,2)$-quad. Let $x \in L$ and let T denote an arbitrary $Q(5,2)$-quad through x which intersects F in a special line. We may suppose that $L \neq T \cap F$. The hex $\mathcal{C}(T,L)$ has at least two $Q(5,2)$ quads through the line $T \cap F$ (namely T and $\mathcal{C}(T \cap F, L)$) and hence is isomorphic to $\mathbb{G}_3$ by Theorem 7.1, Lemma 8.21 and Corollary 8.30. Let T' denote the unique $Q(5,2)$-quad of $\mathcal{C}(T,L)$ through x different from T and $\mathcal{C}(T \cap F, L)$. Then $L \subset T'$ since $T' \cap F$ is a special line.

(b) Let T_1, T_2 and T_3 denote three different $Q(5,2)$-quads through x which are not contained in F. By the proof of (a), we know that T_1 and T_2 are contained in a $\mathbb{G}_3$-hex H_3. Hence T_1 and T_2 intersect in a line M_3. In a similar way one can define hexes H_1 and H_2, and lines M_1 and M_2. Now, $H_1 \cap H_2 \cap H_3 = (H_1 \cap H_2) \cap (H_1 \cap H_3) = T_3 \cap T_2 = M_1$. Similarly $M_2 = M_3 = H_1 \cap H_2 \cap H_3$. Hence, all $Q(5,2)$-quads through x not contained in F have a common line A_x. □

Corollary 8.32. *Let $x \in F$. The $n-1$ $Q(5,2)$-quads through A_x partition the set of lines through x which are not contained in $F \cup A_x$.*

Proof. The $n-1$ $Q(5,2)$-quads through A_x determine $1 + 3(n-1) = 3n-2$ lines through x which are not contained in F. The result now follows since $\delta = 3n-2$. □

Lemma 8.33. *For each $i \in \{1,2\}$, let $\mathcal{S}_i$ be a dense near polygon, let F_i be a big convex subpolygon of $\mathcal{S}_i$ and let x_i be a point of F_i. Suppose that there exists an isomorphism ϕ from F_1 to F_2 mapping x_1 to x_2 and a bijection θ from the set of lines of $\mathcal{S}_1$ through x_1 to the set of lines of $\mathcal{S}_2$ through x_2 such that the following holds for all lines K, L and M through x_1:*

(a) *if K is contained in F_1, then $\theta(K) = \phi(K)$;*

(b) *K, L and M are contained in a quad if and only if $\theta(K)$, $\theta(L)$ and $\theta(M)$ are contained in a quad.*

Then $\mathcal{G}(\mathcal{S}_1, x) \cong \mathcal{G}(\mathcal{S}_2, x_2)$.

Proof. Let A be a convex subpolygon of $\mathcal{S}_1$ through x_1. If A is contained in F_1, then we define $\mu(A) := \phi(A)$. If A is not contained in F_1, then we define $\mu(A) = \mathcal{C}(\theta(K), \phi(A \cap F_1))$ where K is any line of A through x_1 not contained in F_1. This is a good definition. If K' is another line with this property, then K, K' and $\mathcal{C}(K, K') \cap F_1$ are contained in the same quad. By (a) and (b) also $\theta(K)$, $\theta(K')$ and $\phi(\mathcal{C}(K, K') \cap F_1)$ are in the same quad and since $\phi(\mathcal{C}(K, K') \cap F_1) \subseteq \phi(A \cap F_1)$, $\mathcal{C}(\theta(K), \phi(A \cap F_1)) = \mathcal{C}(\theta(K'), \phi(A \cap F_1))$. If A is a near $2i$-gon, $i \in \{1, \ldots, n-1\}$, then also $\mu(A)$ is a near $2i$-gon. Clearly, μ is an incidence preserving bijection between the set of objects of $\mathcal{G}(\mathcal{S}_1, x)$ and the set of objects of $\mathcal{G}(\mathcal{S}_2, x_2)$. □

Lemma 8.34. *For every $x \in F$, $\mathcal{G}(\mathcal{S}, x)$ is isomorphic to $\mathcal{G}(\mathbb{G}_n)$.*

Proof. Let F' denote a convex sub-$2(n-1)$-gon of $\mathbb{G}_n$ isomorphic to $\mathbb{G}_{n-1}$, let $x' \in F'$ and let $A_{x'}$ denote the unique special line through x' not contained in F'. Since $\mathrm{Aut}(\mathbb{G}_{n-1})$ acts transitively on the set of points of $\mathbb{G}_{n-1}$, there exists an isomorphism ϕ from F to F' mapping x to x'. For every line K of F through x, we define $\theta(K) = \phi(K)$. We will now extend θ in such a way that it determines an isomorphism between $\mathcal{L}(\mathcal{S}, x)$ and $\mathcal{L}(\mathbb{G}_n, x')$. Our result then follows from Lemma 8.33.

Extension of θ. We put $\theta(A_x) = A_{x'}$. Let K and K' denote two arbitrary special lines of F through x. Let K, A_x, L_1, L_2 and L_3 denote the five lines of $\mathcal{C}(K, A_x)$ through x. Similarly, let K', A_x, L_1', L_2' and L_3' denote the five lines of $\mathcal{C}(K', A_x)$ through x. Let $\theta(L_1)$ be one of the three lines of $\mathcal{C}(\theta(K), A_{x'})$ through x' different from $\theta(K)$ and $A_{x'}$. Now, let M be an arbitrary line through x not contained in $F \cup \mathcal{C}(K, A_x)$. The quad $\mathcal{C}(L_1, M)$ is a $W(2)$-quad and intersects F in an ordinary line N. The quad $\mathcal{C}(A_x, M)$ is a $Q(5,2)$-quad and intersects F in a special line N'. The hex $\mathcal{C}(A_x, L_1, M)$ is isomorphic to $\mathbb{G}_3$ and intersects F in the $Q(5,2)$-quad $\mathcal{C}(K, N')$. Clearly N is contained in $\mathcal{C}(K, N')$. The hex $\mathcal{C}(A_{x'}, \theta(K), \theta(N'))$ is isomorphic to $\mathbb{G}_3$ and contains the lines $\theta(L_1)$ and $\theta(N)$. The quad $\mathcal{C}(\theta(L_1), \theta(N))$ is isomorphic to $W(2)$ and we put $\theta(M)$ equal to the unique line of $\mathcal{C}(\theta(L_1), \theta(N))$ through x' different from $\theta(L_1)$ and $\theta(M)$. Clearly $\theta(M) \in \mathcal{C}(A_{x'}, \theta(N'))$. We already defined $\theta(L)$ for all lines L through x different from L_2 and L_3. For each $i \in \{2,3\}$, the quad $\mathcal{C}(L_i, L_1')$ is isomorphic to $W(2)$ and intersects F in a line P. Again $\mathcal{C}(\theta(P), \theta(L_1'))$ is a $W(2)$-quad and we put $\theta(L_i)$ equal to the unique line of $\mathcal{C}(\theta(P), \theta(L_1'))$ through x' different from $\theta(P)$ and $\theta(L_1')$. Clearly, $\theta(L_i) \in \mathcal{C}(A_{x'}, \theta(K))$. One easily sees that θ is a bijection between the set of lines of $\mathcal{S}$ through x and the set of lines of $\mathbb{G}_n$ through x'.

A linear space on a certain set of points is completely determined if all lines of size at least 3 are known. The linear spaces $\mathcal{L}(\mathcal{S}, x)$ and $\mathcal{L}(\mathbb{G}_n, x')$ each contain $\frac{n(n-1)}{2}$ lines of size 5 and $\frac{3n(n-1)(n-2)}{2}$ lines of size 3. So, in order to prove that θ determines an isomorphism, it suffices to verify that θ maps lines of size $r \in \{3,5\}$ in $\mathcal{L}(\mathcal{S}, x)$ to lines of size r in $\mathcal{L}(\mathbb{G}_n, x')$. By construction (see above), this holds for the lines of size 5. So, let $\delta = \{M_1, M_2, M_3\}$ denote a line of size 3 in $\mathcal{L}(\mathcal{S}, x)$ and let Q_δ denote the $W(2)$-quad corresponding with it. We will now prove that

$\{\theta(M_1), \theta(M_2), \theta(M_3)\}$ is a line of size 3 in $\mathcal{L}(\mathbb{G}_n, x')$. This trivially holds if $Q_\delta \subset F$. Suppose therefore that M_1, M_2 are outside F and that M_3 is inside F. We may also suppose that $M_1 \neq L_1 \neq M_2$. One of the following cases certainly occurs.

(I) The case $M_1, M_2 \in \{L_2, L_3, L_1', L_2', L_3'\}$.
Let L_1'', L_2'' and L_3'' denote the three lines of $\mathcal{C}(K, K')$ through x different from K and K'. The set $\{L_1, L_2, L_3, L_1', L_2', L_3', L_1'', L_2'', L_3''\}$ together with the subsets $\{L_1, L_2, L_3\}$, $\{L_1', L_2', L_3'\}$, $\{L_1'', L_2'', L_3''\}$, $\{L_i, L_j', \mathcal{C}(L_i, L_j') \cap F\}$, $i, j \in \{1, 2, 3\}$, define an affine plane $\mathcal{A}$ of order 3. In a similar way, an affine plane $\mathcal{A}'$ can be defined on the set $\{\theta(L_1), \ldots, \theta(L_3'')\}$. The set $\{\theta(L_1), \ldots, \theta(L_3'')\}$ also carries the structure of an affine plane $\mathcal{A}^\theta$ if one considers all subsets of the form $\{\theta(P_1), \theta(P_2), \theta(P_3)\}$ where $\{P_1, P_2, P_3\}$ is a line of $\mathcal{A}$. Now, $\mathcal{A}'$ and $\mathcal{A}^\theta$ have the following eight lines in common:

$$\{\theta(L_1), \theta(L_2), \theta(L_3)\}, \{\theta(L_1'), \theta(L_2'), \theta(L_3')\}, \{\theta(L_1''), \theta(L_2''), \theta(L_3'')\},$$

$$\{\theta(L_1), \theta(L_1'), \mathcal{C}(\theta(L_1), \theta(L_1')) \cap F'\}, \{\theta(L_1), \theta(L_2'), \mathcal{C}(\theta(L_1), \theta(L_2')) \cap F'\},$$

$$\{\theta(L_1), \theta(L_3'), \mathcal{C}(\theta(L_1), \theta(L_3')) \cap F'\}, \{\theta(L_2), \theta(L_1'), \mathcal{C}(\theta(L_2), \theta(L_1')) \cap F'\},$$

$$\{\theta(L_3), \theta(L_1'), \mathcal{C}(\theta(L_3), \theta(L_1')) \cap F'\}.$$

So, $\mathcal{A}' = \mathcal{A}^\theta$. This is precisely what we needed to prove.

(II) The case $\{M_1, M_2\} \cap \{L_1, L_2, L_3\} = \emptyset$.
The quad $\mathcal{C}(A_x, M_i)$, $i \in \{1, 2\}$, intersects F in a special line P_i. Clearly, $P_1 \neq P_2$. The $W(2)$-quad $\mathcal{C}(L_1, M_i)$, $i \in \{1, 2\}$, intersects F in an ordinary line N_i which is contained in the $Q(5, 2)$-quad $\mathcal{C}(P_i, K)$. Since N_i is ordinary, $\mathcal{C}(P_i, K)$ is the unique $Q(5, 2)$ quad through N_i. Since $\mathcal{C}(P_1, K) \neq \mathcal{C}(P_2, K)$, $\mathcal{C}(N_1, N_2)$ is not a $Q(5, 2)$-quad. The hex $H = \mathcal{C}(L_1, M_1, M_2)$ intersects F in the quad $\mathcal{C}(N_1, N_2)$. The line M_3 belongs to $\mathcal{C}(N_1, N_2)$ and is different from N_1 and N_2. Hence $\mathcal{C}(N_1, N_2) \cong W(2)$. Since also $\mathcal{C}(\theta(N_1), \theta(N_2)) \cong W(2)$, the lines $\theta(N_1)$, $\theta(N_2)$ and $\theta(M_3)$ are precisely the three lines of $\mathcal{C}(\theta(N_1), \theta(N_2))$ through x'. Since $\mathcal{C}(\theta(L_1), \theta(M_1)) \cap F' = \theta(N_1)$, $\mathcal{C}(\theta(L_1), \theta(M_2)) \cap F' = \theta(N_2)$ and $\mathcal{C}(\theta(L_1), \theta(M_1), \theta(M_2)) \cap F = \mathcal{C}(\theta(N_1), \theta(N_2))$, we necessarily have that $\mathcal{C}(\theta(M_1), \theta(M_2)) \cap F' = \theta(M_3)$. This is precisely what we needed to prove.

(III) The case $\{M_1, M_2\} \cap \{L_1', L_2', L_3'\} = \emptyset$.
By (I) and (II), θ maps the lines $\{L_1', M_1, \mathcal{C}(L_1', M_1) \cap F\}$ and $\{L_1', M_2, \mathcal{C}(L_1', M_2) \cap F\}$ of $\mathcal{L}(\mathcal{S}, x)$ to lines of $\mathcal{L}(\mathbb{G}_n, x')$. With a similar reasoning as in (II), we then derive that also $\{M_1, M_2, \mathcal{C}(M_1, M_2) \cap F\}$ is mapped to a line of $\mathcal{L}(\mathbb{G}_n, x')$. □

Lemma 8.35. *Every point y of $\mathcal{S}$ is contained in a big convex subpolygon isomorphic to $\mathbb{G}_{n-1}$. Hence $\mathcal{G}(\mathcal{S}, y) \cong \mathcal{G}(\mathbb{G}_n)$.*

Proof. We may suppose that $y \notin F$, then y is collinear with a unique point $\pi(y)$ of F. Call a line L through $\pi(y)$ special if it is not contained in a $W(2)$-quad and ordinary otherwise. Since $\mathcal{G}(\mathcal{S}, \pi(y)) \cong \mathcal{G}(\mathbb{G}_n)$, there are precisely n special lines $L_1, \ldots, L_n$ through $\pi(y)$. We may suppose that $y\,\pi(y) \subset \mathcal{C}(L_1, L_2)$. For every $i \in \{2, \ldots, n\}$, we put $F_i := \mathcal{C}(L_1, \ldots, L_i)$. Since $\mathcal{G}(\mathcal{S}, \pi(y)) \cong \mathcal{G}(\mathbb{G}_n)$, we have the following for every $i \in \{2, \ldots, n-1\}$:

(i) F_i is a dense convex subpolygon of order $(2, \frac{3i^2-3i-2}{2})$;

(ii) every quad of F_{i+1} through $\pi(y)$ either is contained in F_i or intersects F_i in a line.

By (i) and Theorem 7.1, $F_2 \cong Q(5,2)$ and $F_3 \cong \mathbb{G}_3$. Suppose now that $F_i \cong \mathbb{G}_i$ for a certain $i \in \{3, \ldots, n-2\}$. By (ii) and Theorem 2.31, F_i is big in F_{i+1}. By the induction hypothesis, F_{i+1} is isomorphic to either $\mathbb{G}_{i+1}$, $\mathbb{G}_i \otimes \mathbb{G}_2$ or $\mathbb{G}_i \times \mathbb{L}_3$. By (i), we have $F_{i+1} \cong \mathbb{G}_{i+1}$. Now, $y \in F_{n-1}$ and $F_{n-1} \cong \mathbb{G}_{n-1}$ is big in $\mathcal{S}$. By Lemma 8.34 applied to F_{n-1} instead of F, $\mathcal{G}(\mathcal{S}, y) \cong \mathcal{G}(\mathbb{G}_n)$. □

Call a line L of $\mathcal{S}$ *special* if it is not contained in a $W(2)$-quad, and *ordinary* otherwise. Since $\mathcal{G}(\mathcal{S}, y) \cong \mathcal{G}(\mathbb{G}_n)$ for every point y of $\mathcal{S}$, every point of $\mathcal{S}$ is incident with n special lines and $\frac{3}{2}n(n-1)$ ordinary lines. Let V_k, $k \in \{1, \ldots, n\}$, denote the set of all convex sub-$2k$-gons generated by k special lines through a fixed point. If $F \in V_k$, $k \in \{1, \ldots, n-1\}$, then a similar reasoning as in the proof of Lemma 8.35 gives that $F \cong \mathbb{G}_k$. From Theorem 7.34, it now follows that $\mathcal{S} \cong \mathbb{G}_n$. □

8.7 Proof of Theorem 8.6

Let Q denote an arbitrary quad intersecting H in a line L. Since $H \cong \mathbb{E}_3$, H has a $W(2)$-quad Q_1 and a $Q(5,2)$-quad Q_2 through L. If Q is a $Q(5,2)$-quad, then the hex $\mathcal{C}(Q, Q_1)$ contains a $Q(5,2)$-quad Q and a big $W(2)$-quad Q_1, contradicting Theorem 7.1. If Q is a $W(2)$-quad, then by Theorem 7.1, the hex $\mathcal{C}(Q, Q_2)$ is isomorphic to either $\mathbb{G}_3$ or $\mathbb{E}_3$. In any case, $\mathcal{C}(Q, Q_2)$ contains a $Q(5,2)$-quad which intersects H in a line, a contradiction. Hence Q is a grid. The theorem now follows from Theorem 4.1. □

8.8 Proof of Theorem 8.7

Let x denote an arbitrary point of F. Since $F \cong \mathcal{S}_1 \times \mathcal{S}_2$, there exist convex subpolygons F_1 and F_2 through x such that

- $F_1 \cong \mathcal{S}_1$ and $F_2 \cong \mathcal{S}_2$;
- $F_1 \cap F_2 = \{x\}$;
- every line through x either is contained in F_1 or in F_2.

If $t_{\mathcal{S}} = t_F + 1$, then $\mathcal{S} \cong F \times L \cong \mathcal{S}_1 \times (\mathcal{S}_2 \times L)$ by Theorem 4.1. Hence, we may suppose that there exist two lines K and L through x which are not contained in F. Since F is big in $\mathcal{S}$, $\mathcal{C}(K, L)$ intersects F in a line L'. By reasons of symmetry, we may suppose that $L' \subseteq F_1$. Put now $F_3 := \mathcal{C}(F_1, K)$. Then $F_1 = F_3 \cap F$ is big in

F_3, $\mathrm{diam}(\mathcal{S}) = \mathrm{diam}(F_3) + \mathrm{diam}(F_2)$ and $F_3 \cap F_2 = \{x\}$. We will show that every line M through x not contained in F is contained in F_3. The theorem then follows from Theorem 4.4. Since $\mathcal{C}(K, L) \subseteq F_3$, we may suppose that M is not contained in $\mathcal{C}(K, L)$. The quad $\mathcal{C}(K, M)$ then intersects F in a line $M' \neq L'$. If M' belongs to F_2, then the grid-quad $\mathcal{C}(L', M')$ is big in the hex $\mathcal{C}(K, L, M)$. From Theorem 7.1 it then follows that $\mathcal{C}(K, L, M)$ is isomorphic to either $\mathbb{L}_3 \times \mathbb{L}_3 \times \mathbb{L}_3$, $W(2) \times \mathbb{L}_3$ or $Q(5,2) \times \mathbb{L}_3$. This contradicts however the fact that none of the quads $\mathcal{C}(K, L)$ and $\mathcal{C}(K, M)$ is a grid. Hence $M' \subseteq F_1$ and $M \subset \mathcal{C}(K, M') \subseteq F_3$. □

8.9 Proof of Theorem 8.8

Lemma 8.36. *The class $\mathcal{D}$ does not contain product near polygons.*

Proof. By Theorem 4.60, it suffices to show that no element of the set $\{DH(2n-1,4) \,|\, n \geq 2\} \cup \{\mathbb{G}_n \,|\, n \geq 3\}$ is a product near polygon. This holds since all modified local spaces of these near polygons are connected (see Theorem 6.31 for the near polygon $\mathbb{G}_n$). □

Definition. If $\mathcal{A}$ is an element of $\mathcal{D}$, then we define $\Omega(\mathcal{A})$ as the set of all pairs $\{T_1, T_2\} \in \Delta_1(\mathcal{A})$ with the property that every element of $T_1 \cup T_2$ belongs to $\mathcal{D}$.

The near polygon $\mathcal{S}$ is a slim dense near $2n$-gon, $n \geq 4$, containing a big convex subpolygon F belonging to $\mathcal{D}$. There are four possibilities. If $F \cong DH(2n-3,4)$ for some $n \geq 3$, then by Theorem 8.2, $\mathcal{S}$ is isomorphic to either $DH(2n-1,4)$, $DH(2n-3,4) \times \mathbb{L}_3$ or $DH(2n-3,4) \otimes Q(5,2)$. If $F \cong \mathbb{G}_{n-1}$ for some $n \geq 3$, then by Theorem 8.5, $\mathcal{S}$ is isomorphic to either $\mathbb{G}_n$, $\mathbb{G}_{n-1} \times \mathbb{L}_3$ or $\mathbb{G}_{n-1} \otimes Q(5,2)$. In view of what we need to prove, we may suppose that

- F is glued, so $\Omega := \Omega(F) \neq \emptyset$;
- $\mathcal{S} \ncong F \times \mathbb{L}_3$;
- no proper convex subpolygon of $\mathcal{S}$ violates Theorem 8.8. (If this were not so, then the reasoning described here would still be applicable to each minimal violating convex subpolygon in $\mathcal{S}$, yielding an obvious contradiction.)

Let V denote the set of all $W(2)$-quads and $Q(5,2)$-quads which intersect F in a line.

Lemma 8.37. *The set V is not empty.*

Proof. Take an arbitrary point x in F. Since $\mathcal{S}$ is not isomorphic to $F \times \mathbb{L}_3$, there exist two lines K and L through x which are not contained in F. By Theorem 1.7, the quad $Q := \mathcal{C}(K, L)$ intersects F in a line. Hence $t_Q + 1 \geq 3$ and Q is isomorphic to either $W(2)$ or $Q(5,2)$. □

Definition. If T is a partition of F in isomorphic convex subpolygons, then $d(T)$ denotes the diameter of an arbitrary element of T. If T_1 and T_2 are two partitions of F in convex subpolygons, then we say that $(T_1, T_2) \in \Omega'$ if $\{T_1, T_2\} \in \Omega$ and if exactly one of the following holds:

(a) $Q \cap F \in \mathcal{L}_1 \cap \mathcal{L}_2$ for every $Q \in V$;

(b) there exists a quad $Q \in V$ such that $Q \cap F \in \mathcal{L}_2 \setminus \mathcal{L}_1$.

Here $\mathcal{L}_i$, $i \in \{1, 2\}$, denotes the set of lines of F which are contained in a subpolygon of T_i. If $\{T_1, T_2\} \in \Omega$, then Ω' contains at least one of the elements (T_1, T_2) and (T_2, T_1). Since $\Omega \neq \emptyset$, also $\Omega' \neq \emptyset$.

Lemma 8.38. *Suppose that $(T_1, T_2) \in \Omega'$.*

(i) *If (T_1, T_2) is of type (a), then for every element $\mathcal{A}$ of T_2, there exists a quad $Q_\mathcal{A} \in V$ such that $Q_\mathcal{A} \cap F \subseteq \mathcal{A}$ and $Q_\mathcal{A} \cap F \in \mathcal{L}_1 \cap \mathcal{L}_2$.*

(ii) *If (T_1, T_2) is of type (b), then for every element $\mathcal{A}$ of T_2, there exists a quad $Q_\mathcal{A} \in V$ such that $Q_\mathcal{A} \cap F \subseteq \mathcal{A}$ and $Q_\mathcal{A} \cap F \in \mathcal{L}_2 \setminus \mathcal{L}_1$.*

(iii) *For every element $\mathcal{A}$ of T_2, there exists a unique convex sub-$2(d(T_2)+1)$-gon $\hat{\mathcal{A}}$ through $\mathcal{A}$ such that $\hat{\mathcal{A}} \cap F = \mathcal{A}$. If $\mathcal{A} \in T_2$ and $\mathcal{B} \in T_1$, then $\hat{\mathcal{A}} \cap \mathcal{B}$ is a line, and every line which intersects $\hat{\mathcal{A}} \cap \mathcal{B}$ in a point either is contained in $\hat{\mathcal{A}}$ or $\mathcal{B}$.*

(iv) *The set $\hat{T}_2 := \{\hat{\mathcal{A}} | \mathcal{A} \in T_2\}$ is a partition of $\mathcal{S}$ in convex subpolygons.*

Proof. (i) Let $\mathcal{A}$ denote an arbitrary element of T_2 and x an arbitrary point of $\mathcal{A}$. Since $\mathcal{S} \not\cong F \times \mathbb{L}_3$, there exist two lines K and L through x not contained in F. Using Theorem 1.7, we see that the quad $\mathcal{C}(K, L)$ satisfies all required properties.

(ii) We will prove that if this property holds for a certain $\mathcal{A} \in T_2$ (with corresponding quad $Q_\mathcal{A}$), then it also holds for any $\mathcal{B} \in T_2$ at distance 1 from $\mathcal{A}$. Property (ii) then follows from the connectedness of F. Put $L_\mathcal{A} := Q_\mathcal{A} \cap F$, let a denote an arbitrary point of $L_\mathcal{A}$ and let b denote the unique point of $\mathcal{B}$ collinear with a. Since $L_\mathcal{A} \in \mathcal{L}_2 \setminus \mathcal{L}_1$, the quad $\mathcal{C}(L_\mathcal{A}, b)$ is a grid and hence contains a unique line $L_\mathcal{B}$ through b disjoint with $L_\mathcal{A}$. Clearly, the line $L_\mathcal{B}$ belongs to $\mathcal{L}_2 \setminus \mathcal{L}_1$ and is contained in $\mathcal{B}$. We will now construct an element $Q_\mathcal{B}$ of V through $L_\mathcal{B}$. By Theorem 1.7, the grid-quad $\mathcal{C}(L_\mathcal{A}, b)$ is big in the hex $\mathcal{C}(Q_\mathcal{A}, b)$. By Theorem 7.1, it then follows that $\mathcal{C}(Q_\mathcal{A}, b) \cong Q_\mathcal{A} \times \mathbb{L}_3$. Hence the hex $\mathcal{C}(Q_\mathcal{A}, b)$ contains a unique quad $Q_\mathcal{B} \cong Q_\mathcal{A}$ through $L_\mathcal{B}$. Clearly $Q_\mathcal{B} \in V$.

(iii) Let $\mathcal{A}$ denote an arbitrary element of T_2. Then there exists a quad $Q \in V$ intersecting $\mathcal{A}$ in a line K with $K \in \mathcal{L}_1 \cap \mathcal{L}_2$ if (T_1, T_2) is of type (a) or $K \in \mathcal{L}_2 \setminus \mathcal{L}_1$ if (T_1, T_2) is of type (b). Let x denote an arbitrary point of K. Clearly, $\mathcal{C}(\mathcal{A}, Q)$ intersects F in $\mathcal{A}$. In order to prove that $\mathcal{C}(\mathcal{A}, Q)$ is the subpolygon which satisfies all required properties, it suffices to show that $\mathcal{C}(\mathcal{A}, L) = \mathcal{C}(\mathcal{A}, Q)$ for any line L through x not contained in F. If (T_1, T_2)

is of type (a), consider then in Q a line L' through x different from K and L. The quad $\mathcal{C}(L, L')$ intersects F in the line K. Hence $L \subseteq \mathcal{C}(L', K) = Q$ and $\mathcal{C}(\mathcal{A}, L) = \mathcal{C}(\mathcal{A}, Q)$. Suppose now that (T_1, T_2) is of type (b) and that L is not contained in Q. By Theorem 1.7 the hex $H := \mathcal{C}(Q, L)$ intersects F in a big quad Q' through K. In H there are at least three quads through K (namely Q, Q' and $\mathcal{C}(K, L)$) and not all these quads are grids. So, H cannot be isomorphic to $\mathbb{L}_3 \times \mathbb{L}_3 \times \mathbb{L}_3$, $W(2) \times \mathbb{L}_3$ or $Q(5, 2) \times \mathbb{L}_3$. By Theorem 7.1, it then follows that H contains no big grid-quad. So, Q' is isomorphic to either $W(2)$ or $Q(5, 2)$. Since $K \in \mathcal{L}_2 \setminus \mathcal{L}_1$, Q' must be contained in $\mathcal{A}$. Hence, $\mathcal{C}(\mathcal{A}, L) = \mathcal{C}(\mathcal{A}, H) = \mathcal{C}(\mathcal{A}, Q)$.

(iv) It suffices to prove that every point x outside F is contained in a unique element of $\hat{T}_2$. If x' denotes the unique point of F nearest to x and $\mathcal{A}$ the unique element of T_2 through x', then $x \in \mathcal{C}(\mathcal{A}, x'x) = \hat{\mathcal{A}}$. If there were two different elements $\mathcal{A}_1$ and $\mathcal{A}_2$ in T_2 such that $x \in \hat{\mathcal{A}}_1 \cap \hat{\mathcal{A}}_2$, then x would be collinear with two different elements of F (notice that $\mathcal{A}_i$, $i \in \{1, 2\}$, is big in $\hat{\mathcal{A}}_i$ by Theorem 1.7), contradicting the fact that F is convex. □

Let (T_1, T_2) be an element of Ω' with $d(T_1)$ as small as possible. Let $\mathcal{G}$ denote an arbitrary near $2\,d(T_1)$-gon of T_1. If $\Omega(\mathcal{G}) \neq \emptyset$, then by Theorem 4.61 and Lemma 8.36, there exists an element $\{\tilde{T}_1, \tilde{T}_2\} \in \Omega(F)$ with $d(\tilde{T}_1) < d(T_1)$ such that T_2 is a refinement of $\tilde{T}_2$. This latter property implies that $(\tilde{T}_1, \tilde{T}_2) \in \Omega'$. But this contradicts our assumption on the minimality of $d(T_1)$. So, $\Omega(\mathcal{G}) = \emptyset$ and $\mathcal{G}$ is isomorphic to an element of the set $\{DH(2m-1, 4) | m \geq 2\} \cup \{\mathbb{G}_m | m \geq 3\}$. We have shown earlier that there exists a unique partition $\hat{T}_2$ of $\mathcal{S}$ in convex sub-$2(d(T_2)+1)$-gons, such that $\hat{A} \cap F \in T_2$ for every $\hat{A} \in \hat{T}_2$. We will now extend the partition T_1 of F to a partition $\hat{T}_1$ of $\mathcal{S}$.

Lemma 8.39. *The partition T_1 of F can be extended to a partition $\hat{T}_1$ of $\mathcal{S}$ such that every element of $\hat{T}_1$ intersects every element of $\hat{T}_2$ in a line.*

Proof. For every point x of F, we define $\hat{F}_1(x)$ as the unique element of T_1 through x. Suppose that x is a point outside F. Let $\hat{F}_2(x)$ denote the unique element of $\hat{T}_2$ through x and let x' denote the unique point of F collinear with x. By Theorem 1.7, $\mathcal{C}(\hat{F}_1(x'), x)$ contains $\hat{F}_1(x')$ as a big convex subpolygon. We distinguish the following cases.

- If $d(T_1) = 2$, then $\hat{F}_1(x') \cong Q(5, 2)$ and $\mathcal{C}(\hat{F}_1(x'), x)$ is isomorphic to either $Q(5, 2) \times \mathbb{L}_3$, $Q(5, 2) \otimes Q(5, 2)$, $\mathbb{G}_3$, $\mathbb{E}_3$ or $DH(5, 4)$ by Theorem 7.1. By property (iii) of Lemma 8.38, every line of $\mathcal{C}(\hat{F}_1(x'), x)$ through x' is contained in one of the quads $\hat{F}_1(x')$ or $\mathcal{C}(\hat{F}_1(x'), x) \cap \hat{F}_2(x)$. So, $\mathcal{C}(\hat{F}_1(x'), x)$ must be isomorphic to either $Q(5, 2) \times \mathbb{L}_3$ or $Q(5, 2) \otimes Q(5, 2)$.
- If $d(T_1) \geq 3$, then $\mathcal{C}(\hat{F}_1(x'), x)$ is isomorphic to either $\mathcal{G} \times \mathbb{L}_3$ or $\mathcal{G} \otimes Q(5, 2)$ by Theorems 8.2 and 8.5.

In any case, there exists a unique convex subpolygon $\hat{F}_1(x)$ in $\mathcal{C}(\hat{F}_1(x'), x)$ satisfying: (i) $x \in \hat{F}_1(x)$, (ii) $\hat{F}_1(x)$ is isomorphic to $\mathcal{G}$ and (iii) $\hat{F}_1(x)$ is disjoint with

$\hat{F}_1(x')$. Clearly, the set $\hat{T}_1$ of all subpolygons $\hat{F}_1(x)$, $x \in \mathcal{S}$, is a partition of $\mathcal{S}$ in subpolygons isomorphic to $\mathcal{G}$. It is also clear that every element of $\hat{T}_1$ intersects every element of $\hat{T}_2$ in a line. □

Lemma 8.40. *If $\hat{\mathcal{A}} \in \hat{T}_1 \cup \hat{T}_2$, then either $\hat{\mathcal{A}}$ belongs to $\mathcal{D}$ or $\hat{\mathcal{A}} \cong \mathbb{E}_3$.*

Proof. Every element of $\hat{T}_1$ is isomorphic to $\mathcal{G}$ and hence belongs to $\mathcal{D}$. Suppose therefore that $\hat{\mathcal{A}} \in \hat{T}_2$. We know that $\mathcal{A} := \hat{\mathcal{A}} \cap F$ is big in $\hat{\mathcal{A}}$, that $t_{\hat{\mathcal{A}}} \geq t_{\mathcal{A}} + 2$ and that $\mathcal{A}$ belongs to $\mathcal{D}$. If $d(T_2) \geq 3$, then $\hat{\mathcal{A}}$ belongs to $\mathcal{D}$ since we assumed that no proper convex subpolygon violates Theorem 8.8. If $d(T_2) = 2$, then $\mathcal{A} \cong Q(5,2)$ and by Theorem 7.1 and the fact that $t_{\hat{\mathcal{A}}} \geq t_{\mathcal{A}} + 2$, $\hat{\mathcal{A}}$ is isomorphic to $Q(5,2) \otimes Q(5,2)$, $\mathbb{G}_3$, $\mathbb{E}_3$ or $DH(5,4)$. Again the lemma holds. □

Lemma 8.41. *$\mathcal{S}$ is not a product near polygon.*

Proof. All modified local spaces of $\mathbb{E}_3$ are connected. By the proof of Lemma 8.36, also all modified local spaces of an arbitrary element of $\mathcal{D}$ are connected. By Lemmas 8.39 and 8.40, it then follows that every modified local space of $\mathcal{S}$ is connected. So, $\mathcal{S}$ is not a product near polygon. □

The following lemma finishes the proof of Theorem 8.8.

Lemma 8.42. (i) *If $\mathcal{G}_1 \in \hat{T}_1$ and $\mathcal{G}_2 \in \hat{T}_2$, then $\mathcal{S}$ is of type $\mathcal{G}_1 \otimes \mathcal{G}_2$.*

(ii) *The near polygons $\mathcal{G}_1$ and $\mathcal{G}_2$ belong to $\mathcal{D}$. As a consequence also $\mathcal{S}$ belongs to $\mathcal{D}$.*

Proof. (i) This follows from Theorem 4.28 and Lemmas 8.39 and 8.41.

(ii) By Theorem 4.46, each element of $\hat{T}_1 \cup \hat{T}_2$ must have a spread of symmetry. By Theorem 6.99, $\mathbb{E}_3$ has no spread of symmetry. So, each element of $\hat{T}_1 \cup \hat{T}_2$ must belong to $\mathcal{D}$ by Lemma 8.40. The lemma now readily follows. □

8.10 Proof of Theorem 8.9

We will first prove that every element of $\mathcal{N}^\times$ has a nice chain of convex subpolygons.

Lemma 8.43. *If $\mathcal{S}$ is an element of $\mathcal{D} \setminus D_2$, then every line of $\mathcal{S}$ is contained in a big convex subpolygon which is isomorphic to an element of $\mathcal{D}$.*

Proof. We will use induction on the diameter of $\mathcal{S}$. Obviously, the lemma holds if $\mathcal{S} \cong DH(2n-1,4)$ or $\mathcal{S} \cong \mathbb{G}_n$ for some $n \geq 3$. So, suppose that $\mathcal{S}$ is of the form $\mathcal{A}_1 \otimes \mathcal{A}_2$ with $\mathcal{A}_1$ and $\mathcal{A}_2$ two elements of $\mathcal{D}$. There exist partitions T_1 and T_2 of $\mathcal{S}$ in convex subpolygons isomorphic to $\mathcal{A}_1$, respectively $\mathcal{A}_2$, such that every element of T_1 intersects every element of T_2 in a line. Now, let L be an arbitrary line of $\mathcal{S}$, then L is contained in an element of $T_1 \cup T_2$. By reasons of symmetry, we may suppose that L is contained in the element F_1 of T_1. If $\mathcal{A}_2$ is a generalized quadrangle, then F_1 is big in $\mathcal{S}$ and we are done. Suppose therefore that $\mathcal{A}_2$ is not a generalized quadrangle. Let F_2 denote an arbitrary element of T_2. By the induction hypothesis we know that the line $F_1 \cap F_2$ of F_2 is contained in a big convex subpolygon F' of F_2 which is isomorphic to an element of $\mathcal{D}$. By Theorem 4.36, $\mathcal{C}(F', F_1)$ is a glued near polygon of type $F_1 \otimes F'$. Since $F_1, F' \in \mathcal{D}$, also $\mathcal{C}(F_1, F') \in \mathcal{D}$. Clearly $\mathcal{C}(F_1, F')$ is big in $\mathcal{S}$ and contains the line L. □

Lemma 8.44. *Every near polygon $\mathcal{S}$ of $\mathcal{N}^\times \setminus \{\mathbb{O}\}$ has a big convex subpolygon which is isomorphic to an element of $\mathcal{N}^\times$.*

Proof. Clearly the lemma holds if $\mathcal{S}$ is isomorphic to an element of $\{\mathbb{L}_3, \mathbb{E}_3\} \cup \{DQ(2n,2) | n \geq 2\} \cup \{\mathbb{H}_n | n \geq 3\} \cup \{\mathbb{I}_n | n \geq 4\}$. Now by Lemma 8.43 the result holds for every $\mathcal{S} \in \mathcal{N}$. Now, if $\mathcal{S}_1$ and $\mathcal{S}_2$ are two dense near polygons and if F_1 is a big convex subpolygon of $\mathcal{S}_1$, then the set of all pairs (x, y) with x a point of F_1 and y a point of $\mathcal{S}_2$ determines a big convex subpolygon of $\mathcal{S}_1 \times \mathcal{S}_2$ isomorphic to $F_1 \times \mathcal{S}_2$. Hence the lemma also holds for every $\mathcal{S} \in \mathcal{N}^\times$. □

Corollary 8.45. *If $\mathcal{S} \in \mathcal{N}^\times$, then $\mathcal{S}$ has a nice chain of convex subpolygons.*

We still need to prove Theorem 8.9 in the other direction.

Theorem 8.46. *Every slim dense near $2n$-gon, $n \geq 1$, containing a nice chain $F_0 \subset F_1 \subset \cdots \subset F_{n-1} \subset F_n$ of convex subpolygons is isomorphic to an element of $\mathcal{N}^\times$.*

Proof. By Theorem 7.1, the theorem holds if $n \leq 3$. Hence the theorem will hold if we can prove that every slim dense near $2n$-gon, $n \geq 4$, containing a big convex sub-$2(n-1)$-gon F which is isomorphic to an element of $\mathcal{N}^\times$, is itself isomorphic to an element of $\mathcal{N}^\times$. By Theorem 8.7, this is true for every $F \in \mathcal{N}^\times$ as soon as it is true for every $F \in \mathcal{N}$. By Theorems 8.1, 8.3, 8.4 and 8.6, it follows that we only need to consider the case $F \in \mathcal{D}$. This case is settled by Theorem 8.8. □

Theorem 8.9 follows from Corollary 8.45 and Theorem 8.46.

Chapter 9

Slim dense near octagons

In this chapter we will classify all slim dense near octagons. It will turn out that there are precisely 24 such near octagons. Recall that by Theorem 2.35, every slim dense near octagon is finite. This chapter is based on the papers [63] and [64].

9.1 Some properties of slim dense near octagons

Lemma 9.1. *For every slim dense near octagon $\mathcal{S}$ with $v_{\mathcal{S}}$ points, there exist constants $a_{\mathcal{S}}$, $b_{\mathcal{S}}$ and $c_{\mathcal{S}}$ such that every point of $\mathcal{S}$ is contained in $a_{\mathcal{S}}$ grid-quads, $b_{\mathcal{S}}$ $W(2)$-quads and $c_{\mathcal{S}}$ $Q(5,2)$-quads. Furthermore, (i) $t_{\mathcal{S}}(t_{\mathcal{S}}+1) = 2a_{\mathcal{S}}+6b_{\mathcal{S}}+20c_{\mathcal{S}}$ and (ii) $|\Gamma_0(x)| = 1$, $|\Gamma_1(x)| = 2(t_{\mathcal{S}}+1)$, $|\Gamma_2(x)| = 4a_{\mathcal{S}}+8b_{\mathcal{S}}+16c_{\mathcal{S}}$, $|\Gamma_3(x)| = \frac{v_{\mathcal{S}}}{3} - 1 - 6t_{\mathcal{S}} + 4a_{\mathcal{S}} + 8b_{\mathcal{S}} + 16c_{\mathcal{S}}$, $|\Gamma_4(x)| = 2(\frac{v_{\mathcal{S}}}{3} - 1 + 2t_{\mathcal{S}} - 4a_{\mathcal{S}} - 8b_{\mathcal{S}} - 16c_{\mathcal{S}})$ for every point x of $\mathcal{S}$.*

Proof. The first part of the lemma is precisely Theorem 7.2.

Let x be an arbitrary point of $\mathcal{S}$. Then $|\Gamma_0(x)| = 1$, $|\Gamma_1(x)| = 2(t_{\mathcal{S}}+1)$ and $|\Gamma_2(x)| = 4a_{\mathcal{S}}+8b_{\mathcal{S}}+16c_{\mathcal{S}}$. Since $\sum_{i=0}^{4}|\Gamma_i(x)| = v_{\mathcal{S}}$ and $\sum_{i=0}^{4}(-\frac{1}{2})^i|\Gamma_i(x)| = 0$ (see Theorem 1.2), we are able to calculate $|\Gamma_3(x)|$ and $|\Gamma_4(x)|$ in terms of $v_{\mathcal{S}}$, $t_{\mathcal{S}}$, $a_{\mathcal{S}}$, $b_{\mathcal{S}}$ and $c_{\mathcal{S}}$. We find the equations mentioned above. □

Lemma 9.2. *Let (x,H) be a point-hex pair of a dense near octagon $\mathcal{S}$ and consider the function $f_x : H \to \mathbb{N}$, $y \mapsto d(x,y) - d(x,H)$. If $d(x,H) \leq 1$, then f_x is a classical valuation of H. If $d(x,H) = 3$, then f_x is an ovoidal valuation of H. If $d(x,H) = 2$, then f_x is a valuation of H which is not classical nor ovoidal.*

Proof. By Theorem 5.5, f_x is a valuation of H.

If $\mathrm{d}(x,H) \leq 1$, then there exists a unique point x' in H nearest to x and $\mathrm{d}(x,y) = \mathrm{d}(x,x') + \mathrm{d}(x',y)$ for every point y of H. Hence, $f_x(y) = \mathrm{d}(x',y)$ for every point y of H. This proves that f_x is classical.

If $\mathrm{d}(x,H)=3$, then $f_x(y)\in\{0,1\}$ for every point y of H. Moreover, every line of H contains a unique point with value 0. This proves that O_{f_x} is an ovoid and that f_x is ovoidal.

Suppose $\mathrm{d}(x,H)=2$. If f_x is classical, then there exists a point $y\in H$ with $f_x(y)=3$. We would then have that $\mathrm{d}(x,y)=5$, a contradiction. Suppose that f_x is ovoidal. Then all points of H are at distance 2 or 3 from x and the points at distance 2 from x form an ovoid O of H. Let y be a point of H not contained in O. Every line of H through y contains a point at distance 2 from x. So, $(\Gamma_1(y)\cap H)\subseteq\mathcal{C}(x,y)$. This implies $H\subseteq\mathcal{C}(x,y)$. This is impossible, since H and $\mathcal{C}(x,y)$ are two different hexes. □

Lemma 9.3 (See Chapter 6). *The nonclassical valuations of the* 11 *slim dense near hexagons are listed in the following table. For each extended valuation f, we mention the type of the quad containing O_f. In the last column we list the possibilities for G_f, where f is the considered valuation.*

near hexagon	*ovoid.*	*semi-cl.*	*extended*	*other*
$\mathbb{L}_3\times\mathbb{L}_3\times\mathbb{L}_3$	*YES*	*YES*	$\mathbb{L}_3\times\mathbb{L}_3$	–
$W(2)\times\mathbb{L}_3$	–	*YES*	$\mathbb{L}_3\times\mathbb{L}_3$, $W(2)$	–
$Q(5,2)\times\mathbb{L}_3$	–	–	$\mathbb{L}_3\times\mathbb{L}_3$	–
$\mathbb{H}_3$	–	–	$W(2)$	$\mathbb{L}_3$, $\mathrm{PG}(2,2)$
$DQ(6,2)$	–	–	$W(2)$	–
$Q(5,2)\otimes Q(5,2)$	–	–	–	$\mathrm{AG}(2,3)$
$\mathbb{G}_3$	–	–	–	$W(2)$
$\mathbb{E}_1$	*YES*	–	–	–
$\mathbb{E}_2$	*YES*	–	–	–
$\mathbb{E}_3$	–	–	–	$\mathrm{PG}(2,4)$
$DH(5,4)$	–	–	–	–

9.2 Existence of big hexes

We label the 11 slim dense near hexagons in the following way: $\mathcal{N}_1=\mathbb{L}_3\times\mathbb{L}_3\times\mathbb{L}_3$, $\mathcal{N}_2=W(2)\times\mathbb{L}_3$, $\mathcal{N}_3=Q(5,2)\times\mathbb{L}_3$, $\mathcal{N}_4=Q(5,2)\otimes Q(5,2)$, $\mathcal{N}_5=\mathbb{H}_3$, $\mathcal{N}_6=DQ(6,2)$, $\mathcal{N}_7=\mathbb{E}_3$, $\mathcal{N}_8=\mathbb{G}_3$, $\mathcal{N}_9=\mathbb{E}_1$, $\mathcal{N}_{10}=\mathbb{E}_2$ and $\mathcal{N}_{11}=DH(5,4)$.

Theorem 9.4. *Let $\mathcal{S}=(\mathcal{P},\mathcal{L},\mathrm{I})$ be a dense near octagon of order $(2,t)$ and let i be the biggest integer such that $\mathcal{S}$ contains a hex isomorphic to $\mathcal{N}_i$. Then every hex of $\mathcal{S}$ isomorphic to $\mathcal{N}_i$ is big in $\mathcal{S}$. As a corollary, every slim dense near octagon contains a big hex.*

Suppose the contrary. Let $H=(\mathcal{P}',\mathcal{L}',\mathrm{I}')$ be a hex isomorphic to $\mathcal{N}_i$ which is not big in $\mathcal{S}$ and let x denote a point of $\mathcal{S}$ at distance 2 from H. Put $f_x:\mathcal{P}'\to\mathbb{N}, y\mapsto\mathrm{d}(x,y)-\mathrm{d}(x,H)$, then $f=f_x$ is a valuation of H. A quad Q of H is called *special* if it is special with respect to the valuation f of H, i.e. if $Q\cap O_f$ is an ovoid of Q. Remark that f is not classical nor ovoidal by Lemma 9.2.

Lemma 9.5. *If Q is a special quad, then $\mathcal{C}(x, Q)$ is a hex.*

Proof. Let y be an arbitrary point of $Q \setminus O_f$. Then $\mathrm{d}(x, y) = 3$; so, $\mathcal{C}(x, y)$ is a hex. Every line of Q through y contains a point of O_f at distance 2 from x and hence is contained in $\mathcal{C}(x, y)$. Hence, $Q \subseteq \mathcal{C}(x, y)$ and $\mathcal{C}(x, Q) = \mathcal{C}(x, y)$. This proves the lemma. □

We will derive a contradiction for each of the possible values of i.

Lemma 9.6. *The case $i \in \{9, 10, 11\}$ cannot occur.*

Proof. By Lemma 9.3 and the fact that f is not classical nor ovoidal, it follows that $i \notin \{9, 10, 11\}$. □

Lemma 9.7. *The case $i = 8$ cannot occur.*

Proof. Since f is not classical, $G_f \cong \overline{W(2)}$ by Lemma 9.3.

Property. $3 \,|\, t + 1$.

Proof. Since $G_f \cong \overline{W(2)}$, there exists a special $W(2)$-quad T. The hex $\mathcal{C}(x, T)$ contains a $W(2)$-quad which is not big. So, $\mathcal{C}(x, T)$ is isomorphic to either $\mathbb{G}_3$ or $\mathbb{E}_3$ by Theorem 7.1. Hence, there exists a $W(2)$-quad Q through x intersecting T in a point z of O_f. The hexes through Q determine a partition of the lines through x which are not contained in Q. By Theorem 2.32, every hex through Q intersects H in at least a line. (Notice that there exists a $Q(5, 2)$-quad through z contained in H.) Now, there are five special quads $R_1, \dots, R_5$ through z and these five quads partition the set of lines of H through z. It follows that the quads $\mathcal{C}(x, R_i)$, $i \in \{1, \dots, 5\}$, are all the quads through Q. If R_i is a grid, then $\mathcal{C}(x, R_i)$ is isomorphic to $\mathbb{G}_3$. If $R_i \cong W(2)$, then $\mathcal{C}(x, R_i)$ is isomorphic to either $\mathbb{G}_3$ or $\mathbb{E}_3$. It follows that

$$t + 1 = 3 + \alpha(t_{\mathbb{G}_3} + 1 - 3) + \beta(t_{\mathbb{E}_3} + 1 - 3),$$

for certain $\alpha, \beta \in \{0, \dots, 5\}$ with $\alpha + \beta = 5$. Hence, $3 \mid t + 1$. □

The previous property leads together with the following property to a contradiction.

Property. $3 \,|\, t$.

Proof. By the proof of the first property, we know that there exists a hex H' through x isomorphic to $\mathbb{G}_3$ intersecting H in a grid-quad T'. By the proof of Theorem 6.45, there exists a unique valuation f' in H' such that $\{x\} \cup (\Gamma_2(x) \cap T') \subseteq O_{f'}$. Since $G_{f'} \cong \overline{W(2)}$, there exists a grid-quad Q' through x intersecting T' in a point z'. Let $R'_1, R'_2, \dots, R'_5$ denote the five special quads of H through z'. As before we know that the quads $\mathcal{C}(x, R'_i)$, $i \in \{1, \dots, 5\}$, partition the set of lines through x not contained in Q'. If R'_i is a grid, then by Theorem 7.1, $\mathcal{C}(x, R'_i)$ is isomorphic to either $\mathbb{G}_3$, $\mathbb{H}_3$ or $Q(5, 2) \otimes Q(5, 2)$. If $R'_i \cong W(2)$, then $\mathcal{C}(x, R'_i) \cong \mathbb{G}_3$. Hence,

$$t+1 = 2+\alpha(t_{\mathbb{G}_3}+1-2)+\beta(t_{\mathbb{H}_3}+1-2)+\gamma(t_{Q(5,2)\otimes Q(5,2)}+1-2) = 2+10\alpha+4\beta+7\gamma,$$

with $\alpha, \beta, \gamma \in \{0, \ldots, 5\}$ such that $\alpha + \beta + \gamma = 5$. The property now immediately follows. □

Lemma 9.8. *The case $i = 7$ cannot occur.*

Proof. By Lemma 9.3, $G_f \cong \mathrm{PG}(2, 4)$ and by Lemmas 9.2 and 9.3, $\Gamma_3(H) = \emptyset$.

Property. Every line through x is contained in a unique quad intersecting H in a point.

Proof. Suppose the contrary and let L be a line through x such that also the remaining points y and z of L have distance 2 from H. Consider a point x' of O_{f_x}. Every $Q(5, 2)$-quad Q of H through x' is classical in $\mathcal{S}$ and we define $L_Q := \pi_Q(L)$. Suppose $L_{Q'} \neq L_Q$ for two different $Q(5, 2)$-quads Q and Q' of H through x'. Put $Q'' := \mathcal{C}(L_Q, L_{Q'})$. Since $|\Gamma_2(y) \cap Q''| \geq 2$ and $|\Gamma_2(z) \cap Q''| \geq 2$, (i) $Q'' \cong W(2)$ and (ii) (y, Q'') and (z, Q'') are ovoidal point-quad pairs. Since any two ovoids of $W(2)$ intersect in a point, there exists a point u in Q'' at distance 2 from y and z. From $\mathrm{d}(y, u) = \mathrm{d}(z, u) = 2$, it follows that $\mathrm{d}(x, u) = 1$, a contradiction. So, the line L_Q must be contained in the intersection of all $Q(5, 2)$-quads of H through x', a contradiction. □

Property. Every quad Q through x intersecting H is isomorphic to $W(2)$.

Proof. Put $Q \cap H = \{x'\}$ and let Q' denote a special quad through x'. The hex $\mathcal{C}(x, Q')$ contains a $W(2)$-quad which is not big and hence is isomorphic to $\mathbb{E}_3$ by Theorem 7.1. It then follows that Q is isomorphic to $W(2)$. □

Corollary. $t + 1 = 63$.

Proof. There are 21 quads through x intersecting H in a point. These quads are isomorphic to $W(2)$ and partition the set of lines through x. □

Property. There are no grid-quads. As a consequence, only hexes isomorphic to $DQ(6, 2)$ or $\mathbb{E}_3$ can occur.

Proof. Suppose the contrary. By Lemma 9.1, there exists a grid-quad G through x. Let Q_1 and Q_2 denote the two $W(2)$-quads through x which intersect H in a point and G in a line. Then $G \subset \mathcal{C}(Q_1, Q_2)$. By the proof of the previous property, we know that $\mathcal{C}(Q_1, Q_2) \cong \mathbb{E}_3$, a contradiction, since $\mathbb{E}_3$ does not contain grid-quads. □

We are now ready to derive a contradiction. Consider a $Q(5, 2)$-quad Q in $\mathcal{S}$ and a point $x' \in Q$. The α hexes through Q determine a partition of the lines through x' not contained in Q. From the previous property, all these hexes are isomorphic to $\mathbb{E}_3$. It follows that $t + 1 = 5 + \alpha(15 - 5)$, contradicting $t + 1 = 63$. □

Lemma 9.9. *The case $i = 6$ cannot occur.*

Proof. By Lemma 9.3, O_f is an ovoid in a $W(2)$-quad Q of H. But then $\mathcal{C}(x, Q)$ is a hex containing a $W(2)$-quad which is not big, contradicting $i = 6$ and Theorem 7.1. □

Lemma 9.10. *The case $i = 5$ cannot occur.*

Proof. Property. There are no $Q(5, 2)$-quads. As a consequence, every hex is isomorphic to either $\mathbb{L}_3 \times \mathbb{L}_3 \times \mathbb{L}_3, W(2) \times \mathbb{L}_3$ or $\mathbb{H}_3$.

Proof. Suppose the contrary, then by Lemma 9.1, there exists a $Q(5, 2)$-quad Q through a point of H. By Theorem 2.32, Q intersects H in a line L. Every hex through Q and a $W(2)$-quad of H through L is isomorphic to $\mathbb{G}_3$ or $\mathbb{E}_3$ by Theorem 7.1, contradicting the fact that $i = 5$. □

Property. Every special quad is a grid. As a consequence, G_f is isomorphic to either $\mathbb{L}_3$ or $\mathrm{PG}(2, 2)$.

Proof. If Q is a special $W(2)$-quad, then $\mathcal{C}(x, Q)$ is a hex containing a $W(2)$-quad which is not big, contradicting Theorem 7.1 and $i = 5$. □

Property. Every quad Q intersecting H in a point is a grid.

Proof. Without loss of generality, we may suppose that $x \in Q$. If Q' is a special quad through the point $Q \cap H$, then Q is not big in the hex $\mathcal{C}(x, Q')$. It follows that $\mathcal{C}(x, Q') \cong \mathbb{H}_3$ and $Q \cong \mathbb{L}_3 \times \mathbb{L}_3$. □

Consider a line L of $\mathcal{S}$ intersecting H in a point y and let $z \in L \setminus \{y\}$. Let G_1, G_2 and G_3 be the grid-quads of H through y and let $H_i := \mathcal{C}(L, G_i)$.

Property. Every line L' through z is contained in at least one of the hexes H_1, H_2, H_3.

Proof. We may suppose that $L' \neq L$. If L' contains a point z' at distance 2 from H, then $G_{f_{z'}}$ is isomorphic to either $\mathbb{L}_3$ or $\mathrm{PG}(2, 2)$. Hence, at least one of the grid-quads G_1, G_2 and G_3 is special with respect to the valuation $f_{z'}$. So, L' is contained in at least one of the hexes H_i, $i = 1, 2, 3$. If L' is contained in $\Gamma_1(H)$, then L' is contained in exactly one of the hexes H_i, $i = 1, 2, 3$. □

Property. $H_i \cap H_j = \{L\}$ for all $i, j \in \{1, 2, 3\}$ with $i \neq j$. As a consequence, $G_f \cong \mathbb{L}_3$.

Proof. Suppose the contrary. Let Q be a quad through L intersecting H in y and suppose that Q is contained in H_j and H_k for certain $j, k \in \{1, 2, 3\}$ with $j \neq k$. Then Q is a grid-quad. Let y' be a point of Q at distance 2 from y. Since the grid-quads G_j and G_k contain three points of $O_{f_{y'}}$, $G_{f_{y'}} \cong \mathrm{PG}(2, 2)$ and it follows that also the third grid-quad G_l $(l \in \{1, 2, 3\} \setminus \{j, k\})$ contains three points of $O_{f_{y'}}$. So, $H_i \cong \mathbb{H}_3$ for every $i \in \{1, 2, 3\}$. Hence $t + 1 = 2 + 3(t_{\mathbb{H}_3} - 1) = 14$ by the previous property. Now, $\mathcal{C}(y', v)$ is a grid-quad for every $v \in O_{f_{y'}}$ and any two of these grid-quads are contained in a unique hex isomorphic to $\mathbb{H}_3$. As a corollary, $(a_S, b_S, c_S) = (7, 28, 0)$. Suppose that y' is contained in α hexes of type $\mathbb{L}_3 \times \mathbb{L}_3 \times \mathbb{L}_3$, β hexes of type $W(2) \times \mathbb{L}_3$ and γ hexes of type $\mathbb{H}_3$. Counting pairs (G, L) where G is a grid-quad through y' and L a line through y' not contained in G yields that $7(t + 1 - 2) = 3\alpha + 6\beta + 12\gamma$ or

$$\alpha + 2\beta + 4\gamma = 28.$$

Counting pairs (W, L) where W is a $W(2)$-quad through y' and L a line through y' not contained in W yields that $28(t + 1 - 3) = \beta + 12\gamma$ or

$$\beta + 12\gamma = 308.$$

Both equalities cannot hold simultaneously. □

We are now ready to derive a contradiction. Since every hex is isomorphic to $\mathbb{L}_3 \times \mathbb{L}_3 \times \mathbb{L}_3$, $W(2) \times \mathbb{L}_3$ or $\mathbb{H}_3$ and since every line through z is contained in at least one of the hexes H_1, H_2, H_3, $7 \leq t + 1 \leq 14$ or $t + 1 = 16$. Consider again a point y' at distance 2 from H such that $y \in O_{f_{y'}}$. By the previous property, $G_{f_{y'}} \cong \mathbb{L}_3$. Let Q denote the unique grid-quad of H containing all points of $O_{f_{y'}}$ and let Q' be a $W(2)$-quad of H disjoint from Q. The point y' is ovoidal with respect to Q'. Let U and V be two hexes through y' intersecting Q' in, say, u and v. Suppose that M is a line through y' contained in U and V. Clearly, M contains a point m at distance 2 from u. But then m is classical with respect to Q', implying that $\mathrm{d}(m, v) = 4$, a contradiction. So, the five hexes through y' intersecting Q' determine at least 15 lines through y'. As a consequence, $t + 1 = 16$ and $H_i \cong \mathbb{H}_3$ for every $i \in \{1, 2, 3\}$. It follows that every quad through L intersecting H in a line is a $W(2)$-quad. If Q'' is a $W(2)$-quad of H through y, then $\mathcal{C}(L, Q'')$ contains at least three $W(2)$-quads through the line L, contradicting Theorem 7.1. □

Lemma 9.11. *The case $i = 4$ cannot occur.*

Proof. Since f is not classical, $G_f \cong \mathrm{AG}(2, 3)$ by Lemma 9.3. Let y be an arbitrary point of O_f, let Q be the quad through x and y and let G_1, G_2, G_3 and G_4 denote the four special grid-quads of H through y. Each of the hexes $\mathcal{C}(x, G_j)$, $1 \leq j \leq 4$, contains a grid-quad which is not big. By Theorem 7.1, all these hexes are isomorphic to $Q(5, 2) \otimes Q(5, 2)$. Let M be a line of Q through y and let M_1, M_2, M_3, respectively M_4, be the lines of G_1, G_2, G_3, respectively G_4, which are contained in a $Q(5, 2)$-quad together with M. Since at least two of these lines are contained in the same $Q(5, 2)$-quad of H, the hex through M and these lines has three $Q(5, 2)$-quads through a point, contradicting $i = 4$. □

Lemma 9.12. *The case $i = 3$ cannot occur.*

Proof. Since f is not classical, O_f is an ovoid in a grid-quad Q by Lemma 9.3. Now, $\mathcal{C}(x, Q)$ is a hex containing a grid-quad which is not big, contradicting $i = 3$. □

Lemma 9.13. *The case $i = 2$ cannot occur.*

Proof. Property. f is a semiclassical valuation.

Proof. Suppose the contrary, then by Lemma 9.3, O_f is an ovoid in a quad Q of H. Now, $\mathcal{C}(x, Q)$ is a hex containing a quad which is not big, contradicting $i = 2$. □

Let $a_{\mathcal{S}}$, respectively $b_{\mathcal{S}}$, denote the constant number of grid-quads, respectively $W(2)$-quads, through a given point of $\mathcal{S}$.

Property. $15 \leq t + 1 \leq 20$ *and* $b_{\mathcal{S}} \geq t + 1 - 15 = t - 14$.

Proof. Let Q be a $W(2)$-quad of H not containing the unique point of O_f. Then $\mathrm{d}(x, Q) = 3$ and $\Gamma_3(x) \cap Q$ is an ovoid $\{x_1, \ldots, x_5\}$ of Q. If there exists a line $\{x, x', x''\}$ through x contained in $\Gamma_3(Q)$, then by Theorem 1.23, the ovoids $\Gamma_3(x) \cap Q$, $\Gamma_3(x') \cap Q$ and $\Gamma_3(x'') \cap Q$ of Q are mutually disjoint. This is impossible since any two ovoids of $W(2)$ have nonempty intersection. As a consequence, every line through x is contained in one of the hexes $H_i := \mathcal{C}(x, x_i)$, $i \in \{1, 2, 3, 4, 5\}$. It follows that $t + 1 \leq 20$. Suppose now that $H_i \cap H_j \neq \{x\}$ for certain $i, j \in \{1, 2, 3, 4, 5\}$ with $i \neq j$ and $\mathrm{d}(x_i, O_f) \neq 1$. Let L be a line through x which is contained in $H_i \cap H_j$ and let y be the unique point of L at distance 2 from x_i. Since $\mathcal{C}(y, x_i) \cap H = \{x_i\}$, $\mathrm{d}(y, H) \neq 1$. So, $\mathrm{d}(y, H) = 2$ and f_y is a semi-classical valuation of H. It follows that $\mathrm{d}(y, x_j) = \mathrm{d}(x, x_i) + \mathrm{d}(x_i, x_j) = 4$, contradicting the fact that y and x_j belong to the same hex H_j. So, $H_i \cap H_j = \{x\}$ for all $i, j \in \{1, \ldots, 5\}$ with $i \neq j$. The property now follows from the fact that every hex is isomorphic to either $\mathbb{L}_3 \times \mathbb{L}_3 \times \mathbb{L}_3$ or $W(2) \times \mathbb{L}_3$. □

Property. $|\Gamma_3(y)| = \frac{v_{\mathcal{S}}}{3} - 1 - 6t + 4a_{\mathcal{S}} + 8b_{\mathcal{S}} = 344 - 216t + 64a_{\mathcal{S}} + 128b_{\mathcal{S}}$ *for every point y of $\mathcal{S}$.*

Proof. Let x' be an arbitrary point of H. For every $i \in \{0, 1, 2, 3\}$, let X_i be the set of all points y of $\mathcal{S}$ such that $\mathrm{d}(y, H) = \mathrm{d}(y, x') = i$. Since H has no ovoids, $X_3 = \emptyset$. Obviously,

- $|X_0| = 1$;
- $|X_1| = 2(t - t_H) = 2t - 6$;
- $|X_2| = |\Gamma_2(x')| - |\Gamma_2(x') \cap H| - 2(t_{H'} + 1)|X_1|$.

Here $|\Gamma_2(x')| = 4a_{\mathcal{S}} + 8b_{\mathcal{S}}$ and $|\Gamma_2(x') \cap H| = 20$. Hence, $|X_i|$, $i \in \{0, 1, 2\}$, is independent from the point x'. Since $|\Gamma_i(y) \cap H| = 1$ for every $i \in \{0, 1, 2\}$ and every $y \in \Gamma_i(H)$, the total number of points of $\mathcal{S}$ is equal to

$$v_{\mathcal{S}} = v_{H'}(|X_0| + |X_1| + |X_2|) = 45(23 - 14t + 4a_{\mathcal{S}} + 8b_{\mathcal{S}}).$$

The property now immediately follows from Lemma 9.1. □

We are now ready to derive a contradiction. Consider a point x' of $\mathcal{S}$ and let A, respectively B, be the number of hexes isomorphic to $\mathbb{L}_3 \times \mathbb{L}_3 \times \mathbb{L}_3$, respectively $W(2) \times \mathbb{L}_3$, through x'. Consider the $a_{\mathcal{S}}$ grid-quads $G_1, \ldots, G_{a_{\mathcal{S}}}$ through x' and suppose that G_i is contained in u_i hexes isomorphic to $\mathbb{L}_3 \times \mathbb{L}_3 \times \mathbb{L}_3$ and v_i hexes isomorphic to $W(2) \times \mathbb{L}_3$. Since each line through x' outside G_i is contained in a unique hex together with G_i, $u_i + 2v_i = t - 1$ for every $i \in \{1, \dots, a_{\mathcal{S}}\}$. Since each hex through x' contains three grid-quads through x', it easily follows that

$$A + 2 \cdot B = \frac{1}{3} \sum_{i=1}^{a_{\mathcal{S}}} (u_i + 2v_i) \ = \ \frac{1}{3} \sum_{i=1}^{a_{\mathcal{S}}} (t-1) \ = \ \frac{1}{3} a_{\mathcal{S}}(t-1).$$

Since each hex through x' isomorphic to $\mathbb{L}_3 \times \mathbb{L}_3 \times \mathbb{L}_3$ (respectively $W(2) \times \mathbb{L}_3$) contains eight (respectively 16) points at distance 3 from x', $|\Gamma_3(x')| = 8A + 16B = \frac{8}{3} a_{\mathcal{S}}(t-1)$. Together with the value of $|\Gamma_3(x')|$ we have calculated earlier, we find that $a_{\mathcal{S}}(t-1) = 129 - 81t + 24a_{\mathcal{S}} + 48b_{\mathcal{S}}$. From Lemma 9.1, $2a_{\mathcal{S}} + 6b_{\mathcal{S}} = t(t+1)$. We can now calculate $b_{\mathcal{S}}$ in terms of t:

$$b_{\mathcal{S}} \ = \ \frac{t^3 - 24t^2 + 137t - 258}{6(t-9)}.$$

Since no two $W(2)$-quads intersect in a line, $3b_{\mathcal{S}} \leq t + 1 \leq 20$. Hence $0 \leq b_{\mathcal{S}} \leq 6$. Since $b_{\mathcal{S}} \in \mathbb{N}$ and $15 \leq t+1 \leq 20$, $t = 17$ and $b_{\mathcal{S}} = 1$, contradicting $b_{\mathcal{S}} \geq t - 14$. □

Theorem 9.4 holds if we can show that also the last case cannot occur.

Lemma 9.14. *The case $i = 1$ cannot occur.*

Proof. In this case, every quad is a grid and every hex is isomorphic to $\mathbb{L}_3 \times \mathbb{L}_3 \times \mathbb{L}_3$. If $x' \in O_{f_x}$, then the grid $\mathcal{C}(x, x')$ intersects H in the point x'. So, $t \geq 4$. The near octagon $\mathcal{S}$ is a regular near octagon. Let A be the collinearity matrix of $\mathcal{S}$. For each possible value of t different from 3, there always exists a multiplicity which is not integral (see Example 2 in Section 3.3 and the remark following this proof). So, also this case cannot occur. □

Remark. Consider again the case $i = 1$. The matrix A has five eigenvalues λ_i, $i \in \{1, \dots, 5\}$, with $-(t+1) = \lambda_1 < \lambda_2 < \cdots < \lambda_5 = 2(t+1)$. The eigenvalues λ_2, λ_3 and λ_4 are the zeros of the polynomial $x^3 - 6x^2 + (21 - 6t)x + (10t - 20)$. The multiplicity of the eigenvalue $2(t+1)$ is equal to 1 and the multiplicity of the eigenvalue $-(t+1)$ is equal to $f(t) := \frac{96+256t-192t^2+128t^3}{48+20t+3t^2+t^3}$. The other multiplicities can also be expressed as functions of t, but these functions are not so nice as f. By Corollary 2.37 and Theorem 2.38, $t \leq 53$. For $4 \leq t \leq 53$, $f(t)$ is only integral if $t \in \{4, 10\}$. To exclude the two remaining possibilities for t, one needs to consider the other eigenvalues or one could use the following argument using valuations.

If there exists a special quad Q in H, then $\mathcal{C}(x, Q)$ is a hex containing a quad which is not big, a contradiction. So, f_x is a semi-classical valuation. There exists a unique point y in H at distance 2 from x and four points x_1, x_2, x_3 and x_4 of $\Gamma_3(y) \cap H$ lie at distance 3 from x. Put now $H_i := \mathcal{C}(x, x_i)$ for every $i \in \{1, 2, 3, 4\}$. Then $H_i \cap H = \{x_i\}$. Suppose that a line L through x is contained in H_i and H_j $(1 \leq i < j \leq 4)$ and let y be the unique point of L at distance 2 from x_i. As in the case $i = 2$ we have that $\mathrm{d}(y, H) = 2$. So, f_y is a semi-classical valuation and since $\mathrm{d}(x_i, x_j) = 2$, we have $\mathrm{d}(y, x_j) = \mathrm{d}(y, x_i) + \mathrm{d}(x_i, x_j) = 4$, contradicting the fact that y and x_j belong to the same hex H_j. Hence, $H_i \cap H_j = \{x\}$ for all $i, j \in \{1, 2, 3, 4\}$ with $i \neq j$. It follows that $t + 1 \geq 12$.

9.3 Classification of the near octagons

Theorem 9.15. *Let $\mathcal{S}$ be a dense near octagon containing a big hex H isomorphic to $\mathbb{E}_1$, then $\mathcal{S}$ is isomorphic to either $\mathbb{E}_1 \times \mathbb{L}_3$ or $\mathbb{E}_1 \otimes Q(5,2)$.*

Proof. If $t_\mathcal{S} = t_{\mathbb{E}_1} + 1$, then $\mathcal{S}$ must be isomorphic to $\mathbb{E}_1 \times \mathbb{L}_3$ by Theorem 4.1. Suppose therefore that $t_\mathcal{S} \geq t_{\mathbb{E}_1} + 2$. Let x denote an arbitrary point of H and let L_1 and L_2 denote two different lines through x. By Theorem 1.7, the quad $Q := \mathcal{C}(L_1, L_2)$ intersects H in a line. So, Q is not a grid. Every hex through Q intersects H in a big grid-quad and hence is isomorphic to either $W(2) \times \mathbb{L}_3$ or $Q(5,2) \times \mathbb{L}_3$, see Theorem 7.1. Since the lines through x not contained in Q are partitioned by the hexes through Q, $t_\mathcal{S} + 1 = t_Q + 12$. From Lemma 9.1, every point y of $\mathcal{S}$ is contained in exactly one quad isomorphic to Q and all other quads through y are grid-quads.

Suppose first that $Q \cong W(2)$. Then $t_\mathcal{S} + 1 = 14$, $b_\mathcal{S} = 1$ and $c_\mathcal{S} = 0$. From Lemma 9.1, $a_\mathcal{S} = 88$. It follows from Theorem 7.1 that only hexes isomorphic to $\mathbb{L}_3 \times \mathbb{L}_3 \times \mathbb{L}_3$, $W(2) \times \mathbb{L}_3$ or $\mathbb{E}_1$ can occur. Consider a point z of $\mathcal{S}$ and let m_1, m_2 and m_3 denote the number of hexes isomorphic to $\mathbb{L}_3 \times \mathbb{L}_3 \times \mathbb{L}_3$, $W(2) \times \mathbb{L}_3$, respectively $\mathbb{E}_1$, through z. Because $b_\mathcal{S} = 1$, $m_2 = 11$. Counting in two ways the number of line - grid-quad pairs intersecting each other in z yields that $3m_1 + 6m_2 + 660m_3 = 88 \cdot 12 = 1056$. Since H is big in $\mathcal{S}$, $\mathcal{S}$ has $v_{\mathbb{E}_1}(1 + 2(t_\mathcal{S} - t_{\mathbb{E}_1})) = 3645$ points. By Lemma 9.1, the number of points at distance 3 from a given point in $\mathcal{S}$ is equal to 1496. Counting the number of points at distance 3 from z yields that $8m_1 + 16m_2 + 440m_3 = 1496$. Together with $3m_1 + 6m_2 + 660m_3 = 1056$ and $m_2 = 12$, this implies that $m_3 = 1$. Hence every point of $\mathcal{S}$ is contained in a unique hex isomorphic to $\mathbb{E}_1$. It is now easy to see that the set of $W(2)$-quads of $\mathcal{S}$ and the set of $\mathbb{E}_1$-hexes of $\mathcal{S}$ determine two partitions of $\mathcal{S}$ satisfying the conditions of Theorem 4.28. So, $\mathcal{S}$ must be a glued near polygon of type $\mathbb{E}_1 \otimes W(2)$. This implies that $W(2)$ has a spread of symmetry, contradicting Theorem 4.42.

Suppose now that $Q \cong Q(5,2)$. Then $t_\mathcal{S} + 1 = 16$, $b_\mathcal{S} = 0$ and $c_\mathcal{S} = 1$. From Lemma 9.1, $a_\mathcal{S} = 110$. It follows from Theorem 7.1 that only hexes isomorphic to $\mathbb{L}_3 \times \mathbb{L}_3 \times \mathbb{L}_3$, $Q(5,2) \times \mathbb{L}_3$ or $\mathbb{E}_1$ can occur. With a similar argument as before, one can show that every point of $\mathcal{S}$ is contained in a unique hex isomorphic to $\mathbb{E}_1$. It follows again that $\mathcal{S}$ is a glued near polygon, this time of type $\mathbb{E}_1 \otimes Q(5,2)$. By Theorem 6.107, there exists a unique glued near hexagon of type $\mathbb{E}_1 \otimes Q(5,2)$. $\square$

Lemma 9.16. *If Q and Q' are two disjoint $W(2)$-quads of $\mathbb{E}_2$, then $\Gamma_1(Q) \cap Q'$ consists of a point of Q' together with all its neighbours in Q'.*

Proof. Let Q_1 be a given $W(2)$-quad of $\mathbb{E}_2$. If Q_2 is a $W(2)$-quad disjoint from Q_1, then by Lemma 7.3, there are three possibilities for $\Gamma_1(Q_1) \cap Q_2$:

(a) $\Gamma_1(Q_1) \cap Q_2$ is the union of three nonconcurrent lines;

(b) $\Gamma_1(Q_1) \cap Q_2$ is a subgrid of Q_2;

(c) $\Gamma_1(Q_1) \cap Q_2 = Q_2$.

Let N_a, N_b, respectively N_c, denote the number of $W(2)$-quads Q_2 for which (a), (b), respectively (c), holds. Since there are 1440 $W(2)$-quads disjoint from Q_1,

$$N_a + N_b + N_c = 1440.$$

Counting in two different ways the number of pairs (x, Q_2) with $x \in \Gamma_2(Q_1)$ and Q_2 a $W(2)$-quad through x disjoint from Q_2, gives

$$8N_a + 6N_b = 384 \cdot 30.$$

It follows that $2N_b + 8N_c = 0$ or $N_b = N_c = 0$. This proves the lemma. □

Lemma 9.17. *Let $\mathcal{S}$ be a dense near polygon, let F be a big convex subpolygon of $\mathcal{S}$ and let x and y be two points outside F. If $\mathcal{C}(x,y)$ is disjoint from F, then $d(x,y) = d(\pi_F(x), \pi_F(y))$.*

Proof. If $y \notin \mathcal{C}(\pi_F(y), x)$, then there exists a shortest path from y to x containing a point of F, a contradiction. So, $y \in \mathcal{C}(\pi_F(y), x)$ and $\mathrm{d}(x,y)$ is equal to either $\mathrm{d}(x, \pi_F(y))$ or $\mathrm{d}(x, \pi_F(y)) - 1$. If $\mathrm{d}(x,y) = \mathrm{d}(x, \pi_F(y))$, then there exists a shortest path between y and x containing a point of the line $y\,\pi_F(y)$. This would imply that the line $y\,\pi_F(y)$ is contained in $\mathcal{C}(x,y)$. In particular, $\pi_F(y) \in \mathcal{C}(x,y)$. This is impossible. Hence, $\mathrm{d}(x,y) = \mathrm{d}(x, \pi_F(y)) - 1 = \mathrm{d}(\pi_F(x), \pi_F(y))$. □

Theorem 9.18. *Let $\mathcal{S}$ be a dense near octagon containing a big convex subhexagon H isomorphic to $\mathbb{E}_2$, then $\mathcal{S}$ is isomorphic to $\mathbb{E}_2 \times \mathbb{L}_3$.*

Proof. (a) Let H' be a hex disjoint with H. The hex H is big; so, we can define a projection π_H from H' on H. By Lemma 9.17, this projection is distance-preserving. So, lines are mapped to lines and quads are mapped to subquadrangles of quads. Suppose that H' is not isomorphic to $\mathbb{E}_2$. By Theorem 9.15 and Section 9.2, we know that H' is not isomorphic to $\mathbb{E}_1$. So, H contains two disjoint big quads Q_1 and Q_2 by Theorem 7.1. Let Q'_i, $i \in \{1,2\}$, denote the unique $W(2)$-quad containing $\pi_H(Q_i)$. Every point of $\pi_H(Q_1)$ has distance 1 from $\pi_H(Q_2)$. Hence, Q'_1 and Q'_2 are disjoint. Now, $|\{x \in Q'_1 \,|\, \mathrm{d}(x, Q'_2) = 1\}| \geq 9$, while by Lemma 9.16, $|\{x \in Q'_1 \,|\, \mathrm{d}(x, Q'_2) = 1\}| = 7$. So, our assumption was wrong and every hex disjoint with H is isomorphic to $\mathbb{E}_2$.

(b) Let x denote a point not contained in H and let H' denote a hex through x disjoint with H. Then $H' \cong \mathbb{E}_2$. Let L denote the line through x intersecting H. Suppose that there exists a line $L' \neq L$ through x not contained in H'. Then L'

is contained in a hex H'' not containing L. Since H'' does not contain L, H'' is disjoint from H and hence is isomorphic to $\mathbb{E}_2$. By Theorem 1.7, the two hexes H' and H'' intersect in a big quad, but that is impossible, since $\mathbb{E}_2$ has no big quads. So, L is the unique line through x not contained in H'. Hence, $t_{\mathcal{S}} = t_{\mathbb{E}_2} + 1$ and $\mathcal{S} \cong \mathbb{E}_2 \times \mathbb{L}_3$. □

Theorem 9.19. *If $\mathcal{S}$ is a slim dense near octagon, then $\mathcal{S}$ is isomorphic to one of the 24 examples given in Table 9.1.*

Proof. Let H be a big hex of $\mathcal{S}$. If $H \cong \mathbb{E}_1$, then $\mathcal{S} \cong \mathbb{E}_1 \times \mathbb{L}_3$ or $\mathcal{S} \cong \mathbb{E}_1 \otimes Q(5,2)$ by Theorem 9.15. If $H \cong \mathbb{E}_2$, then $\mathcal{S}$ is isomorphic to $\mathbb{E}_2 \times \mathbb{L}_3$ by Theorem 9.18. If H is not isomorphic to $\mathbb{E}_1$ or $\mathbb{E}_2$, then H has a big quad. So, $\mathcal{S}$ has a nice chain of convex subpolygons. The theorem now readily follows from Theorem 8.9. □

For each of the possible near octagons $\mathcal{S}$, we have listed all big hexes in Table 9.1. In Table 9.2, we list all the remaining hexes. The number α_H in each table denotes the number of hexes isomorphic to H through a given point of $\mathcal{S}$. In $(Q(5,2) \otimes Q(5,2)) \otimes_b Q(5,2)$, there are two types of points. There are 1458 points which are incident with two hexes isomorphic to $Q(5,2) \otimes Q(5,2)$, 48 hexes isomorphic to $\mathbb{L}_3 \times \mathbb{L}_3 \times \mathbb{L}_3$, eight hexes isomorphic to $Q(5,2) \times \mathbb{L}_3$, and there are 729 points which are incident with three hexes isomorphic to $Q(5,2) \otimes Q(5,2)$, 64 hexes isomorphic to $\mathbb{L}_3 \times \mathbb{L}_3 \times \mathbb{L}_3$, 0 hexes isomorphic to $Q(5,2) \times \mathbb{L}_3$. Using the fact that α_H is constant for all the remaining near octagons, one can show the following results.

Theorem 9.20. *Let $\mathcal{S}$ be a slim dense near polygon. Then, for every slim dense near hexagon H not isomorphic to $\mathbb{L}_3 \times \mathbb{L}_3 \times \mathbb{L}_3$, $Q(5,2) \otimes \mathbb{L}_3$ or $Q(5,2) \otimes Q(5,2)$, there exists a constant α_H such that every point of $\mathcal{S}$ is contained in precisely α_H hexes isomorphic to H.*

Proof. Suppose $\mathcal{S}$ is a near $2d$-gon. Obviously, the property holds if $d \leq 3$. By Tables 9.1 and 9.2, the property also holds if $d = 4$. We will denote by $\alpha_H(O)$ the constant number of H-hexes through a point of a slim dense near octagon O. Now, consider two different collinear points x and y of $\mathcal{S}$. Let μ_H denote the number of H-hexes of $\mathcal{S}$ through the line xy. For every convex suboctagon O through xy, let $\lambda_H(O)$ denote the number of H-hexes of O through xy. Then the total number of H-hexes through x is equal to $\mu_H + \sum(\alpha_H(O) - \lambda_H(O))$, where the summation ranges over all convex suboctagons O through the line xy. By symmetry, the number of H-hexes through y is also equal to $\mu_H + \sum(a_H(O) - \lambda_H(O))$. Hence, every two collinear points of $\mathcal{S}$ are contained in the same number of H-hexes. By connectedness of $\mathcal{S}$, it follows that every point of $\mathcal{S}$ is contained in the same number of H-hexes. □

near octagon $\mathcal{S}$	$v_{\mathcal{S}}$	$t_{\mathcal{S}}$	big hexes H	α_H
$\mathbb{L}_3 \times \mathbb{L}_3 \times \mathbb{L}_3 \times \mathbb{L}_3$	81	3	$\mathbb{L}_3 \times \mathbb{L}_3 \times \mathbb{L}_3$	4
$W(2) \times \mathbb{L}_3 \times \mathbb{L}_3$	135	4	$\mathbb{L}_3 \times \mathbb{L}_3 \times \mathbb{L}_3$	3
			$W(2) \times \mathbb{L}_3$	2
$Q(5,2) \times \mathbb{L}_3 \times \mathbb{L}_3$	243	6	$\mathbb{L}_3 \times \mathbb{L}_3 \times \mathbb{L}_3$	5
			$Q(5,2) \times \mathbb{L}_3$	2
$\mathbb{H}_3 \times \mathbb{L}_3$	315	6	$W(2) \times \mathbb{L}_3$	4
			$\mathbb{I}_3$	1
$DQ(6,2) \times \mathbb{L}_3$	405	7	$W(2) \times \mathbb{L}_3$	7
			$DQ(6,2)$	1
$(Q(5,2) \otimes Q(5,2)) \times \mathbb{L}_3$	729	9	$Q(5,2) \times \mathbb{L}_3$	2
			$Q(5,2) \otimes Q(5,2)$	1
$\mathbb{G}_3 \times \mathbb{L}_3$	1215	12	$Q(5,2) \times \mathbb{L}_3$	3
			$\mathbb{G}_3$	1
$\mathbb{E}_1 \times \mathbb{L}_3$	2187	12	$\mathbb{E}_1$	1
$\mathbb{E}_2 \times \mathbb{L}_3$	2277	15	$\mathbb{E}_2$	1
$\mathbb{E}_3 \times \mathbb{L}_3$	1701	15	$Q(5,2) \times \mathbb{L}_3$	6
			$\mathbb{E}_3$	1
$DH(5,4) \times \mathbb{L}_3$	2673	21	$Q(5,2) \times \mathbb{L}_3$	21
			$DH(5,4)$	1
$W(2) \times W(2)$	225	5	$W(2) \times \mathbb{L}_3$	6
$Q(5,2) \times W(2)$	405	7	$W(2) \times \mathbb{L}_3$	5
			$Q(5,2) \times \mathbb{L}_3$	3
$Q(5,2) \times Q(5,2)$	729	9	$Q(5,2) \times \mathbb{L}_3$	10
$(Q(5,2) \otimes Q(5,2)) \otimes_a Q(5,2)$	2187	12	$Q(5,2) \otimes Q(5,2)$	3
$(Q(5,2) \otimes Q(5,2)) \otimes_b Q(5,2)$	2187	12	$Q(5,2) \otimes Q(5,2)$	2–3
$\mathbb{G}_3 \otimes Q(5,2)$	3645	15	$Q(5,2) \otimes Q(5,2)$	2
			$\mathbb{G}_3$	1
$\mathbb{E}_1 \otimes Q(5,2)$	6561	15	$\mathbb{E}_1$	1
$DH(5,4) \otimes Q(5,2)$	8019	24	$Q(5,2) \otimes Q(5,2)$	5
			$DH(5,4)$	1
$\mathbb{G}_4$	8505	21	$\mathbb{G}_3$	4
$\mathbb{H}_4$	945	9	$\mathbb{H}_3$	5
$\mathbb{I}_4$	2025	13	$DQ(6,2)$	8
$DQ(8,2)$	2295	14	$DQ(6,2)$	15
$DH(7,4)$	114939	84	$DH(5,4)$	85

Table 9.1: Slim dense near octagons and their big hexes

near octagon	other hexes H	α_H
$\mathbb{L}_3 \times \mathbb{L}_3 \times \mathbb{L}_3 \times \mathbb{L}_3$	–	–
$W(2) \times \mathbb{L}_3 \times \mathbb{L}_3$	–	–
$Q(5,2) \times \mathbb{L}_3 \times \mathbb{L}_3$	–	–
$\mathbb{H}_3 \times \mathbb{L}_3$	$\mathbb{L}_3 \times \mathbb{L}_3 \times \mathbb{L}_3$	3
$DQ(6,2) \times \mathbb{L}_3$	–	–
$(Q(5,2) \otimes Q(5,2)) \times \mathbb{L}_3$	$\mathbb{L}_3 \times \mathbb{L}_3 \times \mathbb{L}_3$	16
$\mathbb{G}_3 \times \mathbb{L}_3$	$\mathbb{L}_3 \times \mathbb{L}_3 \times \mathbb{L}_3$	9
	$W(2) \times \mathbb{L}_3$	9
$\mathbb{E}_1 \times \mathbb{L}_3$	$\mathbb{L}_3 \times \mathbb{L}_3 \times \mathbb{L}_3$	66
$\mathbb{E}_2 \times \mathbb{L}_3$	$W(2) \times \mathbb{L}_3$	35
$\mathbb{E}_3 \times \mathbb{L}_3$	$W(2) \times \mathbb{L}_3$	15
$DH(5,4) \times \mathbb{L}_3$	–	–
$W(2) \times W(2)$	–	–
$Q(5,2) \times W(2)$	–	–
$Q(5,2) \times Q(5,2)$	–	–
$(Q(5,2) \otimes Q(5,2)) \otimes_a Q(5,2)$	$\mathbb{L}_3 \times \mathbb{L}_3 \times \mathbb{L}_3$	64
$(Q(5,2) \otimes Q(5,2)) \otimes_b Q(5,2)$	$\mathbb{L}_3 \times \mathbb{L}_3 \times \mathbb{L}_3$	48–64
	$Q(5,2) \times \mathbb{L}_3$	8–0
$\mathbb{G}_3 \otimes Q(5,2)$	$\mathbb{L}_3 \times \mathbb{L}_3 \times \mathbb{L}_3$	24
	$W(2) \times \mathbb{L}_3$	36
	$Q(5,2) \times \mathbb{L}_3$	7
$\mathbb{E}_1 \otimes Q(5,2)$	$\mathbb{L}_3 \times \mathbb{L}_3 \times \mathbb{L}_3$	220
	$Q(5,2) \times \mathbb{L}_3$	11
$DH(5,4) \otimes Q(5,2)$	$Q(5,2) \times \mathbb{L}_3$	64
$\mathbb{G}_4$	$W(2) \times \mathbb{L}_3$	36
	$Q(5,2) \times \mathbb{L}_3$	18
	$\mathbb{H}_3$	27
$\mathbb{H}_4$	$W(2) \times \mathbb{L}_3$	10
$\mathbb{I}_4$	$\mathbb{H}_3$	7
$DQ(8,2)$	–	–
$DH(7,4)$	–	–

Table 9.2: Other hexes of the slim dense near octagons

In a completely similar way, one can show the following.

Theorem 9.21. *Let $\mathcal{S}$ be a slim dense near polygon not containing a convex suboctagon isomorphic to $(Q(5,2) \otimes Q(5,2)) \otimes_b Q(5,2)$ and let H be one of the 11 slim dense near hexagons. Then there exists a constant α_H such that every point of $\mathcal{S}$ is contained in precisely α_H hexes isomorphic to H.*

Chapter 10

Nondense slim near hexagons

In this chapter, we discuss slim near hexagons which are not necessarily dense. This chapter is based on the articles [42] and [43].

10.1 A few lemmas

Let $\mathcal{S}$ be a slim near hexagon. For every point x of $\mathcal{S}$, let $t_x + 1$ denote the total number of lines through x. If K and L are two distinct intersecting lines of $\mathcal{S}$ such that $\mathcal{C}(K, L) = K \cup L$, then $\mathcal{C}(K, L)$ is called a *degenerate quad.* If Q is a degenerate quad, then we define $t_Q := 0$. In this section, a quad can be either degenerate or nondegenerate. A nondegenerate quad is isomorphic to either the (3×3)-grid, $W(2)$ or $Q(5, 2)$. Recall that a (possibly degenerate) quad is called *big* if every point has distance at most 1 from it.

Lemma 10.1. *If $\mathcal{S}$ is a slim near polygon and if x and y are two points of $\mathcal{S}$ at distance 2 from each other, then $|\Gamma_1(x) \cap \Gamma_1(y)| \in \{1, 2, 3, 5\}$.*

Proof. If $|\Gamma_1(x) \cap \Gamma_1(y)| \geq 2$, then x and y are contained in a nondegenerate quad by Theorem 1.20. This quad is isomorphic to either $\mathbb{L}_3 \times \mathbb{L}_3$, $W(2)$ or $Q(5, 2)$. The lemma now immediately follows. $\square$

Lemma 10.2. *Let Q be a big quad, let x be an arbitrary point outside Q, let x' denote the unique point of Q collinear with x, let $K_1, \dots, K_k$ denote all the lines of Q through x' and let $Q_i := \mathcal{C}(xx', K_i)$ for every $i \in \{1, \dots, k\}$. Then $t_x = \sum_{i=1}^{k} t_{Q_i}$. As a consequence, $t_x \leq \max\{t_{Q_i} | 1 \leq i \leq k\} \cdot k$.*

Proof. Every line L through x different from xx' is contained in a unique quad of the form $\mathcal{C}(xx', K_i)$. [If $L \subset \mathcal{C}(xx', K_i)$, then K_i necessarily is the projection of L on Q.] Hence $t_x = \sum_{i=1}^{k} t_{Q_i}$. $\square$

Lemma 10.3. *Let Q and R be two different intersecting quads and suppose that Q is big.*

- *If R is nondegenerate, then $Q \cap R$ is a line.*
- *If R is degenerate, i.e. the join of two lines zx and zy, then $Q \cap R$ is either equal to $\{z\}$, zx or zy.*

Proof. Let u denote a common point of Q and R. Suppose that there exists a point v in R at distance 2 from u. Let v' denote the projection of v on Q. Then $2 = \mathrm{d}(v, u) = \mathrm{d}(v, v') + \mathrm{d}(v', u) = 1 + \mathrm{d}(v', u)$, so u and v' are collinear. Since v' is on a shortest path between v and u, $v' \in R$. Hence $Q \cap R$ is the line uv'. If there exists no point v in R at distance 2 from u, then R is degenerate (i.e. $R = zx \cup zy$) and $u = z$. Obviously, $Q \cap R$ is equal to either $\{z\}$, zx or zy. □

Lemma 10.4. *Let x and y denote two points at distance 3 from each other. Put $\Gamma_1(x) \cap \Gamma_2(y) = \{u_1, u_2, \ldots, u_{t_x+1}\}$ and $\Gamma_1(y) \cap \Gamma_2(x) = \{v_1, v_2, \ldots, v_{t_y+1}\}$. For every $i \in \{1, \ldots, t_x + 1\}$, let $Q_i = \mathcal{C}(u_i, y)$ and for every $j \in \{1, \ldots, t_y + 1\}$, let $R_j = \mathcal{C}(v_j, x)$. If A denotes the number of paths of length 3 connecting x and y, then $A = \sum_{i=1}^{t_x+1}(t_{Q_i} + 1) = \sum_{j=1}^{t_y+1}(t_{R_j} + 1)$. As a consequence, $t_x + 1 \leq A \leq \max\{t_{Q_i} + 1 | 1 \leq i \leq t_x + 1\} \cdot (t_x + 1)$ and $t_y + 1 \leq A \leq \max\{t_{R_j} + 1 | 1 \leq j \leq t_y + 1\} \cdot (t_y + 1)$.*

Proof. Count in two different ways the pairs (u, v) with $x \sim u \sim v \sim y$. □

10.2 Slim near hexagons with special points

10.2.1 Special points

Definition. A point x of a slim near hexagon is called *special* if it has distance at most 2 from any other point.

Suppose that $\mathcal{S}$ is a slim near hexagon. Let X denote the set of all special points of $\mathcal{S}$.

Theorem 10.5. *If x is a special point, then every nondegenerate quad contains x.*

Proof. Let L_1 and L_2 denote two disjoint lines of a nondegenerate quad Q. Since x has distance at most 2 from any point of L_i, $i \in \{1, 2\}$, x has distance at most 1 from L_i. It then immediately follows that $x \in Q$ since Q is convex. □

Theorem 10.6. *If $X \neq \emptyset$, then one of the following possibilities occurs:*

(I) *X consists of one point;*

(II) *X is a line L;*

(III) *$X = Q \setminus \Gamma_2(x)$ for some nondegenerate quad Q and some point $x \in Q$. In this case $\mathcal{S}$ consists of the quad Q and some additional lines through x.*

Proof. Suppose that $|X| \geq 2$.

- Suppose that there are two special points x_1 and x_2 at distance 2 from each other. Put $Q := \mathcal{C}(x_1, x_2)$. If $z \in \Gamma_2(Q)$, then $\mathrm{d}(z, x_1) = 2$; so z and x_1 have a common neighbour u. Now, $\mathrm{d}(u, x_2) = \mathrm{d}(u, x_1) + \mathrm{d}(x_1, x_2) = 3$, a contradiction. Hence $\Gamma_2(Q) = \emptyset$. If $z \in \Gamma_1(Q)$, then z is collinear with a unique point $z' \in Q$. Since $\mathrm{d}(z, x_1) = 1 + \mathrm{d}(z', x_1) \leq 2$ and $\mathrm{d}(z, x_2) = 1 + \mathrm{d}(z', x_2) \leq 2$, z' is a common neighbour of x_1 and x_2. If L is a line through z different from zz', then $L \subseteq \Gamma_1(Q)$ and hence L projects to a line L' which is completely contained in $\Gamma_1(x_1) \cap \Gamma_1(x_2)$, a contradiction. Hence, every point $z \in \Gamma_1(Q)$ is contained in a unique line. If there exist two points z_1 and z_2 outside Q such that $z_1' \neq z_2'$, then $\mathrm{d}(z_1, z_2)$ would be equal to 4, a contradiction. So, there exists a unique point $x \in Q$ which is collinear with every point outside Q. Clearly, Q must be nondegenerate. So, case (III) occurs.

- Suppose now that every two special points are collinear, so X is part of a line L. Let $x_1, x_2 \in X$ with $x_1 \neq x_2$ and let x_3 denote the other point of L. Every point z of $\mathcal{S}$ has distance at most 2 from x_1 and x_2, and hence has distance at most 2 from x_3. This proves that X is the whole line L. So, case (II) occurs. □

10.2.2 Slim near hexagons of type (III)

To every slim near hexagon $\mathcal{S}$ of type (III), we can associate two parameters λ and λ':

- $\lambda = t_Q$, where Q denotes the unique nondegenerate quad of $\mathcal{S}$,
- λ' is the number of lines not contained in Q.

For each value of $\lambda \in \{1, 2, 4\}$ and for each value of $\lambda' > 0$, there exists, up to isomorphism, a unique near hexagon of type (III) with parameters λ and λ'.

10.2.3 Slim near hexagons of type (II)

If $\mathcal{S}$ is of type (II), then there is a unique line $L = \{x_1, x_2, x_3\}$ of special points. By Theorem 10.5, every nondegenerate quad contains L. Now, let x denote an arbitrary point not incident with L. Since x has distance at most 2 from each point of L, it has distance 1 from L. Hence, x is contained in a unique line L_x meeting L. Any other line M through x is contained in the quad $\mathcal{C}(L_x, L)$ which is necessarily nondegenerate. So, every line of $\mathcal{S}$ disjoint with L is contained in a unique nondegenerate quad.

We can now associate six parameters $(\alpha_1, \alpha_2, \alpha_4, \beta_1, \beta_2, \beta_3)$ to $\mathcal{S}$:

- α_i, $i \in \{1, 2, 4\}$, denotes the number of quads of order $(2, i)$ through L.
- β_i, $i \in \{1, 2, 3\}$, denotes the number of lines K through x_i for which $\mathcal{C}(K, L) = K \cup L$.

If $\alpha_1, \alpha_2, \alpha_4, \beta_1, \beta_2, \beta_3$ are natural numbers such that $\alpha_1+\alpha_2+\alpha_4+\beta_1'+\beta_2'+\beta_3' \geq 2$ (put β_i' equal to 0 if $\beta_i = 0$ and equal to 1 otherwise), then there exists, up to isomorphism, a unique near hexagon $\mathcal{H}(\alpha_1, \alpha_2, \alpha_4, \beta_1, \beta_2, \beta_3)$ of type (II) with parameters $(\alpha_1, \alpha_2, \alpha_4, \beta_1, \beta_2, \beta_3)$. For every permutation σ of $\{1, 2, 3\}$, we have $\mathcal{H}(\alpha_1, \alpha_2, \alpha_4, \beta_1, \beta_2, \beta_3) \cong \mathcal{H}(\alpha_1, \alpha_2, \alpha_4, \beta_{\sigma(1)}, \beta_{\sigma(2)}, \beta_{\sigma(3)})$.

10.2.4 Slim near hexagons of type (I)

The following theorem can be used to construct many near hexagons of type (I).

Theorem 10.7 (Section 2 of [53]). *Let $\mathcal{L}$ denote a connected partial linear space with the property that every two different lines meet in a point and that no point is incident with all lines. Put $A = \{T \,|\, \mathcal{L}$ has a line of size $T + 1\}$. Then there exists a near hexagon $\mathcal{A}$ of type (I) such that:*

(i) *$\mathcal{L}(\mathcal{A}, x^*) \cong \mathcal{L}$ with x^* the unique special point of $\mathcal{A}$;*

(ii) *every line of $\mathcal{A}$ is incident with $s + 1$ points ($s \in \mathbb{N} \setminus \{0\}$),*

if and only if there exists a generalized quadrangle of order (s, T) for every $T \in A$.

10.3 Slim near hexagons without special points

10.3.1 Examples

Let $\mathcal{S}$ be a finite slim near hexagon without special points. If every two points at distance 2 from each other have at least two common neighbours, then $\mathcal{S}$ is dense and hence isomorphic to either $\mathbb{L}_3 \times \mathbb{L}_3 \times \mathbb{L}_3$, $W(2) \times \mathbb{L}_3$, $Q(5, 2) \times \mathbb{L}_3$, $\mathbb{H}_3$, $DQ(6, 2)$, $Q(5, 2) \otimes Q(5, 2)$, $\mathbb{G}_3$, $\mathbb{E}_1$, $\mathbb{E}_2$, $\mathbb{E}_3$ or $DH(5, 4)$. If every two points at distance 2 have a unique common neighbour, then $\mathcal{S}$ is a generalized hexagon of order $(2, t)$. As we have already mentioned in Section 3.5, there are four such generalized hexagons: one of order $(2, 1)$, one of order $(2, 8)$ and two nonisomorphic ones of order 2. Besides the dense near hexagons and the generalized hexagons, there are many other examples (of mixed type) as we will show now.

(1) Suppose that H is a slim dense near hexagon with a big quad Q' which is not isomorphic to $Q(5, 2)$. Embed Q' properly as a subquadrangle in a generalized quadrangle Q (i.e. the (3×3)-grid in $W(2)$ or $Q(5, 2)$, $W(2)$ in $Q(5, 2)$). Then the resulting incidence structure is a slim near hexagon. The following possibilities for (H, Q', Q) give rise to such a near hexagon: $(\mathbb{L}_3 \times \mathbb{L}_3 \times \mathbb{L}_3, \mathbb{L}_3 \times \mathbb{L}_3, W(2))$, $(\mathbb{L}_3 \times \mathbb{L}_3 \times \mathbb{L}_3, \mathbb{L}_3 \times \mathbb{L}_3, Q(5, 2))$, $(\mathbb{L}_3 \times W(2), \mathbb{L}_3 \times \mathbb{L}_3, W(2))$, $(\mathbb{L}_3 \times W(2), \mathbb{L}_3 \times \mathbb{L}_3, Q(5, 2))$, $(\mathbb{L}_3 \times W(2), W(2), Q(5, 2))$, $(\mathbb{L}_3 \times Q(5, 2), \mathbb{L}_3 \times \mathbb{L}_3, W(2))$, $(\mathbb{L}_3 \times Q(5, 2), \mathbb{L}_3 \times \mathbb{L}_3, Q(5, 2))$, $(\mathbb{H}_3, W(2), Q(5, 2))$ and $(DQ(6, 2), W(2), Q(5, 2))$.

(2) Let Δ be one of the dual polar spaces $DW(5, 2)$ or $DH(5, 4)$. Let Q_1, Q_2 and Q_3 be disjoint quads of Δ such that $Q_3 = \mathcal{R}_{Q_2}(Q_1)$ and let A be a set of quads intersecting Q_1 and Q_2 in a line. Put $V := Q_1 \cup Q_2 \cup Q_3 \cup (\bigcup_{Q \in A} Q)$. If V is a

subspace, then the points and lines of Δ which are contained in V determine a slim subhexagon $\mathcal{S}_V$. Let X denote the set of lines of Q_1 which are contained in a quad of A. In [56], all slim near hexagons $\mathcal{S}_V$ of the above type were classified. If $\Delta \cong DW(5,2)$, then we have the following possibilities for X:

(a) X is a (possibly empty) set of lines through a given point of Q_1;

(b) X is a regulus of Q_1;

(c) X consists of a regulus of Q_1 together with one line of its opposite regulus;

(d) X consists of the six lines which are contained in a (3×3)-grid of Q_1;

(e) X consists of the lines of Q_1 which meet a given line of Q_1;

(f) X consists of the lines of Q_1 which intersect a given line of Q_1 in a unique point;

(g) X consists of the whole set of lines of Q_1;

(h) X is the complement of a spread of Q_1.

If $\Delta \cong DH(5,4)$, then we have the following possibilities for X:

(a) X is a (possibly empty) set of lines through a given point of Q_1;

(b) X is a regulus of Q_1;

(c) X consists of a regulus of Q_1 together with one line of its opposite regulus;

(d) X consists of the lines contained in a (3×3)-grid of Q_1;

(e) X is a regular spread of Q_1;

(f) X consists of the lines of Q_1 having nonempty intersection with a given line of Q_1;

(g) X consists of the lines of Q_1 which intersect a given line of Q_1 in a unique point;

(h) there exists a subquadrangle $R \cong Q(4,2)$ in Q_1 and X' consists of all lines of R having nonempty intersection with a given line of R;

(i) there exists a subquadrangle $R \cong Q(4,2)$ in Q_1 and X consists of all lines of R intersecting a given line of R in a unique point;

(j) X consists of all lines contained in a subquadrangle $R \cong Q(4,2)$ of Q_1;

(k) X is the whole set of lines of Q_1;

(l) X consists of all 18 lines which are contained in three mutually disjoint grids of Q_1;

(m) there exists a subquadrangle $R \cong Q(4,2)$ in Q_1 and a spread S in R such that X consists of all lines of R which are not contained in S.

10.3.2 Upper bounds for the number of lines through a point

Theorem 10.8 (Section 10.4). *Let $\mathcal{S}$ be a (possibly infinite) slim near hexagon without special points. Suppose $\mathcal{S}$ contains at least one $Q(5,2)$-quad. Let $t+1$ denote the maximal number of lines through a given point of $\mathcal{S}$.*

- *If there exist at least two $Q(5,2)$-quads, then $t+1 \leq 21$.*
- *If there exists a unique $Q(5,2)$-quad Q, then $t+1 \leq 25$ and every point outside Q is incident with at most 21 lines.*

Theorem 10.9 (Section 10.5). *Let $\mathcal{S}$ be a (possibly infinite) slim near hexagon without special points. Suppose $\mathcal{S}$ contains at least one $W(2)$-quad, but no $Q(5,2)$-quad. For every point x of $\mathcal{S}$, let t_x+1 denote the total number of lines through x and let $t+1$ denote the maximal value attained by t_x+1. We have:*

- *$t+1 \leq 45$;*
- *if $\mathcal{S}$ contains a big $W(2)$-quad, then $t+1 \leq 15$;*
- *if the point x is contained in a $W(2)$-quad, then $t_x+1 \leq 33$;*
- *if the point x has distance 2 from a $W(2)$-quad, then $t_x+1 \leq 15$;*
- *if the point x has distance 1 from a big $W(2)$-quad, then $t_x+1 \leq 7$.*

Theorem 10.10 (Section 10.6). *Let $\mathcal{S}$ be a finite slim near hexagon without special points, $W(2)$-quads and $Q(5,2)$-quads. Then the number of lines through a given point is at most equal to 76. If $\mathcal{S}$ has an order, then the number of lines through a given point is at most equal to 33.*

From Theorems 10.8, 10.9 and 10.10, we immediately obtain:

Corollary 10.11. *There are finitely many finite slim near hexagons without special points.*

10.4 Proof of Theorem 10.8

In this section, we suppose that $\mathcal{S}$ is a slim near hexagon having at least one $Q(5,2)$-quad, but no special points. Since $Q(5,2)$ contains no ovoid, every $Q(5,2)$-quad is big.

Lemma 10.12. *Every line K of $\mathcal{S}$ is contained in a nondegenerate quad.*

Proof. Let Q denote an arbitrary $Q(5,2)$-quad of $\mathcal{S}$.

(i) If K is contained in Q, then we are done.

(ii) Suppose that K is disjoint with Q. Let x and y be arbitrary points of K and let x' and y' be their respective projections on Q. Since x and y' have at least two common neighbours (namely x' and y), the quad $\mathcal{C}(x,y')$ is nondegenerate. Clearly, $K \subseteq \mathcal{C}(x,y')$.

(iii) Suppose that K intersects Q in a point u. If $t_x = 0$ for every point x of $K \setminus \{u\}$, then $\mathrm{d}(x,y) = 1 + \mathrm{d}(u,y)$ for every point x of $K \setminus \{u\}$ and for every point y not incident with K. If we take for y an arbitrary point at distance 3 from u, then we obtain a contradiction. Hence, there exists a line L which intersects K in a point different from u. By (ii), K and L are contained in a nondegenerate quad. □

Let I be the intersection of all $Q(5,2)$-quads. By Lemma 10.2, $t_x + 1 \leq 21$ for every point x outside I.

Lemma 10.13. *If there exists a line K which is contained in at least four $Q(5,2)$-quads, then $I = \emptyset$.*

Proof. Suppose that there exists a point x in I, then x lies on K. Let y denote an arbitrary point at distance 3 from x and let z denote the unique point of K at distance 2 from y. Now, by Lemma 10.3, $\mathcal{C}(y,z)$ has a line in common with each $Q(5,2)$-quad through K. Since there are at least 4 such quads, $\mathcal{C}(y,z)$ must be a $Q(5,2)$-quad. But $\mathcal{C}(y,z)$ does not contain x, a contradiction. Hence, $I \neq \emptyset$. □

Lemma 10.14. *Let K be a line of $\mathcal{S}$ which is incident with three $Q(5,2)$-quads Q_1, Q_2 and Q_3, let x be an arbitrary point of K and suppose that there exists a $Q(5,2)$-quad Q_4 through x not containing K. Then $t_x + 1 \leq 21$.*

Proof. If $x \notin I$, then we are done. So, we may suppose that $x \in I$. So, $I = \{x\}$ by Lemma 10.13. Every nondegenerate quad through K intersects Q_4 in a line. So, there are at most five nondegenerate quads through K.

- *First case*: There are precisely five nondegenerate quads through K.
  Let A denote the set of lines through x which are contained in one of the five nondegenerate quads through K. Since $I \neq \emptyset$, there are at most three $Q(5,2)$-quads through K; hence, $|A| \leq 17$. Suppose that there exists a line U through x not belonging to A. By Lemma 10.12, U is contained in a nondegenerate quad Q_U. Since Q_U intersects each of the quads Q_1, Q_2 and Q_3 in a line different from K, $t_{Q_U} \geq 3$. So, $Q_U \cong Q(5,2)$. But then Q_U intersects each of the five nondegenerate quads through K in a line and $t_{Q_U} \geq 5$, a contradiction. As a consequence every line through x is contained in A, implying that $t_x + 1 = |A| \leq 17$.

- *Second case*: There are precisely four nondegenerate quads through K.
  Let R denote the fourth quad through K. By Lemma 10.13, R is either a grid-quad or a $W(2)$-quad. If U is a line through x which is not contained in one of the four nondegenerate quads through K, then as before, U is contained in a $Q(5,2)$-quad Q_U. The quad Q_U uniquely determines the line U since U is the unique line of Q_U through x which is not contained in Q_1, Q_2, Q_3 and R. Hence, $t_x + 1 \leq 15 + N$, where N is the total number of $Q(5,2)$-quads through x different from Q_1, Q_2 and Q_3. By Lemma 10.13, there are at most $3 \cdot t_R \leq 6$ $Q(5,2)$-quads meeting R in a line through x different from K. Hence, $N \leq 6$ and $t_x + 1 \leq 21$.

- *Third case*: There are precisely three nondegenerate quads through K.
If N denotes the total number of $Q(5,2)$-quads through x different from Q_1, Q_2 and Q_3, then with a similar reasoning as in the second case, we have that $t_x+1 \leq 13+2N$. If $N \leq 4$, then we are done. Suppose therefore that $N \geq 5$. There then exists a line K' in Q_1 through the point x which is contained in at least three $Q(5,2)$-quads. By Lemma 10.13 it follows that K' is contained in precisely three $Q(5,2)$-quads. Let Q_1, R_1 and R_2 denote these quads. By Lemma 10.13, each of the lines $R_1 \cap Q_1$, $R_1 \cap Q_2$ and $R_1 \cap Q_3$ is contained in at most three $Q(5,2)$-quads. Hence, the number of $Q(5,2)$-quads on x which intersect R_1 in a line outside $Q_1 \cup Q_2 \cup Q_3$ is at least $N-4$. A similar conclusion holds for R_2. We can now improve the bound $t_x+1 \leq 13+2N$ to $t_x+1 \leq 13+2N-(N-4)-(N-4)=21$. □

Lemma 10.15. *If there exist $Q(5,2)$-quads Q_1, Q_2 and Q_3 through x such that $Q_1 \cap Q_2 \cap Q_3 = \{x\}$, then $t_x+1 \leq 21$.*

Proof. Define $K_i := Q_j \cap Q_k$ if $\{i,j,k\} = \{1,2,3\}$. We distinguish two cases.

- There exists a fourth $Q(5,2)$-quad R through x.
Suppose that R intersects Q_i, $i \in \{1,2,3\}$, in a line L_i. If there exists a line through x which is contained in at least three $Q(5,2)$-quads, then we are done by Lemma 10.14. So, suppose that there exist no line through x which is contained in at least three $Q(5,2)$-quads. If there are precisely i $Q(5,2)$-quads through x, then all these quads cover $5i - \frac{i(i-1)}{2}$ lines through x. Clearly, $5i - \frac{i(i-1)}{2} \leq 15$. Now, let U denote an arbitrary line through x not contained in $Q_1 \cup Q_2 \cup Q_3 \cup R$, then U is contained in a nondegenerate quad Q_U. Since Q_U meets each of the quads Q_1, Q_2, Q_3 and R in a line, it must be a $Q(5,2)$-quad or coincide with one of the quads $\mathcal{C}(K_i, L_i)$, $i \in \{1,2,3\}$. Since none of the quads $\mathcal{C}(K_i, L_i)$ is a $Q(5,2)$-quad, we have $t_x+1 \leq 15+3=18$.

- Q_1, Q_2 and Q_3 are the only $Q(5,2)$-quads through x.
Every line U through x not contained in Q_1, Q_2 and Q_3 is contained in a nondegenerate quad Q_U. The quad Q_U is not isomorphic to $Q(5,2)$ and intersects each of the quads Q_1, Q_2 and Q_3 in a line. Hence, there exists an $i \in \{1,2,3\}$ such that Q_U contains K_i and intersects Q_i in a line. So, the number of possibilities for Q_U is at most $9 = 3 \times 3$. Hence, at most 9 lines through x are not contained in Q_1, Q_2 or Q_3 and $t_x+1 \leq 12+9=21$. □

Corollary 10.16. *If I is the empty set or a point, then $t_x+1 \leq 21$ for every point x of $\mathcal{S}$.*

Lemma 10.17. *If I is a line K, then $t_x+1 \leq 19$ for every point x of K.*

Proof. Let x denote an arbitrary point of K, let y be a point at distance 3 from x and let z denote the unique point of K at distance 2 from y. Let i denote the number of $Q(5,2)$-quads through K. By Lemma 10.13, $i \in \{2,3\}$. Let A be the set of lines through x which are contained in one of the $Q(5,2)$-quads through K, then $|A| = 1 + 4i$. Since y is not contained in a $Q(5,2)$-quad, we have $t_y \leq 10$ by Lemma 10.2. Put $\Gamma_1(y) \cap \Gamma_2(x) := \{u_1, u_2, \ldots, u_{t_y+1}\}$. Take an arbitrary point u_j in $\Gamma_1(y) \cap \Gamma_2(x)$, then one of the following possibilities occurs.

- The point u_j is contained in a $Q(5,2)$-quad through K. There are precisely i such points, one for each $Q(5,2)$-quad through K. All these points are contained in the quad $\mathcal{C}(z,y)$.
- The point u_j is not contained in the quad $\mathcal{C}(z,y)$. There are precisely $t_y - \delta$ such points ($\delta := t_{\mathcal{C}(y,z)}$). For each such point u_j, the quad $\mathcal{C}(x,u_j)$ is not isomorphic to $Q(5,2)$. Since the quad $\mathcal{C}(x,u_j)$ intersects each $Q(5,2)$-quad, it contains at most $3 - i$ lines through x which are not contained in A.
- The point u_j is contained in the quad $\mathcal{C}(z,y)$ and $\mathcal{C}(x,u_j)$ is not isomorphic to $Q(5,2)$. The number of such points is equal to $\epsilon := \delta + 1 - i$ which is equal to 1 if $\delta = i = 2$ and 0 otherwise. For each such point u_j, $\mathcal{C}(x,u_j)$ contains K and hence there are at most two lines through x in $\mathcal{C}(x,u_j)$ which are not contained in A.

Since each line through x is contained in one of the sets $\mathcal{C}(x,u_j)$, we have $t_x + 1 \leq (1+4i) + (t_y - \delta)(3-i) + 2\epsilon$. Taking into account the above-mentioned restrictions on i, t_y, δ and ϵ, we find that $t_x + 1 \leq 19$. □

Lemma 10.18. *If I is a quad Q, then $t_x + 1 \leq 25$ for every point x of Q.*

Proof. Let y be a point at distance 3 from x and let y' denote the unique point of Q collinear with y. Since $y \notin Q$ and since Q is the only $Q(5,2)$-quad, we have $t_y \leq 10$ by Lemma 10.2. Now, count pairs (u,v) with $u, v \notin Q$ and $x \sim u \sim v \sim y$. We have the following.

- On each of the $t_x - 4$ lines through x not contained in Q, there exists a unique point u at distance 2 from y and for each such point u there exists a point v in $(\Gamma_1(u) \cap \Gamma_1(y)) \setminus Q$.
- On each of the t_y lines through y different from yy', there exists a unique point v at distance 2 from x and for each such point v there are at most two points $u \in (\Gamma_1(x) \cap \Gamma_1(v)) \setminus Q$. [Notice that the projection of v on Q is collinear with x.]

Hence $t_x - 4 \leq 2t_y \leq 20$. So, $t_x + 1 \leq 25$. □

10.5 Proof of Theorem 10.9

In this section, we suppose that $\mathcal{S}$ is a slim near hexagon having at least one $W(2)$-quad, but no $Q(5,2)$-quads and no special points.

Lemma 10.19. *If Q is a big $W(2)$-quad of $\mathcal{S}$, then $t_x + 1 \leq 7$ for every point $x \in \Gamma_1(Q)$ and $t_x + 1 \leq 15$ for every point $x \in Q$.*

Proof. If $x \in \Gamma_1(Q)$, then $t_x + 1 \leq 7$ by Lemma 10.2. Suppose now that $x \in Q$. Let y denote a point at distance 3 from x, then $y \in \Gamma_1(Q)$. Let y' denote the unique point of Q collinear with y. Now, count pairs (u, v) with $u, v \notin Q$ and $x \sim u \sim v \sim y$. We have the following.

- On each of the $t_x - 2$ lines through x not contained in Q, there exists a unique point u at distance 2 from y and for each such point u there exists a point v in $(\Gamma_1(u) \cap \Gamma_1(y)) \setminus Q$.
- On each of the t_y lines through y different from yy', there exists a unique point v at distance 2 from x and for each such point v there are at most two points $u \in (\Gamma_1(x) \cap \Gamma_1(v)) \setminus Q$. [Notice that the projection of v on Q is collinear with x.]

Hence $t_x - 2 \leq 2t_y \leq 12$. So, $t_x + 1 \leq 15$. □

From now on we assume that no $W(2)$-quad is big in $\mathcal{S}$. Let Q denote an arbitrary $W(2)$-quad of $\mathcal{S}$.

Lemma 10.20. *For every ovoid O of Q, there exists a point $y \in \Gamma_2(Q)$ such that $\Gamma_2(y) \cap Q = O$.*

Proof. Let x denote an arbitrary point of $\Gamma_2(Q)$ and consider the ovoid $O' := \Gamma_2(x) \cap Q$. If $O' = O$, then we are done. Suppose therefore that $O' \neq O$. Then O and O' intersect in a point u. Let v denote a common neighbour of u and x and let y denote the unique point on vx different from v and x. Clearly, $\Gamma_2(y) \cap Q$ coincides with O. □

Lemma 10.21. *If $x \in \Gamma_2(Q)$, then $t_x + 1 \leq 15$.*

Proof. Since Q has no partition in ovoids, every line through x must meet $\Gamma_1(Q)$. Now, consider the ovoid $\Gamma_2(x) \cap Q = \{x_1, x_2, x_3, x_4, x_5\}$. Then every line through x is contained in one of the quads $\mathcal{C}(x, x_i)$, $i \in \{1, \ldots, 5\}$. Hence $t_x + 1 \leq 3 \cdot 5 = 15$. □

Lemma 10.22. *If $x \in Q$, then $t_x + 1 \leq 33$.*

Proof. Consider an ovoid O in Q not containing x and let y be a point of $\Gamma_2(Q)$ such that $\Gamma_2(y) \cap Q = O$. Let x_1, x_2, x_3 denote the three points of O collinear with x and let x_4 and x_5 denote the two other points of O. Put $\alpha_i := t_{\mathcal{C}(y,x_i)} + 1$. Every line through y contains a unique point of $\Gamma_1(Q)$ and hence is contained in at least one of the quads $\mathcal{C}(y, x_i)$, $i \in \{1, 2, 3, 4, 5\}$. If a line K through y were contained in two such quads, then the unique point of $K \cap \Gamma_1(Q)$ would be collinear with two points of Q, contradicting the fact that Q is geodetically closed. As a consequence, $t_y + 1 = \alpha_1 + \alpha_2 + \alpha_3 + \alpha_4 + \alpha_5$. If $\alpha_4 = 3$ or $\alpha_5 = 3$, then by Lemma 10.21, $t_x + 1 \leq 15$. Suppose therefore that that $\alpha_4 \neq 3 \neq \alpha_5$. Counting pairs (u, v) such that $x \sim u \sim v \sim y$ yields $3(\alpha_1 + \alpha_2 + \alpha_3 + \alpha_4 + \alpha_5) = 3(t_y + 1) \geq (t_x - 2) + (\alpha_1 + \alpha_2 + \alpha_3)$ or $t_x + 1 \leq 2(\alpha_1 + \alpha_2 + \alpha_3) + 3(\alpha_4 + \alpha_5) + 3$. Now, $\alpha_1, \alpha_2, \alpha_3 \leq 3$ and $\alpha_4, \alpha_5 \leq 2$ and so $t_x + 1 \leq 33$. □

Lemma 10.23. *For every point $x \in \Gamma_1(Q)$, there exists a point $y \in \Gamma_3(x)$ such that $t_y + 1 \leq 15$. As a consequence, $t_x + 1 \leq 45$ for every point $x \in \Gamma_1(Q)$.*

Proof. Let x' denote the unique point of Q collinear with x, let O be an ovoid of Q through x' and let z be a point of $\Gamma_2(Q)$ for which $\Gamma_2(z) \cap Q = O$. If $x \notin \mathcal{C}(z, x')$, then we define $y := z$. Clearly, $\mathrm{d}(x, y) = 3$ and $t_y + 1 \leq 15$. If $x \in \mathcal{C}(z, x')$, then we distinguish the following two cases.

- There exists a point $u \in \mathcal{C}(z, x') \cap \Gamma_2(x') \cap \Gamma_2(x)$. Notice that $u \in \Gamma_2(Q)$. Let y be any point in $\Gamma_2(Q)$ collinear with u and not contained in $\mathcal{C}(z, x') = \mathcal{C}(u, x')$. Clearly, $y \in \Gamma_3(x)$ and $t_y + 1 \leq 15$.

- $\mathcal{C}(z, x') \cap \Gamma_2(x') \cap \Gamma_2(x) = \emptyset$. Then $x \sim z$ and $\mathcal{C}(z, x')$ is the degenerate quad $xx' \cup xz$. If $t_z + 1 \geq 6$, then because every line through z is contained in a quad $\mathcal{C}(z, y')$ for some $y' \in O$, there exists a nondegenerate quad through z intersecting Q and the required point y can be taken in this quad. If $t_z + 1 = 5$, then by Lemma 10.4, any point y of $Q \cap \Gamma_3(z) \cap \Gamma_3(x) = (Q \cap \Gamma_2(x')) \setminus O$ satisfies the required properties ($t_y + 1 \leq A \leq 3(t_z + 1) = 15$).

If $y \in \Gamma_3(x)$ and $t_y + 1 \leq 15$, then again by Lemma 10.4, $t_x + 1 \leq 3(t_y + 1) = 45$. $\square$

10.6 Proof of Theorem 10.10

In this section, we suppose that $\mathcal{S} = (\mathcal{P}, \mathcal{L}, \mathrm{I})$ is a finite slim near hexagon which satisfies the following properties.

- Every two points at distance 2 have 1 or 2 common neighbours. So, there are no $W(2)$-quads and no $Q(5, 2)$-quads.

- $\mathcal{S}$ has no special points.

For every point x of $\mathcal{S}$, $t_x + 1$ denotes the total number of lines through x. Put $t + 1 := \max\{t_x + 1 \,|\, x \in \mathcal{P}\}$ and let x^* denote a point of $\mathcal{S}$ for which $t_{x^*} = t$. Since $\mathcal{S}$ has no special points, $\Gamma_3(x^*) \neq \emptyset$.

10.6.1 Upper bound for $|\Gamma_3(x^*)|$

Lemma 10.24. *Let η denote the total number of grid-quads through x^*.*

(i) *We have $|\Gamma_3(x^*)| \leq 8t^2 - \frac{16\eta}{t+1}$. In particular, $|\Gamma_3(x^*)| \leq 8t^2$.*

(ii) *If $\mathcal{S}$ has an order $(2, t)$, then η and $|\Gamma_3(x^*)|$ are independent of the chosen point x^*, and $|\Gamma_3(x^*)| = 8t^2 - 8\eta$.*

(iii) *If $x \in \Gamma_3(x^*)$, then $t_x + 1 \geq \frac{t+1}{2}$.*

Proof. (i) Put $n_i := |\Gamma_i(x^*)|$, $i \in \{0, 1, 2, 3\}$. Then $n_0 = 1$ and $n_1 = 2(t + 1)$. We now derive some inequalities.

(a) There are 4η points x in $\Gamma_2(x^*)$ for which $|\Gamma_1(x) \cap \Gamma_1(x^*)| = 2$ and $n_2 - 4\eta$ points in $\Gamma_2(x^*)$ for which $|\Gamma_1(x) \cap \Gamma_1(x^*)| = 1$. Counting pairs (x_1, x_2) of points satisfying $x_1 \in \Gamma_1(x^*)$, $x_2 \in \Gamma_2(x^*)$ and $x_1 \sim x_2$, gives $(n_2 - 4\eta) \cdot 1 + (4\eta) \cdot 2 \leq 2(t+1) \cdot 2t$ or $n_2 \leq 4t(t+1) - 4\eta$.

(b) Counting pairs (y, x) of points satisfying $y \in \Gamma_2(x^*)$, $x \in \Gamma_3(x^*)$ and $x \sim y$, gives $\sum_{x \in \Gamma_3(x^*)} (t_x + 1) \leq (n_2 - 4\eta) \cdot 2t + 4\eta \cdot 2(t-1) = 2tn_2 - 8\eta$.

(c) For every point $x \in \Gamma_3(x^*)$, let α_x denote the number of grid-quads Q through x^* containing a point at distance 1 from x. Obviously, $\alpha_x \leq t_x + 1$. Counting pairs (x, Q) with x a point of $\Gamma_3(x^*)$ and Q a grid-quad through x^* such that $\mathrm{d}(x, Q) = 1$, gives $\sum_{x \in \Gamma_3(x^*)} \alpha_x \leq \eta \cdot 4 \cdot 2(t-1) = 8\eta(t-1)$.

(d) Let x be a fixed point of $\Gamma_3(x^*)$. Counting pairs (y, z) of points satisfying $x^* \sim y \sim z \sim x$, gives $(t_x + 1 - \alpha_x) \cdot 1 + \alpha_x \cdot 2 \geq t + 1$ or $t_x + 1 \geq t + 1 - \alpha_x$.

Combining the above inequalities, we find that $(t+1)n_3 - 8\eta(t-1) \leq (t+1)n_3 - \sum \alpha_x \leq \sum (t_x + 1) \leq 2tn_2 - 8\eta \leq 8t^2(t+1) - 8\eta t - 8\eta$. So, $(t+1)n_3 \leq 8t^2(t+1) - 16\eta$ or $n_3 \leq 8t^2 - \frac{16\eta}{t+1}$.

(ii) If $\mathcal{S}$ has an order $(2, t)$, then we have equality in (a) and (b). So, we have $n_2 = 4t(t+1) - 4\eta$, $n_3(t+1) = 2tn_2 - 8\eta$ and $n_2 + n_3 = v - 1 - 2(t+1)$, with v the total number of points of $\mathcal{S}$. Property (ii) now follows from an easy calculation.

(iii) If $x \in \Gamma_3(x^*)$, then by (d), $2(t_x + 1) \geq t_x + 1 + \alpha_x \geq t + 1$, proving property (iii). □

10.6.2 Some classes of paths in $\Gamma_3(x^*)$

Definitions.

(a) For every path $\gamma = (x_0, \ldots, x_k)$ in $\Gamma_3(x^*)$, we define $s(\gamma) = x_0$, $e(\gamma) = x_k$, $l(\gamma) = k$, $\epsilon(\gamma) := (-1)^k$ and $\Omega(\gamma) := \bigcup_{0 \leq i \leq k-1} S(x^*, x_i \diamond x_{i+1})$. Since $\mathrm{d}(x^*, x_i) = \mathrm{d}(x^*, x_{i+1}) = 3$, $\mathrm{d}(x^*, x_i \diamond x_{i+1}) = 2$, $|S(x^*, x_i \diamond x_{i+1})| \leq 2$ and $|\Omega(\gamma)| \leq 2 \cdot l(\gamma)$.

(b) Let y^* be a fixed point of $\Gamma_3(x^*)$. Let $\mathcal{V}$ denote the set of all paths $\gamma = (x_0, \ldots, x_k)$ in $\Gamma_3(x^*)$ which satisfy the following properties:

- $s(\gamma) = x_0 = y^*$,
- the k sets $S(x^*, x_i \diamond x_{i+1})$, $i \in \{0, \ldots, k-1\}$, are mutually disjoint,
- $|\Omega(\gamma)| \leq \frac{t}{2}$.

For every path $\gamma \in \mathcal{V}$, we have $l(\gamma) \leq |\Omega(\gamma)| \leq 2 \cdot l(\gamma)$. If $\mathcal{S}$ is a generalized hexagon of order $(2, t)$, then $|\Omega(\gamma)| = l(\gamma)$.

(c) For every path $\gamma = (y^*, x_1, x_2, \ldots, x_k) \in \mathcal{V}$ with $k = l(\gamma) \geq 2$, we define $\tilde{\gamma} := (x_2, \ldots, x_k)$. We put $\mathcal{W} := \{\tilde{\gamma} \,|\, \gamma \in \mathcal{V} \text{ and } l(\gamma) \geq 2\}$.

Lemma 10.25. *Let L be a line through x^*. For every $x_1, x_2 \in L \setminus \{x^*\}$, we define $\epsilon_L(x_1, x_2) := +1$ if $x_1 = x_2$ and $\epsilon_L(x_1, x_2) := -1$ if $x_1 \neq x_2$.*

(a) *If y_1 and y_2 are two different collinear points of $\Gamma_3(x^*)$, then L is contained in $S(x, y_1 \diamond y_2)$ if and only if $\epsilon_L(p_L(y_1), p_L(y_2)) = 1$.*

(b) *If γ is a path of $\mathcal{V}$, then $L \in \Omega(\gamma)$ if and only if $\epsilon_L(p_L(y^*), p_L(e(\gamma))) = -\epsilon(\gamma)$.*

Proof. (a) The point $y_1 \diamond y_2$ is the unique point of y_1y_2 at distance 2 from x^*. Now, $L \subseteq S(x^*, y_1 \diamond y_2) \Leftrightarrow \exists u \in L : \mathrm{d}(u, y_1 \diamond y_2) = 1 \Leftrightarrow \exists u \in L : \mathrm{d}(u, y_1) = \mathrm{d}(u, y_2) = 2 \Leftrightarrow \exists u \in L : u = p_L(y_1) = p_L(y_2) \Leftrightarrow \epsilon_L(p_L(y_1), p_L(y_2)) = 1$.

(b) If $\gamma = (x_0, \ldots, x_k)$, then $\epsilon_L(p_L(y^*), p_L(e(\gamma))) = \Pi_{i=0}^{k-1} \epsilon_L(p_L(x_i), p_L(x_{i+1}))$. If $L \notin \Omega(\gamma)$, then $\epsilon_L(p_L(x_i), p_L(x_{i+1})) = -1$ for all $i \in \{0, \ldots, k-1\}$ and hence $\epsilon_L(p_L(y^*), p_L(e(\gamma))) = (-1)^k = \epsilon(\gamma)$. If $L \in \Omega(\gamma)$, then $\epsilon_L(p_L(x_i), p_L(x_{i+1})) = -1$ for all but one value of i and hence $\epsilon_L(p_L(y^*), p_L(e(\gamma))) = (-1)^{k-1} = -\epsilon(\gamma)$. □

Lemma 10.26. *If γ_1 and γ_2 are two paths in $\mathcal{V}$ with $e(\gamma_1) = e(\gamma_2)$, then $\Omega(\gamma_1) = \Omega(\gamma_2)$ and $\epsilon(\gamma_1) = \epsilon(\gamma_2)$.*

Proof. By (b) of Lemma 10.25, it follows that:

- if $\epsilon(\gamma_1) = \epsilon(\gamma_2)$, then $\Omega(\gamma_1) = \Omega(\gamma_2)$;
- if $\epsilon(\gamma_1) \neq \epsilon(\gamma_2)$, then $\Omega(\gamma_1) = \overline{\Omega(\gamma_2)}$ (i.e. the complement of $\Omega(\gamma_2)$ in the set of lines through x^*).

Now, $|\Omega(\gamma_1)|, |\Omega(\gamma_2)| \leq \frac{t}{2}$ and so the situation $\Omega(\gamma_1) = \overline{\Omega(\gamma_2)}$ cannot occur. □

Definitions. Put $E := \{e(\gamma) | \gamma \in \mathcal{V}\}$. For every point x of E, we define $\Omega(x) := \Omega(\gamma)$ and $\epsilon(x) := \epsilon(\gamma)$ where γ is any path of $\mathcal{V}$ with $e(\gamma) = x$. For all $a, l \in \mathbb{N}$, we define $E(a, l)$ as the set of all points $x \in E$ for which there exists a path $\gamma \in \mathcal{V}$ satisfying $e(\gamma) = x$, $l(\gamma) = l$ and $|\Omega(\gamma)| = a$. For every nonempty subset A of $\mathbb{N}$ and every $l \in \mathbb{N}$, we define $E(A, l) := \bigcup_{a \in A} A(a, l)$ and $E_l := E(\mathbb{N}, l)$.

Lemma 10.27. *Let $l, l', a, a' \in \mathbb{N}$.*

(a) *$E(a, l) = \emptyset$ if $a > \frac{t}{2}$ or if $a \notin \{l, l+1, \ldots, 2l\}$. Hence, $E_l := E(\{l, l+1, \ldots, 2l\}, l)$.*

(b) *If $E(a, l) \cap E(a', l') \neq \emptyset$, then $a = a'$ and $l - l'$ is even.*

(c) *If $E_l \cap E_{l'} \neq \emptyset$, then $l' - l$ is even, $l' \leq 2l$ and $l \leq 2l'$.*

Proof. (a) This follows from the fact that $|\Omega(\gamma)| \leq \frac{t}{2}$ and $l(\gamma) \leq |\Omega(\gamma)| \leq 2 \cdot l(\gamma)$ for every $\gamma \in \mathcal{V}$.

(b) If $x \in E(a, l) \cap E(a', l')$, then $a = |\Omega(x)| = a'$ and $(-1)^l = \epsilon(x) = (-1)^{l'}$.

(c) If $E_l \cap E_{l'} \neq \emptyset$, then $E(a,l) \cap E(a',l') \neq \emptyset$ for a certain $a \in \{l, \ldots, 2l\}$ and a certain $a' \in \{l', \ldots, 2l'\}$. But then $l - l'$ is even, $l' \leq a' = a \leq 2l$ and $l \leq a = a' \leq 2l'$. $\square$

Lemma 10.28. *Let x_1 and x_2 be two different collinear points of E.*

(i) *If $\epsilon(x_1) = -\epsilon(x_2)$, then $\Omega(x_2) = \Omega(x_1) \,\Delta\, S(x^*, x_1 \diamond x_2)$ (i.e. $\Omega(x_2)$ is the symmetric difference of $\Omega(x_1)$ and $S(x^*, x_1 \diamond x_2)$).*

(ii) *If $\epsilon(x_1) = \epsilon(x_2)$, then $\Omega(x_2) = \overline{\Omega(x_1) \,\Delta\, S(x^*, x_1 \diamond x_2)}$. Moreover, $|\Omega(x_1)| = |\Omega(x_2)| = n$ if $t = 2n+1$ and $|\Omega(x_1)|, |\Omega(x_2)| \in \{n-1, n\}$ if $t = 2n$.*

Proof. Let L be an arbitrary line through x^*.

(i) If $\epsilon(x_1) = -\epsilon(x_2)$, then by Lemma 10.25, we have

$$\begin{aligned} L \in \Omega(x_2) &\Leftrightarrow \epsilon_L(p_L(y^*), p_L(x_2)) = -\epsilon(x_2) \\ &\Leftrightarrow (\epsilon_L(p_L(y^*), p_L(x_2)) = -\epsilon(x_2) \text{ and } p_L(x_2) = p_L(x_1)) \\ &\qquad \text{or } (\epsilon_L(p_L(y^*), p_L(x_2)) = -\epsilon(x_2) \text{ and } p_L(x_2) \neq p_L(x_1)) \\ &\Leftrightarrow (\epsilon_L(p_L(y^*), p_L(x_1)) = \epsilon(x_1) \text{ and } p_L(x_2) = p_L(x_1)) \\ &\qquad \text{or } (\epsilon_L(p_L(y^*), p_L(x_1)) = -\epsilon(x_1) \text{ and } p_L(x_2) \neq p_L(x_1)) \\ &\Leftrightarrow (L \notin \Omega(x_1) \text{ and } L \in S(x^*, x_1 \diamond x_2)) \\ &\qquad \text{or } (L \in \Omega(x_1) \text{ and } L \notin S(x^*, x_1 \diamond x_2)) \\ &\Leftrightarrow L \in \Omega(x_1) \,\Delta\, S(x^*, x_1 \diamond x_2). \end{aligned}$$

Hence, $\Omega(x_2) = \Omega(x_1) \,\Delta\, S(x^*, x_1 \diamond x_2)$.

(ii) If $\epsilon(x_1) = \epsilon(x_2)$, then we have

$$\begin{aligned} L \in \Omega(x_2) &\Leftrightarrow \epsilon_L(p_L(y^*), p_L(x_2)) = -\epsilon(x_2) \\ &\Leftrightarrow (\epsilon_L(p_L(y^*), p_L(x_2)) = -\epsilon(x_2) \text{ and } p_L(x_2) = p_L(x_1)) \\ &\qquad \text{or } (\epsilon_L(p_L(y^*), p_L(x_2)) = -\epsilon(x_2) \text{ and } p_L(x_2) \neq p_L(x_1)) \\ &\Leftrightarrow (\epsilon_L(p_L(y^*), p_L(x_1)) = -\epsilon(x_1) \text{ and } p_L(x_2) = p_L(x_1)) \\ &\qquad \text{or } (\epsilon_L(p_L(y^*), p_L(x_1)) = \epsilon(x_1) \text{ and } p_L(x_2) \neq p_L(x_1)) \\ &\Leftrightarrow (L \in \Omega(x_1) \text{ and } L \in S(x^*, x_1 \diamond x_2)) \\ &\qquad \text{or } (L \notin \Omega(x_1) \text{ and } L \notin S(x^*, x_1 \diamond x_2)) \\ &\Leftrightarrow L \in \overline{\Omega(x_1) \,\Delta\, S(x^*, x_1 \diamond x_2)}. \end{aligned}$$

Hence, $\Omega(x_2) = \overline{\Omega(x_1) \,\Delta\, S(x^*, x_1 \diamond x_2)}$. Now, $|\Omega(x_1)| \leq \frac{t}{2}$, $|\Omega(x_2)| \leq \frac{t}{2}$, $|S(x^*, x_1 \diamond x_2)| \leq 2$ and so $|\Omega(x_1)| = |\Omega(x_2)| = n$ if $t = 2n+1$ and $|\Omega(x_1)|, |\Omega(x_2)| \in \{n-1, n\}$ if $t = 2n$. $\square$

Definitions. For every $a, l \in \mathbb{N}$, we define $N(a,l) := |E(a,l)|$ and $N_l := |E_l|$. Obviously, $N(0,0) = 1$ and $N(1,1) + N(2,1) = t_{y^*} + 1$ if $t \geq 4$. In the following section we will derive inequalities involving the values $N(a,l)$ and N_l.

10.6.3 Some inequalities involving the values $N(a,l)$ and N_l

Definitions.

(a) Suppose that x is a point of E. Let $a = |\Omega(x)|$, let $K_1, \dots, K_a$ denote the a elements of $\Omega(x)$ and let u_i, $i \in \{1, \dots, a\}$, denote the unique point of K_i at distance 2 from x. Then we define V_x as the set of lines through x which are not contained in $\bigcup_{i=1}^{a} S(x, u_i)$.

(b) For every $l \in \mathbb{N}$ and every $a \in \{l, l+1, \dots, 2l\}$, we will now define a number $g(t,l,a)$. If $\mathcal{S}$ is a generalized hexagon of order $(2,t)$, then we define $g(t,l,a) := t+1-l$. If $\mathcal{S}$ has order $(2,t)$ but is not a generalized hexagon, then we define $g(t,l,a) := t+2-2a$ if $a = 2l \neq 0$ and $g(t,l,a) := t+1-2a$ otherwise. If $\mathcal{S}$ has no order, then we define $g(t,l,a) := \frac{t+3-3a}{2}$ if $a = 2l \neq 0$ and $g(t,l,a) := \frac{t+1-3a}{2}$ otherwise.

Lemma 10.29. *If $x \in E(a,l)$, then $|V_x| \geq g(t,l,a)$.*

Proof. Let $u_1, \dots, u_a$ be as above. Suppose that $|\Gamma_1(u_i) \cap \Gamma_1(x)| = 1$ for precisely a_1 values of $i \in \{1, \dots, a\}$. Then

$$|V_x| \geq t_x + 1 - a_1 - 2(a - a_1).$$

If $a = 2l \neq 0$, then we can improve this lower bound. Then there exists a path $(y^*, \dots, x', x) \in \mathcal{V}$ with $x' \in E(a-2, l-1)$. Without loss of generality, we may suppose that $\Omega(x) \setminus \Omega(x') = \{u_{a-1}, u_a\}$. Since $xx' \in S(x, u_a) \cap S(x, u_{a-1})$, we then have that $|V_x| \geq t_x + 2 - a_1 - 2(a - a_1)$.

If $\mathcal{S}$ is a generalized hexagon of order $(2,t)$, then $a = l = a_1$ and hence $|V_x| \geq t+1-l = g(t,l,a)$. If $\mathcal{S}$ has order $(2,t)$, then $t_x+1-a_1-2(a-a_1) \geq t+1-2a$. In the general case, we count pairs (y,z) with $x^* \sim y \sim z \sim x$ and we obtain $a_1 \cdot 1 + (a-a_1) \cdot 2 + (t+1-a) \cdot 1 \leq (t_x+1) \cdot 2$. Hence, $t_x + 1 - a_1 - 2(a-a_1) \geq \frac{t+1}{2} - \frac{3}{2}a + \frac{a_1}{2} \geq \frac{t+1-3a}{2}$. The lemma now immediately follows. □

Lemma 10.30. *Let $x \in E(a,l)$ with $a \leq \frac{t}{2} - 2$ if $\mathcal{S}$ is not a generalized hexagon and $a = l = \frac{t}{2} - 1$ if $\mathcal{S}$ is a generalized hexagon of order $(2,t)$. Let $K \in V_x$ and let x_K denote the unique point of $K \setminus \{x\}$ at distance 3 from x^*. Then $x_K \in E(a+1, l+1) \cup E(a+2, l+1)$ if $\mathcal{S}$ is not a generalized hexagon and $x_K \in E(l+1, l+1)$ if $\mathcal{S}$ is a generalized hexagon. Moreover, $S(x^*, x \diamond x_K)$ is disjoint with $\Omega(x)$ and $\Omega(x_K) = \Omega(x) \cup S(x^*, x \diamond x_K)$.*

Proof. Suppose that L is a common line of $S(x^*, x \diamond x_K)$ and $\Omega(x)$. The unique point of L collinear with $x \diamond x_K$ has distance 2 from x and so it coincides with one of the points u_i, $i \in \{1, \dots, a\}$, which we defined above. But then the line K would be contained in $S(x, u_i)$, a contradiction. So, $S(x^*, x \diamond x_K) \cap \Omega(x) = \emptyset$. Let γ denote a path of $\mathcal{V}$ with $l(\gamma) = l$ and $e(\gamma) = x$. If we add x_K to this path, then we obtain a path γ' with $l(\gamma') = l+1$ and $e(\gamma') = x_K$. Since $S(x^*, x \diamond x_K) \cap \Omega(x) = \emptyset$ and $|\Omega(\gamma')| = |\Omega(x)| + |S(x^*, x \diamond x_K)| = a + |S(x^*, x \diamond x_K)| \leq \frac{t}{2}$, γ' belongs to $\mathcal{V}$. So, $\Omega(x_K) = \Omega(\gamma') = \Omega(x) \cup S(x^*, x \diamond x_K)$, $x_K \in E(l+1, l+1)$ if $\mathcal{S}$ is a generalized hexagon and $x_K \in E(a+1, l+1) \cup E(a+2, l+1)$ otherwise. □

Definitions.

(a) For every $l \in \mathbb{N}$, put $g(t,l) := \min\{g(t,l,a) \,|\, l \leq a \leq 2l\} = g(t,l,2l)$. So, $g(t,l) = t+1-l$ if $\mathcal{S}$ is a generalized hexagon of order $(2,t)$, $g(t,l) = t+2-4l$ if $\mathcal{S}$ has order $(2,t)$ but is not a generalized hexagon, and $g(t,l) = \frac{t+3-6l}{2}$ if $\mathcal{S}$ has no order.

(b) If $\mathcal{S}$ is a generalized hexagon, then we define $l^* := \lfloor \frac{t}{2} \rfloor$. If $\mathcal{S}$ has an order $(2,t)$, but is not a generalized hexagon, then we define $l^* := \lfloor \frac{t}{4} \rfloor$. If $\mathcal{S}$ has no order, then we define $l^* := \min\{\lfloor \frac{t}{4} \rfloor, \lfloor \frac{t+8}{6} \rfloor\}$.

Lemma 10.31. (i) *For every $l \in \{0,\ldots,l^*\}$, $E_l \neq \emptyset$.*

(ii) *Suppose that $l^* \geq 2$. Then for every point x in E_2 and every $l \in \{2,\ldots,l^*\}$, there are at least $\prod_{i=2}^{l-1} g(t,i)$ paths $\gamma \in \mathcal{W}$ such that $s(\gamma) = x$ and $e(\gamma) \in E_l$.*

Proof. Let γ denote a path of $\mathcal{V}$ with $l(\gamma) \leq l^* - 1$. Put $x = e(\gamma)$ and $a = |\Omega(x)|$. If $\mathcal{S}$ is a generalized hexagon of order $(2,t)$, then $a = l(\gamma) \leq l^* - 1 \leq \frac{t}{2} - 1$. If $\mathcal{S}$ is not a generalized hexagon, then $a \leq 2 \cdot l(\gamma) \leq 2l^* - 2 \leq \frac{t}{2} - 2$. Since $g(t,l(\gamma),a) \geq g(t,l(\gamma)) \geq g(t,l^*-1) > 0$, there exists a line $K \in V_x$ by Lemma 10.29. By Lemma 10.30, the unique point x_K in $K \setminus \{x\}$ at distance 3 from x^* extends γ to a path $\gamma' \in \mathcal{V}$ whose length is equal to $l(\gamma)+1$. Since $g(t,l(\gamma),a) \geq g(t,l(\gamma))$, we can extend γ in at least $g(t,l(\gamma))$ ways. The lemma now easily follows. □

Definitions. For every point x of $E(a,l)$, $l \geq 2$, let $\delta_1(x)$, respectively $\delta_2(x)$, denote the number of points of $E(a-1,l-1)$, respectively $E(a-2,l-2)$, collinear with x.

Lemma 10.32. (i) *Suppose that $\mathcal{S}$ is a generalized hexagon. If $x \in E(l,l)$, $l \geq 2$, then $\delta_1(x) \leq l$ and $\delta_2(x) = 0$. If $x \in E(2,2)$, then $\delta_1(x) = 1$.*

(ii) *Suppose that $\mathcal{S}$ is not a generalized hexagon. If $x \in E(a,l)$, $l \geq 2$, then $\delta_1(x) + 2\delta_2(x) \leq 2a$. As a consequence, $\delta_1(x) \leq 2a$ and $\delta_2(x) \leq a$. If $x \in E(a,2)$, then $\delta_1(x) \leq 2$ and $\delta_2(x) \leq 2$.*

Proof. Put $\Omega(x) = \{K_1,\ldots,K_a\}$ (with $a = l$ in the case of generalized hexagons) and let u_i, $i \in \{1,\ldots,a\}$, denote the unique point of K_i at distance 2 from x.

Suppose that x' is a point of $E(a-1,l-1) \cup E(a-2,l-1)$ collinear with x. Then $\epsilon(x) = -\epsilon(x')$ and so $\Omega(x) = \Omega(x') \cup S(x^*, x \diamond x')$ by Lemma 10.28. Since $\Omega(x) > \Omega(x')$ and $|S(x^*, x \diamond x')| \leq 2$, we have $\Omega(x') \cap S(x^*, x \diamond x') = \emptyset$ and $S(x^*, x \diamond x') \subseteq \{K_1,\ldots,K_a\}$. If $K_i \in S(x^*, x \diamond x')$, then the line xx' is contained in $S(x,u_i)$.

Now, counting pairs (K,u) satisfying (i) $u \in \{u_1,\ldots,u_a\}$, (ii) K is a line through x containing a point of $E(a-1,l-1) \cup E(a-2,l-1)$ and (iii) $K \subseteq S(x,u)$ gives $2a \geq \delta_1(x) + 2\delta_2(x)$. If $\mathcal{S}$ is a generalized hexagon, then we have that $a \geq \delta_1(x)$ and $\delta_2(x) = 0$.

If $l = 2$, then any point of $E(a-1,1) \cup E(a-2,1)$ collinear with x is a common neighbour of x and y^*. Since there are at most two such common neighbours, we have $\delta_1(x) \leq 2$ and $\delta_2(x) \leq 2$. In the case of generalized hexagons, we have $\delta_1(x) = 1$. □

Lemma 10.33. *If $t \geq 2(a+2)$, then*

- $N(a,1) \cdot g(t,1,a) \leq 2(N(a+1,2) + N(a+2,2))$,
- $N(a,l) \cdot g(t,l,a) \leq N(a+1,l+1) \cdot (2a+2) + N(a+2,l+1) \cdot (a+2)$ *for every* $l \geq 2$.

If $\mathcal{S}$ is a generalized hexagon of order $(2,t)$, then

- $N(1,1) \cdot g(t,1) \leq N(2,2)$ *if* $t \geq 4$,
- $N(l,l) \cdot g(t,l) \leq N(l+1,l+1) \cdot (l+1)$ *if* $l \geq 2$ *and* $t \geq 2l+2$.

Proof. This follows from Lemmas 10.29, 10.30 and 10.32. □

Definitions.

(a) Let $a, l \in \mathbb{N}$ with $l \geq 2$. If $E(a,l) \neq \emptyset$, then we put $M(a,l)$ equal to $\max_{x \in E(a,l)} \lambda_x$, where λ_x denotes the number of paths $\gamma \in \mathcal{W}$ with $e(\gamma) = x$. If $E(a,l) = \emptyset$, then we put $M(a,l)$ equal to 0. If $\mathcal{S}$ is a generalized hexagon of order $(2,t)$, then $M(a,l) = 0$ if $a \neq l$.

(b) If $\mathcal{S}$ is a generalized hexagon of order $(2,t)$, then we define $\Delta_l := \frac{l!}{2}$ for every $l \geq 3$. If $\mathcal{S}$ is not a generalized hexagon, then we define

$$\Delta_l := \max\left\{ \frac{(2a_l)(2a_{l-1})\cdots(2a_3)}{2^{(a_l-a_2)-(l-2)}} \;\middle|\; \begin{array}{l} a_2 = \{2,3,4\} \text{ and} \\ a_i - a_{i-1} \in \{1,2\} \text{ if } 3 \leq i \leq l \end{array} \right\}$$

for every $l \geq 3$. If $\mathcal{S}$ is not a generalized hexagon, then one calculates that $\Delta_3 = 10$, $\Delta_4 = 120$, $\Delta_5 = 1680$, $\Delta_6 = 26880$, $\Delta_7 = 483840$, $\Delta_8 := 10644480$, etc.

(c) We define

- $f_0(t) := 1$;
- $f_1(t) := t+1$ if $\mathcal{S}$ has order $(2,t)$ and $f_1(t) := \frac{t+1}{2}$ otherwise;
- $f_2(t) := (t+1)t$ if $\mathcal{S}$ is a generalized hexagon of order $(2,t)$, $f_2(t) := \frac{(t+1)(t-2)}{2}$ if $\mathcal{S}$ is a near hexagon of order $(2,t)$ and $f_2(t) := \frac{(t+1)(t-3)}{8}$ if $\mathcal{S}$ has no order;
- for every $i \geq 3$, we define $f_i(t) := \frac{f_2(t) \cdot \prod_{j=2}^{i-1} g(t,j)}{\Delta_i}$.

Lemma 10.34. (i) *If $E(a,2) = \emptyset$, then $M(a,2) = 0$. If $E(a,2) \neq \emptyset$, then $M(a,2) = 1$. If $l \geq 3$ and $a \geq 2$, then $M(a,l) \leq (2a) \max\{\frac{M(a-2,l-1)}{2}, M(a-1,l-1)\}$.*

(ii) *Let $\mathcal{S}$ be a generalized hexagon of order $(2,t)$. If $l^* \geq 2$, then $M(2,2) = 1$ and $M(l,l) \leq l \cdot M(l-1,l-1)$ for every $l \geq 3$.*

(iii) $M(a,l) \le \Delta_l$ *for every* $l \ge 3$.

Proof. (i) We still need to prove that $M(a,l) \le (2a)\max\{\frac{M(a-2,l-1)}{2}, M(a-1,l-1)\}$ for every $l \ge 3$ and every $a \ge 2$. Obviously, this holds if $E(a,l) = \emptyset$. Suppose therefore that $E(a,l) \ne \emptyset$. For every point $x \in E(a,l)$,

$$\begin{aligned}\lambda_x &\le \delta_1(x) \cdot M(a-1,l-1) + \delta_2(x) \cdot M(a-2,l-1) \\ &= \delta_1(x) \cdot M(a-1,l-1) + 2\delta_2(x) \cdot \tfrac{M(a-2,l-1)}{2} \\ &\le (\delta_1(x) + 2\delta_2(x)) \cdot \max\{\tfrac{M(a-2,l-1)}{2}, M(a-1,l-1)\} \\ &\le (2a)\max\{\tfrac{M(a-2,l-1)}{2}, M(a-1,l-1)\}.\end{aligned}$$

(The latter inequality follows from Lemma 10.32.) So, $M(a,l) = \max\{\lambda_x \mid x \in E(a,l)\} \le (2a)\max\{\frac{M(a-2,l-1)}{2}, M(a-1,l-1)\}$.

(ii) If $x \in E(l,l)$ with $l \ge 3$, then $\lambda_x \le \delta_1(x) \cdot M(l-1,l-1) \le l \cdot M(l-1,l-1)$.

(iii) This follows from (i), (ii) and the definition of Δ_l. □

Lemma 10.35. *For every* $l \in \{0,\dots,l^*\}$*, we have* $N_l \ge f_l(t)$.

Proof. Clearly, $N_0 = f_0(t) = 1$. If $l^* \ge 1$, then $N_1 = t_{y^*} + 1$ which is at least equal to $f_1(t)$ by Lemma 10.24. If $l^* \ge 2$, then $N_2 \ge \frac{N_1 \cdot g(t,1)}{\kappa}$, where $\kappa = 1$ if $\mathcal{S}$ is a generalized hexagon of order $(2,t)$ and $\kappa = 2$ otherwise. So, $N_2 \ge \frac{f_1(t)\cdot g(t,1)}{\kappa} = f_2(t)$. If $l^* \ge 3$, then $N_l \ge f_l(t)$ for every $l \in \{3,\dots,l^*\}$ by Lemmas 10.31, 10.34 and the definition of $f_l(t)$. □

10.6.4 The proof of Theorem 10.10

We will use the following lemma to derive an upper bound for t.

Lemma 10.36. *If the sets* $E_{l_1},\dots,E_{l_k}$ ($k \ge 1$, $0 \le l_1 < l_2 < \cdots < l_k \le l^*$ *and* $l_k \ge 3$) *are mutually disjoint, then* $t \le \lfloor R \rfloor$*, where* R *is the greatest real root of the polynomial* $f(X) := f_{l_1}(X) + f_{l_2}(X) + \cdots + f_{l_k}(X) - 8X^2 \in \mathbb{Q}[X]$.

Proof. By Lemmas 10.24 and 10.35, $f_{l_1}(t) + f_{l_2}(t) + \cdots + f_{l_k}(t) \le |E_{l_1}| + \cdots + |E_{l_k}| = |E_{l_1} \cup \cdots \cup E_{l_k}| \le |\Gamma_3(x^*)| \le 8t^2$ or $f(t) \le 0$. Now, $\deg(f) = l_k \ge 3$ and the coefficient of X^{l_k} in $f(X)$ is strictly positive. So, $\lim_{T\to+\infty} f(T) = +\infty$ and $t \le R$. □

(A) The case of generalized hexagons

Suppose that $t \ge 8$. Then $l^* = \lfloor \frac{t}{2} \rfloor \ge 4$ and $f_0(t) = 1$, $f_1(t) = t+1$, $f_2(t) = (t+1)t$, $f_3(t) = \frac{(t+1)t(t-1)}{3}$, $f_4(t) = \frac{(t+1)t(t-1)(t-2)}{12}$. Since $E_l = E(l,l)$ for every $l \in \{0,\dots,4\}$, the sets E_0, E_1, E_2, E_3, E_4 are mutually disjoint by Lemma 10.27. So, we can apply Lemma 10.36 and we find that $t \le \lfloor R \rfloor = \lfloor 8.10\cdots \rfloor = 8$.

As a consequence, any finite generalized hexagon of order $(2,t)$ satisfies $t \le 8$. This is precisely the bound predicted by the Haemers-Roos inequality, see Theorem 1.28.

(B) The case of near hexagons with an order

We will prove that $t \leq 33$. Suppose that $t \geq 30$. Then $l^* = \lfloor \frac{t}{4} \rfloor \geq 7$. By Lemma 10.27, the sets E_0, E_1, E_2, E_3, E_6 and E_7 are disjoint. One calculates that $f_0(t) = 1$, $f_1(t) = t+1$, $f_2(t) = \frac{(t+1)(t-2)}{2}$, $f_3(t) = \frac{(t+1)(t-2)(t-6)}{20}$, $f_6(t) = \frac{(t+1)(t-2)(t-6)(t-10)(t-14)(t-18)}{53760}$, $f_7(t) = \frac{(t+1)(t-2)(t-6)(t-10)(t-14)(t-18)(t-22)}{967680}$. By Lemma 10.36, $t \leq \lfloor R \rfloor = \lfloor 33.77 \cdots \rfloor = 33$.

The maximal known value for t is 11, attained for the near hexagon $\mathbb{E}_1$.

(C) The case of near hexagons without an order

We will prove that $t \leq 76$. Suppose that $t \geq 50$. Then $l^* = \min\{\lfloor \frac{t}{4} \rfloor, \lfloor \frac{t+8}{6} \rfloor\} \geq 9$. By Lemma 10.27, the sets E_0, E_1, E_2, E_3, E_7 and E_8 are disjoint. One calculates that $f_0(t) = 1$, $f_1(t) = \frac{t+1}{2}$, $f_2(t) = \frac{(t+1)(t-3)}{8}$, $f_3(t) = \frac{(t+1)(t-3)(t-9)}{160}$, $f_7(t) = \frac{(t+1)(t-3)(t-9)(t-15)(t-21)(t-27)(t-33)}{123863040}$, $f_8(t) = \frac{1}{5449973760}(t+1)(t-3)(t-9)(t-15)(t-21)(t-27)(t-33)(t-39)$. By Lemma 10.36, $t \leq \lfloor R \rfloor = \lfloor 76.80 \cdots \rfloor = 76$.

We have considered other possibilities for the sets $E_{l_1}, \ldots, E_{l_k}$, but we have not found better upper bounds than the ones given above.

10.7 Slim near hexagons with an order

In this section $\mathcal{S} = (\mathcal{P}, \mathcal{L}, \mathrm{I})$ denotes a near hexagon of order $(2,t)$. Since $\mathcal{S}$ has two intersecting lines, we necessarily have $t+1 \geq 2$. Let v denote the total number of points of $\mathcal{S}$. Define $n_0 := 1$, $n_1 := 2(t+1)$, $n_2 := \frac{1}{3}v + 2t - 1$ and $n_3 := \frac{2}{3}v - 4t - 2$. By Theorem 1.2, we have

Lemma 10.37. *For every $i \in \{0,1,2,3\}$, $|\Gamma_i(x)| = n_i$.*

Lemma 10.38. *No point of $\mathcal{S}$ is special.*

Proof. Suppose that $\mathcal{S}$ has a special point x. Let y be a point at distance 2 from x and put $Q := \mathcal{C}(x,y)$. Every line through y contains a unique point nearest to and hence collinear with x. So, $t_Q + 1 = t_y + 1 = t + 1 \geq 2$ and Q is a nondegenerate quad. Now, since $t = t_Q$, no point u of Q is collinear with a point outside Q, which is impossible. □

Lemma 10.39. *$\mathcal{S}$ is dense if and only if every local space is linear.*

Proof. Suppose that the local space at a point x is not linear. Then there exist two lines K_1 and K_2 incident with x such that $\mathcal{C}(K,L)$ is a degenerate quad. If x_i, $i \in \{1,2\}$, is a point of $K_i \setminus K_{3-i}$, then $K_1 \cap K_2$ is the unique common neighbour of x_1 and x_2. So, $\mathcal{S}$ is not dense.

Conversely, suppose that $\mathcal{S}$ is not dense. Then there exist two points x_1 and x_2 at distance 2 from each other which have exactly 1 common neighbour z. Since $\mathcal{C}(zx_1, zx_2)$ is a degenerate quad, the local space at z is not linear. □

Theorem 10.40. *If $\mathcal{S}$ contains a $Q(5,2)$-quad, then it is dense and hence isomorphic to either $Q(5,2) \times \mathbb{L}_3$, $Q(5,2) \otimes Q(5,2)$, $\mathbb{G}_3$, $\mathbb{E}_3$ or $DH(5,4)$.*

Proof. Since every $Q(5,2)$-quad is big, we have

$$v = 27[1 + 2(t-4)] = 54t - 189. \tag{10.1}$$

By Lemma 10.37, we then have

$$n_2 = 20t - 64, \tag{10.2}$$

$$n_3 = 32t - 128. \tag{10.3}$$

Let x denote an arbitrary point of $\mathcal{S}$. Let x be contained in α_x grid-quads, β_x $W(2)$-quads and γ_x $Q(5,2)$-quads. The number of points $y \in \Gamma_2(x)$ for which $\mathcal{C}(x,y)$ is a grid-quad, a $W(2)$-quad, a $Q(5,2)$-quad or a degenerate quad is respectively equal to $4\alpha_x$, $8\beta_x$, $16\gamma_x$ or $n_2 - 4\alpha_x - 8\beta_x - 16\gamma_x$. Counting pairs (z,y) with $y \in \Gamma_2(x)$ and $x \sim z \sim y$ yields $4t(t+1) = (n_2 - 4\alpha_x - 8\beta_x - 16\gamma_x) + (4\alpha_x)\cdot 2 + (8\beta_x)\cdot 3 + (16\gamma_x)\cdot 5$. Together with (10.2) this implies that:

$$\alpha_x + 4\beta_x + 16\gamma_x = t^2 - 4t + 16. \tag{10.4}$$

Counting pairs $\{K, L\}$ of lines through x, we obtain:

$$\alpha_x + 3\beta_x + 10\gamma_x \leq \frac{t(t+1)}{2}, \tag{10.5}$$

with equality if and only if the local space at x is linear.

(a) First, suppose that $\gamma_x = 0$. From (10.4) and (10.5), it then follows that $(t-7)^2 \leq 1$. So, $t \in \{6,7,8\}$. If $t \in \{6,8\}$, then we must have equality in (10.5) which implies that $\mathcal{S}_x$ is linear. If $t = 7$, then $\alpha_x = 1$ and $\beta_x = 9$ by (10.4) and (10.5). Again we have equality in (10.5) and so $\mathcal{S}_x$ is linear.

(b) Next, suppose that $\gamma_x \geq 1$. From equations (10.4) and (10.5), we obtain

$$\beta_x \geq \frac{t^2 - 9t + 32}{2} - 6\gamma_x. \tag{10.6}$$

We now derive an upper bound for β_x. Let Q denote an arbitrary $Q(5,2)$-quad through x. By Lemma 10.3, every $W(2)$-quad or $Q(5,2)$-quad through x different from Q intersects Q in a line. Now, counting pairs (K,L) of lines through x not contained in Q, we find that $2\beta_x + 12(\gamma_x - 1) \leq (t-4)(t-5)$, or that

$$\beta_x \leq \frac{t^2 - 9t + 32}{2} - 6\gamma_x. \tag{10.7}$$

From inequalities (10.6) and (10.7), we see that we must have equality in (10.5). So, $\mathcal{S}_x$ is linear.

Since every local space is linear, $\mathcal{S}$ is dense. The theorem now follows from Theorem 7.1. □

Theorem 10.41. *If $\mathcal{S}$ contains a big $W(2)$-quad, then it is dense and hence isomorphic to either $W(2) \times \mathbb{L}_3$, $DQ(6,2)$ or $\mathbb{H}_3$.*

Proof. Since none of the near hexagons $Q(5,2) \times \mathbb{L}_3$, $Q(5,2) \otimes Q(5,2)$, $\mathbb{G}_3$, $\mathbb{E}_3$, $DH(5,4)$ contains a big $W(2)$-quad, $\mathcal{S}$ cannot contain a $Q(5,2)$-quad by Theorem 10.40. Since there is a big $W(2)$-quad, we have

$$v = 15[1 + 2(t-2)] = 30t - 45. \tag{10.8}$$

By Lemma 10.37, we then have

$$n_2 = 12t - 16, \tag{10.9}$$

$$n_3 = 16t - 32. \tag{10.10}$$

Let x denote an arbitrary point of $\mathcal{S}$. Let the point x be contained in α_x grid-quads and β_x $W(2)$-quads. The number of points $y \in \Gamma_2(x)$ for which $\mathcal{C}(x,y)$ is a grid-quad, a $W(2)$-quad or a degenerate quad is respectively equal to $4\alpha_x$, $8\beta_x$ or $n_2 - 4\alpha_x - 8\beta_x$. Counting pairs (z,y) with $y \in \Gamma_2(x)$ and $x \sim z \sim y$ yields $4t(t+1) = (n_2 - 4\alpha_x - 8\beta_x) + (4\alpha_x)\cdot 2 + (8\beta_x)\cdot 3$. Together with (10.9) this implies that:

$$\alpha_x + 4\beta_x = t^2 - 2t + 4. \tag{10.11}$$

Counting pairs $\{K, L\}$ of lines through x, we obtain

$$\alpha_x + 3\beta_x \leq \frac{t(t+1)}{2}. \tag{10.12}$$

From (10.11) and (10.12), it then follows that

$$\beta_x \geq \frac{t^2 - 5t + 8}{2}. \tag{10.13}$$

So, $\beta_x \neq 0$ and there exists a $W(2)$-quad Q through x. Since $|\Gamma_0(Q)| + |\Gamma_1(Q)| = 15[1 + 2(t-2)] = v$, Q is big. By Lemma 10.3, every other $W(2)$-quad through x intersects Q in a line. Now, counting pairs (K, L) of lines through x not contained in Q, we find that $2(\beta_x - 1) \leq (t-2)(t-3)$, i.e.

$$\beta_x \leq \frac{t^2 - 5t + 8}{2}. \tag{10.14}$$

From inequalities (10.13) and (10.14), we see that we must have equality in (10.12). So, $\mathcal{S}_x$ is linear. Since every local space is linear, $\mathcal{S}$ is dense. The theorem now follows from Theorem 7.1. □

Theorem 10.42. *If $\mathcal{S}$ contains a big grid-quad, then $\mathcal{S}$ is isomorphic to either $\mathbb{L}_3 \times \mathbb{L}_3 \times \mathbb{L}_3$, $W(2) \times \mathbb{L}_3$ or $Q(5,2) \times \mathbb{L}_3$.*

Proof. Let Q be a big grid-quad of $\mathcal{S}$. We have

$$v = 9[1 + 2(t-1)] = 18t - 9.$$

Suppose R is a $W(2)$-quad of $\mathcal{S}$. If R is disjoint from Q, then $\pi_Q(R)$ is a subquadrangle of Q isomorphic to $W(2)$, a contradiction. Hence, R intersects Q in a line and $t+1 \geq 4$. Since

$$18t - 9 = v \geq |R| + |\Gamma_1(R)| = 30t - 45,$$

$t+1 \leq 4$. Hence, $t+1 = 4$ and R is big. By Theorem 10.41, $\mathcal{S} \cong W(2) \times \mathbb{L}_3$.

Suppose R is a $Q(5,2)$-quad of $\mathcal{S}$. If R is disjoint from Q, then $\pi_Q(R)$ is a subquadrangle of Q isomorphic to $Q(5,2)$, a contradiction. Hence, R intersects Q in a line and $t+1 \geq 6$. Since

$$18t - 9 = v \geq |R| + |\Gamma_1(R)| = 54t - 189,$$

$t+1 \leq 6$. Hence, $t+1 = 6$ and R is big. By Theorem 10.40, $\mathcal{S} \cong Q(5,2) \times \mathbb{L}_3$.

Suppose now that $\mathcal{S}$ does not contain $W(2)$-quads nor $Q(5,2)$-quads. Then $t+1 \leq 3$ by Lemma 10.2. Hence, $t+1 = 3$ and $v = 27$. By Lemma 10.37, $n_2 = 12$. Now, let x denote an arbitrary point of $\mathcal{S}$. Suppose x is contained in α_x grid-quads. Counting pairs of points (z, y) with $y \in \Gamma_2(x)$ and $x \sim z \sim y$ yields $4t(t+1) = (n_2 - 4\alpha_x) + 4\alpha_x \cdot 2$. Hence $\alpha_x = 3$. So, the local space at x is a linear space. By Lemma 10.39, $\mathcal{S}$ is dense. By Theorem 7.1, $\mathcal{S} \cong \mathbb{L}_3 \times \mathbb{L}_3$. $\square$

Appendix A

Dense near polygons of order $(3, t)$

In this appendix, we will briefly discuss the known results regarding the classification of dense near $2d$-gons of order $(3, t)$. In the following sections, we will discuss the generalized quadrangles, the near hexagons, the near octagons and the near $2d$-gons with a nice chain of convex subpolygons. We refer to [31], [47], [50], [51], [52] and [82] for more details.

A.1 Generalized quadrangles of order $(3, t)$

Definition. Let ζ denote a symplectic polarity of $PG(3, q)$. The totally isotropic points and lines of $PG(3, q)$ define a generalized quadrangle $W(q)$. For every point x of $PG(3, q)$, we can define the incidence structure $\mathcal{P}(W(q), x)$ with points the points of $PG(3, q) \setminus x^\zeta$ and with lines the lines of $PG(3, q)$ not contained in x^ζ which either contain x or are totally isotropic (natural incidence). $\mathcal{P}(W(q), x)$ is a generalized quadrangle of order $(q - 1, q + 1)$, see [80] or [82]. If q is odd, then $\mathcal{P}(W(q), x)$ is the so-called *Ahrens-Szekeres generalized quadrangle* $AS(q)$ ([1]).

Theorem A.1 ([67], [10], [82, 6.2]). *Every (possibly infinite) generalized quadrangle of order* $(3, t)$ *is isomorphic to either the* (4×4)*-grid* $\mathbb{L}_4 \times \mathbb{L}_4$, $W(3)$, $Q(4, 3)$, $Q(5, 3)$ *or* $\mathcal{P}(W(4), x)$.

We will now discuss the ovoids of the five generalized quadrangles of order $(3, t)$.

- $\mathbb{L}_4 \times \mathbb{L}_4$ has 24 ovoids. Fans and rosettes of ovoids do exist.
- By [94] (see also [82, 1.8.4]), $W(q)$, q odd, has no ovoid. In particular, $W(3)$ does not have ovoids.

- If π is a hyperplane of $PG(4, q)$ intersecting $Q(4, q)$ in a nonsingular elliptic quadric, then $\pi \cap Q(4, q)$ is an ovoid of $Q(4, q)$. If q is prime, then every ovoid of $Q(4, q)$ is obtained in this way, see [3]. In particular, every ovoid of $Q(4, 3)$ is an elliptic quadric. It follows that $Q(4, 3)$ has rosettes of ovoids, but no fans of ovoids.

- Let Π_∞ be a $PG(2, 4)$ which is embedded as a hyperplane in a projective space Π and let $\mathcal{H}$ be a hyperoval of Π_∞. The points of $\Pi \setminus \Pi_\infty$ and the lines L of Π for which $|L \cap \Pi_\infty| = 1$ and $L \cap \Pi_\infty \subseteq \mathcal{H}$ define a generalized quadrangle $T_2^*(\mathcal{H})$ isomorphic to $\mathcal{P}(W(4), x)$, see [82]. If α is a plane of Π intersecting Π_∞ in a line disjoint with $\mathcal{H}$, then the affine points in α form an ovoid of $T_2^*(\mathcal{H})$. By [81], every ovoid of $T_2^*(\mathcal{H})$ is obtained in this way. As a consequence, $T_2^*(\mathcal{H})$ has fans of ovoids, but no rosettes of ovoids.

- By [88] or [94] (see also [82, 1.8.3]), no generalized quadrangle of order (s, s^2), $s \geq 2$, has ovoids. In particular, $Q(5, 3)$ has no ovoid.

A.2 Dense near hexagons of order $(3, t)$

For the finite dense near hexagons of order $(3, t)$, we have an "almost complete" classification.

Theorem A.2 (Main Theorem of [31]). *Let $\mathcal{S}$ be a finite dense near hexagon of order $(3, t)$. If $\mathcal{S}$ is classical or glued, then it is isomorphic to either $\mathbb{L}_4 \times \mathbb{L}_4 \times \mathbb{L}_4$, $W(3) \times \mathbb{L}_4$, $Q(4, 3) \times \mathbb{L}_4$, $T_2^*(\mathcal{H}) \times \mathbb{L}_4$, $Q(5, 3) \times \mathbb{L}_4$, $DW(5, 3)$, $DQ(6, 3)$, $DH(5, 9)$, the unique glued near hexagon of type $T_2^*(\mathcal{H}) \otimes T_2^*(\mathcal{H})$ or the unique glued near hexagon of type $Q(5, 3) \otimes Q(5, 3)$. If $\mathcal{S}$ is not classical nor glued, then only quads isomorphic to the (4×4)-grid or to $Q(4, 3)$ occur. Moreover, there exist constants a and b such that every point of $\mathcal{S}$ is contained in a grids and b $Q(4, 3)$-quads. If v denotes the total number of points of $\mathcal{S}$, then (v, t, a, b) is equal to either $(5848, 19, 160, 5)$, $(6736, 21, 171, 10)$, $(8320, 27, 120, 43)$ or $(20608, 34, 595, 0)$.*

For an alternative description of the glued near hexagon $T_2^*(\mathcal{H}) \otimes T_2^*(\mathcal{H})$, we refer to [54].

Definition. A non-classical and non-glued near hexagon of order $(3, t)$ is called *exceptional of type* (I), (II), (III), respectively (IV), if (v, t, a, b) is equal to (5848,19, 160,5), (6736,21,171,10), (8320,27,120,43), respectively (20608,34,595,0).

Conjecture. Every dense near hexagon of order $(3, t)$ is classical or glued and hence is isomorphic to one of the 10 examples mentioned in Theorem A.2.

The valuations of the 10 known dense near hexagons of order $(3, t)$ have been classified in [51]. They are either classical, semi-classical, ovoidal, extended or semi-diagonal. In the following table we list for each of the near hexagons the number of nonisomorphic valuations.

NEAR HEXAGON	SEMI-CL.	OVOID.	EXT.	SEMI-DIAG.
$\mathbb{L}_4 \times \mathbb{L}_4 \times \mathbb{L}_4$	1	1	1	–
$W(3) \times \mathbb{L}_4$	–	–	1	–
$Q(4,3) \times \mathbb{L}_4$	1	–	2	–
$T_2^*(\mathcal{H}) \times \mathbb{L}_4$	–	1	2	–
$Q(5,3) \times \mathbb{L}_4$	–	–	1	–
$DW(5,3)$	–	–	1	–
$DQ(6,3)$	–	–	–	–
$DH(5,9)$	–	–	–	–
$Q(5,3) \otimes Q(5,3)$	–	–	–	1
$T_2^*(\mathcal{H}) \otimes T_2^*(\mathcal{H})$	–	1	1	1

In the following table, we list the number of nonisomorphic spreads of symmetry for each classical and glued near hexagon of order $(3, t)$ ([50]).

NEAR HEXAGON	SPR. OF SYM.
$\mathbb{L}_4 \times \mathbb{L}_4 \times \mathbb{L}_4$	1
$W(3) \times \mathbb{L}_4$	1
$Q(4,3) \times \mathbb{L}_4$	1
$T_2^*(\mathcal{H}) \times \mathbb{L}_4$	2
$Q(5,3) \times \mathbb{L}_4$	2
$DW(5,3)$	–
$DQ(6,3)$	–
$DH(5,9)$	1
$Q(5,3) \otimes Q(5,3)$	2
$T_2^*(\mathcal{H}) \otimes T_2^*(\mathcal{H})$	2

Regarding the existence of spreads of symmetry in the exceptional near hexagons of order $(3, t)$, one can say the following.

Theorem A.3 ([50]). *An exceptional near hexagon of order $(3, t)$ has no spreads of symmetry.*

A.3 Dense near octagons of order $(3, t)$

Theorem A.4 ([52]). *Let $\mathcal{S}$ be a finite dense near octagon of order $(3, t)$ with the property that every hex is classical or glued, then $\mathcal{S}$ is isomorphic to one of the following 28 near octagons:* $\mathbb{L}_4 \times \mathbb{L}_4 \times \mathbb{L}_4 \times \mathbb{L}_4$, $\mathbb{L}_4 \times \mathbb{L}_4 \times W(3)$, $\mathbb{L}_4 \times \mathbb{L}_4 \times Q(4,3)$, $\mathbb{L}_4 \times \mathbb{L}_4 \times T_2^*(\mathcal{H})$, $\mathbb{L}_4 \times \mathbb{L}_4 \times Q(5,3)$, $\mathbb{L}_4 \times DW(5,3)$, $\mathbb{L}_4 \times DQ(6,3)$, $\mathbb{L}_4 \times DH(5,9)$, $\mathbb{L}_4 \times (T_2^*(\mathcal{H}) \otimes T_2^*(\mathcal{H}))$, $\mathbb{L}_4 \times (Q(5,3) \otimes Q(5,3))$, $W(3) \times W(3)$, $W(3) \times Q(4,3)$, $W(3) \times T_2^*(\mathcal{H})$, $W(3) \times Q(5,3)$, $Q(4,3) \times Q(4,3)$, $Q(4,3) \times T_2^*(\mathcal{H})$, $Q(4,3) \times Q(5,3)$, $T_2^*(\mathcal{H}) \times T_2^*(\mathcal{H})$, $T_2^*(\mathcal{H}) \times Q(5,3)$, $Q(5,3) \times Q(5,3)$, $DH(5,9) \otimes Q(5,3)$, $(Q(5,3) \otimes Q(5,3)) \otimes_a Q(5,3)$, $(Q(5,3) \otimes Q(5,3)) \otimes_b Q(5,3)$, $(T_2^*(\mathcal{H}) \otimes T_2^*(\mathcal{H})) \otimes_a T_2^*(\mathcal{H})$, $(T_2^*(\mathcal{H}) \otimes T_2^*(\mathcal{H})) \otimes_b T_2^*(\mathcal{H})$, $DW(7,3)$, $DQ(8,3)$, $DH(7,9)$.

Theorem A.5 ([50]). *Let $\mathcal{S}$ be a finite dense near octagon of order $(3, t)$ having a big hex which is classical or glued, then $\mathcal{S}$ is isomorphic to one of the 28 near octagons mentioned in Theorem A.4.*

Corollary A.6. *Let $\mathcal{S}$ be a finite dense near octagon of order $(3, t)$ with a nice chain of convex subpolygons, then $\mathcal{S}$ is isomorphic to one of the 28 near octagons mentioned in Theorem A.4*

Theorem A.7 ([50]). *Let $\mathcal{S}$ be a finite dense near octagon of order $(3, t)$ having a big hex H which not classical nor glued, then $\mathcal{S} \cong H \times \mathbb{L}_4$.*

Theorem A.8 ([52]). *Let $\mathcal{S}$ be a finite dense near octagon of order $(3, t)$ without big hexes. Then all quads are isomorphic to $\mathbb{L}_4 \times \mathbb{L}_4$ and all hexes are isomorphic to $\mathbb{L}_4 \times \mathbb{L}_4 \times \mathbb{L}_4$ or to an exceptional near hexagon of type (IV).*

A.4 Some properties of dense near $2d$-gons of order $(3, t)$

Theorem A.9 ([52]). *Let $\mathcal{S}$ be a finite dense near polygon of order $(3, t)$ and let Q be one of the five generalized quadrangles of order $(3, t)$. Then there exists a constant α such that every point of $\mathcal{S}$ is contained in α quads isomorphic to Q.*

Theorem A.10 ([52]). *Let $\mathcal{S}$ be a finite dense near polygon of order $(3, t)$ and let H be a dense near hexagon of order $(3, t)$ isomorphic to either $W(3) \times \mathbb{L}_4$, $Q(4, 3) \times \mathbb{L}_4$, $T_2^*(\mathcal{H}) \times \mathbb{L}_4$, $DW(5, 3)$, $DQ(6, 3)$, $DH(5, 9)$, $T_2^*(\mathcal{H}) \otimes T_2^*(\mathcal{H})$ or an exceptional near hexagon of type (I), (II) or (III). Then there exists a constant α such that every point of $\mathcal{S}$ is contained in α hexes isomorphic to H.*

Theorem A.11 ([52]). *Let $\mathcal{S}$ be a finite dense near polygon of order $(3, t)$. If $\mathcal{S}$ does not contain convex suboctagons of type $(Q(5, 3) \otimes Q(5, 3)) \otimes Q(5, 3)$, then every point of $\mathcal{S}$ is contained in the same number of $\mathbb{L}_4 \times \mathbb{L}_4 \times \mathbb{L}_4$-hexes, the same number of $Q(5, 3) \times \mathbb{L}_4$-hexes and the same number of $Q(5, 3) \otimes Q(5, 3)$-hexes.*

Theorem A.12 ([52]). *Let $\mathcal{S}$ be a finite dense near polygon of order $(3, t)$. Then there exists a constant α such that every point of $\mathcal{S}$ is contained in α exceptional hexes of type (IV).*

A.5 Dense near polygons of order $(3,t)$ with a nice chain of convex subpolygons

Corollary A.6 can be generalized. Define

$$\begin{aligned}
\mathcal{C}_2 &:= \{T_2^*(\mathcal{H})\}, \\
\mathcal{C}_i &:= \bigcup_{2\le j\le i-1} \mathcal{C}_j \otimes \mathcal{C}_{i+1-j} \qquad \text{if } i \ge 3, \\
\mathcal{C} &:= \mathcal{C}_2 \cup \mathcal{C}_3 \cup \cdots, \\
\mathcal{D}_2 &:= \{DH(5,9)\}, \\
\mathcal{D}_i &:= \{DH(2i-1,9)\} \cup \bigcup_{2\le j\le i-1} \mathcal{D}_j \otimes \mathcal{D}_{i+1-j} \qquad \text{if } i \ge 3, \\
\mathcal{D} &:= \mathcal{D}_2 \cup \mathcal{D}_3 \cup \cdots, \\
\mathcal{N} &:= \{\mathbb{O}, \mathbb{L}_4\} \cup \{DW(2n-1,3) \,|\, n \ge 2\} \cup \{DQ(2n,3) \,|\, n \ge 2\} \cup \mathcal{C} \cup \mathcal{D}.
\end{aligned}$$

Let $\mathcal{N}^\times$ denote the set of all near polygons which are the direct product of $k \ge 1$ elements of $\mathcal{N}$.

Theorem A.13 ([47]). *A dense near polygon of order $(3,t)$ has a nice chain of convex subpolygons if and only if it is isomorphic to an element of $\mathcal{N}^\times$.*

Conjecture. Every dense near polygon of order $(3,t)$ has a nice chain of convex subpolygons and hence is isomorphic to an element of $\mathcal{N}^\times$.

Bibliography

[1] R. W. Ahrens and G. Szekeres. On a combinatorial generalization of 27 lines associated with a cubic surface. *J. Austral. Math. Soc.* **10** (1969), 485–492.

[2] M. Aschbacher. Flag structures on Tits geometries. *Geom. Dedicata* **14** (1983), 21–32.

[3] S. Ball, P. Govaerts and L. Storme. On ovoids of parabolic quadrics. *Des. Codes Cryptogr.* **38** (2006), 131–145.

[4] C. T. Benson. On the structure of generalized quadrangles. *J. Algebra* **15** (1970), 443–454.

[5] N. Biggs. *Algebraic graph theory.* Cambridge Tracts in Mathematics **67**, Cambridge University Press, 1974.

[6] R. C. Bose and D. M. Mesner. On linear associative algebras corresponding with association schemes of partially balanced designs. *Ann. Math. Statist.* **30** (1959), 21–38.

[7] A. E. Brouwer. The uniqueness of the near hexagon on 729 points. *Combinatorica* **2** (1982), 333–340.

[8] A. E. Brouwer. The uniqueness of the near hexagon on 759 points. pp. 47–60 in *Finite Geometries* (N. L. Johnson, M. J. Kallaher and C. T. Long, eds.), Lecture Notes in Pure and Appl. Math. **82**, Marcel Dekker, 1982.

[9] A. E. Brouwer. The classification of near hexagons with lines of size three. Report R85-16, Instit. f. Elekroniski Systemer (Aalborg, Denmark), 1985.

[10] A. E. Brouwer. A nondegenerate generalized quadrangle with lines of size four is finite. pp. 47–49 in *Advances in Finite Geometries and Designs*, Oxford Univ. Press, New York, 1991.

[11] A. E. Brouwer and A. M. Cohen. Local recognition of Tits geometries of classical type. *Geom. Dedicata* **20** (1986), 181–199.

[12] A. E. Brouwer, A. M. Cohen, J. I. Hall and H. A. Wilbrink. Near polygons and Fischer spaces. *Geom. Dedicata* **49** (1994), 349–368.

[13] A. E. Brouwer, A. M. Cohen and A. Neumaier. *Distance-regular graphs.* Ergebnisse der mathematik und ihrer grenzgebiete, 3. Folge, Band 18, Springer-Verlag, 1989.

[14] A. E. Brouwer, A. M. Cohen and H. A. Wilbrink. Near polygons with lines of size 3 and Fisher spaces. Math. Centr. Report ZW 191 (Amsterdam), 1983.

[15] A. E. Brouwer and E. W. Lambeck. An inequality on the parameters of distance regular graphs and the uniqueness of a graph related to M_{23}. *Ann. of Discrete Math.* **34** (1987), 101–106.

[16] A. E. Brouwer and H. A. Wilbrink. The structure of near polygons with quads. *Geom. Dedicata* **14** (1983), 145–176.

[17] A. E. Brouwer and H. A. Wilbrink. Two infinite sequences of near polygons. Math. Centr. Report ZW 194 (Amsterdam), 1983.

[18] F. Buekenhout. La géometrie des groupes de Fisher. preprint, 1974.

[19] F. Buekenhout and X. Hubaut. Locally polar spaces and related rank 3 groups. *J. Algebra* **45** (1977), 391–434.

[20] P. J. Cameron. Orbits of permutation groups on ordered sets, II. *J. London Math. Soc. (2)* **23** (1981), 249–264.

[21] P. J. Cameron. Dual polar spaces. *Geom. Dedicata* **12** (1982), 75–86.

[22] P. J. Cameron and J. H. van Lint. *Graph theory, coding theory and block designs.* London Mathematical Society Lecture Notes **19**, Cambridge University Press, 1975.

[23] A. M. Cohen. Finite quaternionic reflection groups. *J. Algebra* **64** (1980), 293–324.

[24] A. M. Cohen. Geometries arising from certain distance-regular graphs. pp. 81–87 in *Finite Geometries and Designs, Proc. Second Isle of Thorns Conference* (P. J. Cameron, J. W. P. Hirschfeld and D. R. Hughes, eds.), London Mathematical Society Lecture Notes **49**, Cambridge University Press, 1981.

[25] A. M. Cohen and J. Tits. On generalized hexagons and a near octagon whose lines have three points. *European J. Combin.* **6** (1985), 13–27.

[26] J. H. Conway. Three lectures on exceptional groups. pp. 215–247 in *Finite simple groups* (M. B. Powell and G. Higman, eds.), Academic Press, 1971.

[27] J. H. Conway, R. A. Wilson, R. T. Curtis, S. P. Norton and R. P. Parker. *Atlas of finite groups.* Clarendon Press, 1985.

[28] H. S. M. Coxeter. Twelve points in PG(5, 3) with 95040 self-transformations. *Proc. Roy. Soc. London Ser. A* 247 (1958), 279–293.

[29] B. De Bruyn. Generalized quadrangles with a spread of symmetry. *European J. Combin.* **20** (1999), 759–771.

[30] B. De Bruyn. On near hexagons and spreads of generalized quadrangles. *J. Algebraic Combin.* **11** (2000), 211–226.

[31] B. De Bruyn. Near hexagons with four points on a line. *Adv. Geom.* **1** (2001), 211–228.

[32] B. De Bruyn. Recent results on near polygons: a survey. pp. 39–59 in *Proceedings of the Academy Contact Forum 20 October 2000* (F. De Clerck, L. Storme, J. A. Thas and H. Van Maldeghem, eds.), Universa Press, 2001.

[33] B. De Bruyn. Glued near polygons. *European J. Combin.* **22** (2001), 973–981.

[34] B. De Bruyn. On the uniqueness of near polygons with three points on every line. *European J. Combin.* **23** (2002), 523–528.

[35] B. De Bruyn. The glueing of near polygons. *Bull. Belg. Math. Soc. Simon Stevin* **9** (2002), 621-630.

[36] B. De Bruyn. Near polygons having a big sub near polygon isomorphic to $\mathbb{H}_n$. *Ann. Comb.* **6** (2002), 285–294.

[37] B. De Bruyn. On the finiteness of near polygons with 3 points on every line. *J. Algebraic Combin.* **18** (2003), 41–46.

[38] B. De Bruyn. Characterizations of near hexagons by means of one local space. *J. Combin. Theory Ser. A* **102** (2003), 283–292.

[39] B. De Bruyn. New near polygons from Hermitian varieties. *Bull. Belg. Math. Soc. Simon Stevin* **10** (2003), 561–577.

[40] B. De Bruyn. Near polygons having a big sub near polygon isomorphic to $\mathbb{G}_n$. *Bull. Belg. Math. Soc. Simon Stevin* **11** (2004), 321–341.

[41] B. De Bruyn. Decomposable near polygons. *Ann. Comb.* **8** (2004), 251–267.

[42] B. De Bruyn. Slim near polygons. *Des. Codes Cryptogr.* **37** (2005), 263–280.

[43] B. De Bruyn. Bounding the size of near hexagons with lines of size 3. *J. Graph Theory*, to appear.

[44] B. De Bruyn. A note on near hexagons with lines of size 3. preprint 2005.

[45] B. De Bruyn. Dense near polygons with two types of quads and three types of hexes. *J. Combin. Math. Combin. Comput.*, to appear.

[46] B. De Bruyn. Compatible spreads of symmetry in near polygons. *J. Algebraic Combin.*, to appear.

[47] B. De Bruyn. A general theory for dense near polygons with a nice chain of subpolygons. *European J. Combin.*, to appear.

[48] B. De Bruyn. Valuations of glued near hexagons. *J. Combin. Des.*, to appear.

[49] B. De Bruyn. A coordinatization of generalized quadrangles with a regular spread. preprint, 2005.

[50] B. De Bruyn. Dense near octagons with four points on each line, I. preprint, 2005.

[51] B. De Bruyn. Dense near octagons with four points on each line, II. preprint, 2005.

[52] B. De Bruyn. Dense near octagons with four points on each line, III. preprint, 2005.

[53] B. De Bruyn and F. De Clerck. Near polygons from partial linear spaces. *Geom. Dedicata* **75** (1999), 287–300.

[54] B. De Bruyn and F. De Clerck. On linear representations of near hexagons. *European J. Combin.* **20** (1999), 45–60.

[55] B. De Bruyn and S. E. Payne. Some notes on generalized quadrangles with a span of regular points. preprint, 2005.

[56] B. De Bruyn and H. Pralle. The hyperplanes of $DW(5, q)$ with no ovoidal quads. *Glasg. Math. J.* **48** (2006), 75–82.

[57] B. De Bruyn and H. Pralle. The hyperplanes of $DH(5, q^2)$. preprint, 2005.

[58] B. De Bruyn and P. Vandecasteele. Two conjectures regarding dense near polygons with three points on each line. *European J. Combin.* **24** (2003), 631–647.

[59] B. De Bruyn and P. Vandecasteele. Near polygons with a nice chain of sub near polygons. *J. Combin. Theory Ser. A* **108** (2004), 297–311.

[60] B. De Bruyn and P. Vandecasteele. Valuations of near polygons. *Glasg. Math. J.* **47** (2005), 347–361.

[61] B. De Bruyn and P. Vandecasteele. Valuations and hyperplanes of dual polar spaces. *J. Combin. Theory Ser. A* **112** (2005), 194–211.

[62] B. De Bruyn and P. Vandecasteele. Near polygons having a big sub near polygon isomorphic to $H^D(2n-1, 4)$. *Ars Combin.*, to appear.

[63] B. De Bruyn and P. Vandecasteele. The valuations of the near hexagons related to the Witt designs. *J. Combin. Des.*, to appear.

[64] B. De Bruyn and P. Vandecasteele. The classification of the slim dense near octagons. *European J. Combin.*, to appear.

[65] F. De Clerck and H. Van Maldeghem. Some classes of rank 2 geometries. pp. 433–475 in *Handbook of Incidence Geometry* (F. Buekenhout, ed.), North-Holland, 1995.

[66] P. Delsarte and J. M. Goethals. Unrestricted codes with the Golay parameters are unique. *Discr. Math.* **12** (1975), 211–224.

[67] S. Dixmier and F. Zara. Etude d'une méthode d'étude de certains graphes liés aux groupes classiques. preprint, 1976.

[68] J. D. Dixon and B. Mortimer. *Permutations Groups.* Graduate Texts in Mathematics **163**, Springer-Verlag, 1996.

[69] A. Dress and R. Scharlau. Gated sets in metric spaces. *Aequationes Math.* **34** (1987), 112–120.

[70] P. Erdös, Chao Ko and R. Rado. Intersection theorems for systems of finite sets. *Quart. J. Math. Oxford (2)* **12** (1961), 313–320.

[71] W. Feit and G. Higman. The non-existence of certain generalized polygons. *J. Algebra* **1** (1964), 114–131.

[72] B. Fisher. Finite groups generated by 3-transpositions. *Invent. Math.* **13** (1971), 232–246.

[73] H. Freudenthal. Une étude des quelques quadrangles généralisés. *Ann. Mat. Pura Appl.* **102** (1975), 109–133.

[74] W. Haemers and C. Roos. An inequality for generalized hexagons. *Geom. Dedicata* **10** (1981), 219–222.

[75] D. G. Higman. Invariant relations, coherent configurations and generalized polygons. pp. 27–43 in *Proceedings of the Advanced Study Institute on Combinatorics, Breukelen* (M. Hall Jr. and J. H. van Lint, eds.), Mathematical Centre Tracts **57**, 1974.

[76] R. Mathon. On primitive association schemes with three classes. preprint.

[77] A. Neumaier. Krein conditions and regular near polygons. *J. Combin. Theory Ser. A* **54** (1990), 201–209.

[78] D. Pasechnik. Extending polar spaces of rank at least 3. *J. Combin. Theory Ser. A* **72** (1995), 232–242.

[79] A. Pasini. *Diagram Geometries.* Oxford Science Publications, The Clarendon Press, Oxford University Press, 1994.

[80] S. E. Payne. Nonisomorphic generalized quadrangles. *J. Algebra* **18** (1971), 201–212.

[81] S. E. Payne. The generalized quadrangle with $(s,t) = (3,5)$. In "Proceedings of the Twenty-first Southeastern Conference on Combinatorics, Graph Theory, and Computing (Boca Raton, FL, 1990)". *Congr. Numer.* **77** (1990), 5–29.

[82] S. E. Payne and J. A. Thas. *Finite Generalized Quadrangles.* Research Notes in Mathematics **110**, Pitman, Boston, 1984.

[83] V. Pless. On the uniqueness of the Golay codes. *J. Comb. Theory* **5** (1968), 215–228.

[84] H. Pralle and S. Shpectorov. The ovoidal hyperplane of a dual polar space of rank 4. preprint, 2004.

[85] L. L. Scott, Jr. A condition on Higman's parameters. *Notices Amer. Math. Soc.* **20** (1973), 97.

[86] J. J. Seidel. Strongly regular graphs with $(-1, 1, 0)$ adjacency matrix having eigenvalue 3. *Linear Algebra Appl.* **1** (1968), 281–298.

[87] S. Shad. *Characterizations of geometries related to polar spaces.* Ph.D. thesis, Kansas State University, 1980.

[88] S. A. Shad and E. E. Shult. The near n-gon geometries. preprint, 1979.

[89] E. E. Shult. Characterization of certain classes of graphs. *J. Combin. Theory Ser. B* **13** (1972), 142–167.

[90] E. E. Shult and J. A. Thas. Constructions of polygons from buildings. *Proc. London Math. Soc.* **71** (1995), 397-440.

[91] E. E. Shult and A. Yanushka. Near n-gons and line systems. *Geom. Dedicata* **9** (1980), 1–72.

[92] J. J. Sylvester. *Collected Mathematical Papers I.* Cambridge University Press, 1904.

[93] J. A. Thas. On 4-gonal configurations with parameters $r = q^2 + 1$ and $k = q + 1$, part I. *Geom. Dedicata* **4** (1974), 51–59.

[94] J. A. Thas. Some results on quadrics and a new class of partial geometries. *Simon Stevin* **55** (1981), 129–139.

[95] J. A. Thas. Generalized polygons. pp. 383–431 in *Handbook of Incidence Geometry* (F. Buekenhout, ed.), North-Holland, 1995.

[96] J. Tits. Sur la trialité et certains groupes qui s'en déduisent. *Inst. Hautes Etudes Sci. Publ. Math.* **2** (1959), 14–60.

[97] J. Tits. *Buildings and BN-pairs of spherical type.* Lecture Notes in Mathematics **386**, Springer-Verlag, 1974.

[98] J. Tits. Quaternions over $\mathbb{Q}(\sqrt{5})$, Leech's lattice and the sporadic group of Hall-Janko. *J. Algebra* **63** (1980), 56–75.

[99] J. A. Todd. On representations of Mathieu groups as collineation groups. *J. London Math. Soc.* **34** (1959), 406–416.

[100] H. Van Maldeghem. *Generalized Polygons.* Monographs in Mathematics **93**, Birkhäuser, 1998.

[101] R. A. Wilson. The geometry of the Hall-Janko group as a quaternionic reflection group. *Geom. Dedicata* **20** (1986), 157–173.

[102] E. Witt. Uber Steinersche Systeme. *Abh. Math. Sem. Univ. Hamburg* **12** (1938), 265–275.

Index